Introduction to Engineering Experimentation

Second Edition

Anthony J. Wheeler
Ahmad R. Ganji

School of Engineering
San Francisco State University

PEARSON
Prentice
Hall

Upper Saddle River, New Jersey 07458

Library of Congress Cataloging-in-Publication Data

Wheeler, Anthony J.
  Introduction to engineering experimentation / Anthony J. Wheeler and Ahmad R. Ganji.—2nd ed.
    p. cm.
  Includes bibliographical references and index.
  ISBN 0-13-065844-8
  1. Engineering—Experiments. 2. Experimental design. I. Ganji, A.R. (Ahmad Reza) II.
  Title
  TA153.W47 2003
  6208 .0078—dc21   2003043374

Vice President and Editorial Director, ECS: *Marcia J. Horton*
Acquisitions Editor: *Laura Fischer*
Vice President and Director of Production and Manufacturing, ESM: *David W. Riccardi*
Executive Managing Editor: *Vince O'Brien*
Managing Editor: *David A. George*
Production Editor: *Craig Little*
Director of Creative Services: *Paul Belfanti*
Creative Director: *Carole Anson*
Art Director: *Jayne Conte*
Cover Designer: *Bruce Kenselaar*
Art Editor: *Greg Dulles*
Manufacturing Manager: *Trudy Pisciotti*
Manufacturing Buyer: *Lisa McDowell*
Marketing Manager: *Holly Stark*

© 2004 by Pearson Education, Inc.
Upper Saddle River, New Jersey 07458

The authors and publisher of this book have used their best efforts in preparing this book. These efforts include the development, research, and testing of the theories and programs to determine their effectiveness. The authors and publisher make no warranty of any kind, expressed or implied, with regard to these programs or the documentation contained in this book. The authors and publisher shall not be liable in any event for incidental or consequential damages in connection with, or arising out of, the furnishing, performance, or use of these programs.

Printed in the United States of America

10 9 8

ISBN 0-13-065844-8

Pearson Education Ltd., *London*
Pearson Education Australia Pty. Ltd., *Sydney*
Pearson Education Singapore, Pte. Ltd.
Pearson Education—North Asia Ltd., *Hong Kong*
Pearson Education Canada, Inc., *Toronto*
Pearson Educación de Mexico, S.A. de C.V.
Pearson Education—Japan, *Tokyo*
Pearson Education Malaysia, Pte. Ltd.
Pearson Education, Upper Saddle River, *New Jersey*

Contents

Preface

This book is an introduction to many of the topics that an engineer needs to master in order to successfully design experiments and measurement systems. In addition to descriptions of common measurement systems, the book describes computerized data acquisition systems, common statistical techniques, experimental uncertainty analysis, and guidelines for planning and documenting experiments. It should be noted that this book is introductory in nature. Many of the subjects covered in a chapter or a few pages here are the subjects of complete books or major technical papers. Only the most common measurement systems are included—there exist many others that are used in practice. More comprehensive studies of available literature and consultation with product vendors are appropriate when engaging in a significant real-world experimental program. It is to be expected that the skills of the experimenter will be enhanced by more advanced courses in experimental and measurement systems design and practical experience.

The design of an experimental or measurement system is inherently an interdisciplinary activity. For example, the instrumentation and control system of a process plant might require the skills of chemical engineers, mechanical engineers, electrical engineers, and computer engineers. Similarly, the specification of the instrumentation used to measure the earthquake response of a large structure will involve the skills of civil, electrical, and computer engineers. Based on these facts, the topics presented in this book have been selected to prepare engineering students and practicing engineers of different disciplines to design experimental projects and measurement systems.

This book was conceived when a decision was made at San Francisco State University to upgrade the laboratory of our first experimental course from using primarily mechanical instruments to using electrical output devices, and to introduce the students to the acquisition and processing of the data with computer systems. The lecture was upgraded at the same time to include the new topics. A survey was made of available texts, and none was found to provide complete coverage of the material in the revised course. The primary deficiencies were in the coverage of computerized data acquisition systems, statistics, and the design and documentation of experiments. Consequently, we created a course reader, which was subsequently expanded to become this book.

The book first introduces the essential general characteristics of instruments, electrical measurement systems, and computerized data acquisition systems. This introduction gives the students a foundation for the laboratory associated with the course.

The theory of discretely sampled systems is introduced next. The book then moves into statistics and experimental uncertainty analysis, which are both considered central to a modern course in experimental methods. It is not anticipated that the remaining chapters will necessarily be covered either in their entirety or in the presented sequence in lectures—the instructor will select appropriate subjects. Descriptions and theory are provided for a wide variety of measurement systems. There is an extensive discussion of dynamic measurement systems with applications. Finally, guidance for planning experiments, including scheduling, cost estimation, and outlines for project proposals and reports, are presented in the last chapter.

There are some subjects included in the introductory chapters that are frequently of interest, but are often not considered vital for an introductory experimental methods course. These subjects include the material on circuits using operational amplifiers (Sections 3.2.2, 3.2.5 and 3.2.6), details on various types of analog-to-digital converters (Section 4.3.3), and the material on Fourier transforms (Section 5.3). Any or all of these sections can be omitted without significant impact on the remainder of the text.

The book has been designed for a semester course of up to three lectures with one laboratory per week. Depending on the time available, it is expected that only selected topics will be covered. The material covered depends on the number of lectures per week, the prior preparation of students in the area of statistics, and the scope of included design project(s). The book can serve as a reference for subsequent laboratory courses.

Our introductory course in engineering experimentation is presented to all undergraduate engineers in civil, electrical, and mechanical engineering. The one-semester format includes two lectures per week and one three-hour laboratory. In our two-lecture-per-week format, the course content is broken down as follows:

1. General aspects of measurement systems (2 lectures)
2. Electrical output measurement systems (2 lectures)
3. Computerized data acquisition systems (3 lectures)
4. Fourier analysis and the sampling rate theorem (4 lectures)
5. Statistical methods and uncertainty analysis (10 lectures)
6. Selected measurement devices (4 lectures)
7. Dynamic measurement systems (3 lectures)

Additional measurement systems and the material on planning and documenting experiments are covered in the laboratory. The laboratory also includes an introduction to computerized data acquisition systems and applicable software; basic measurements such as temperature, pressure, and displacement; statistical analysis of data; the sampling rate theorem; and a modest design project. A subsequent laboratory-only course expands on the introductory course and includes a significant design project.

There is sufficient material for a one-semester, three-lecture-per-week course even if the students have taken a prior course in statistics. Areas that can be covered in greater detail include operational amplifiers, analog-to-digital converters, spectral analysis, uncertainty analysis, measurement devices, dynamic measurements, and experiment design.

In this second edition, Chapter 6 on statistics has been significantly enhanced to include the Poisson distribution, multiple and polynomial regression, outlier analysis for x–y data sets, and linear functions of random variables. Chapter 7 on uncertainty analysis has been extensively modified to make it compatible with the latest ASME standard and to provide a simpler path through the material for large data samples. Chapter 5 has been modified to include the folding diagram for predicting alias frequencies and to make the nomenclature for Fourier series consistent with common current usage. Numerous lesser alterations have been made throughout the book to clarify, update, or enhance the material. Finally, the number of homework problems has been increased by 50%.

ACKNOWLEDGMENTS

The authors would like to acknowledge the many individuals who reviewed all or portions of the book. We would like to thank Sergio Franco, Sung Hu, and V. Krishnan of San Francisco State University; Howard Skolnik of Intelligent Instrumentation; Ali Rejali of Isfahan University of Technology (Iran) and Ronald Diek of Pratt & Whitney, each of whom reviewed portions of the book. Particular thanks go to reviewers of the complete book: Charles Edwards of the University of Missouri, Rolla, and David Bogard of the University of Texas, Austin.

ANTHONY J. WHEELER

AHMAD R. GANJI

SAN FRANCISCO, CALIFORNIA

CHAPTER 1

Introduction

Experimentation is the backbone of modern physical science. In engineering, carefully designed experiments are needed to conceive and verify theoretical concepts, develop new methods and products, commission sophisticated new engineering systems, and evaluate the performance and behavior of existing products. Experimentation and the design of measurement systems are major engineering activities. In this chapter we give an overview of the applications of experiments and measurement systems and describe briefly how this book will prepare the reader for professional activities in these areas.

1.1 APPLICATIONS OF ENGINEERING EXPERIMENTATION AND MEASUREMENT

Engineering measurement applications can broadly be broken into two categories. The first of these is measurement in engineering experimentation, in which new information is being sought, and the second is measurement in operational devices for monitoring and control purposes.

1.1.1 Measurement in Engineering Experimentation

Engineering experimentation, which in a general sense involves using the measurement process to seek new information, ranges in scope from experiments to establish new concepts all the way to testing of existing products to determine maintenance requirements. Such experimentation falls broadly into three categories:

1. Research experimentation
2. Development experimentation
3. Performance testing

The primary difference between research and development is that in the former, concepts for new products or processes are being sought (often unsuccessfully), while in the latter, known concepts are being used to establish potential commercial products.

Carbon-fiber composites represent a relatively recent example of the research and development process. Carbon-fiber composites are now used commercially for such diverse products as golf clubs and aircraft control surfaces. In the research phase, methods were suggested and evaluated to produce carbon fibers in small quantities and tests were performed to determine the physical properties of samples. The results of the research activities were so promising that many development activities were initiated. These activities included development of large-scale fiber manufacturing processes and development of methods to fabricate fiber composite parts. Although there are now many products using carbon fibers, developmental activities in this area continue.

Research experiments are frequently highly uncertain and often lead to dead ends. The risk is high, either because the experiment itself may be unsuccessful or because the experimental result may not be as wanted. Research experimentation is usually performed in universities or special research organizations. On the other hand, development programs usually have better defined goals than research programs and frequently result in an operational product. Sometimes, however, the product will not be deemed competitive and will never be produced in quantity. Development programs are usually performed by product manufacturers.

Although the instrumentation must function properly during the research or development program, it may be delicate and require considerable attention. Special measurement techniques may be created. Experimental measuring systems whose characteristics are not completely defined may also be suitable for such testing programs. The engineers and scientists performing such tests are generally sophisticated in the fine points of the instruments and can compensate for deficiencies.

Performance testing is somewhat different from research and development experimental activities. Performance testing is done on products that have been developed and in many cases are already on the market. Performance testing may be carried out to demonstrate applicability for a particular application, to assess reliability, or to determine product lifetime. This testing may be done either by the manufacturer, the supplier, the customer, or an independent laboratory. As an example, a performance test might be used to demonstrate that an electronic device which functions satisfactorily in a stationary environment will also function in an aircraft application with high levels of vibration.

Another type of performance testing is the periodic testing of operating devices to determine needs for maintenance. Utilities normally perform this type of testing in power plants to make sure that the efficiencies of large pumps, heat exchangers, and other components are adequate. Poor performance will lead to maintenance actions. Instruments may be in place for such tests, but they may need repair, and supplementary instruments may be required at the time of the tests. Commissioning of process plants may also involve extensive but standardized testing to demonstrate conformance to design specifications.

Measuring systems for performance testing are generally established devices with well-defined characteristics. The systems need to be reliable, and significant interpretation of ambiguities in the measured values should not be required since the people performing the tests are often technicians.

1.1.2   Measurement in Operational Systems

Many dynamic systems are instrumented for monitoring or control purposes. Such systems range from simple furnaces for home heating to extremely complex jet aircraft. One very sophisticated but everyday measurement and control system is the engine control system of modern automobiles. These systems have sensors to measure variables such as airflow, engine speed, water temperature, and exhaust gas composition and use a computer to determine the correct fuel flow rate. These engine control systems are very compact and are specially engineered for the particular application.

Elaborate measurement and control systems are needed in complex process plants such as oil refineries, steam power plants, and sewage treatment facilities. Such systems may have hundreds of sensors and use computers to collect and interpret the data and control the process. This particular class of applications is so large that it is a specialized field in its own right, called *process control*. While the complete measuring systems for such applications are specifically engineered, the components are generally modular and standardized.

Instrumentation for operating systems must be very durable and reliable. Sensors that need to be calibrated very frequently would present major problems in these applications. In many cases, the measuring systems have to be designed such that by redundancy or other techniques, a failed component can be readily identified so that the operating system can continue to operate correctly or at least be safely shut down.

1.2   OBJECTIVE AND OVERVIEW

The objective of this book is to provide the reader with the skills necessary to perform an engineering experiment systematically—from the definition of the experimental need to the completion of the final report. A systematic approach includes careful planning and analytical design of the experiment before it is constructed, demonstration of the validity of the test apparatus, analysis of the test results, and reporting of the final results. The emphasis is on the design of the experiment and the analysis of the results; however, guidance is given on other activities. Chapters 2 through 11 provide the technical information necessary to design an experimental system and interpret the results. This information is also applicable to the design of the measurement (but not control) systems of process plants. Chapter 12 provides an overview of the overall experimental design process and provides guidelines on planning, designing, scheduling, and documenting experimental projects.

1.3   DIMENSIONS AND UNITS

The International System of Units (SI) is the most widely used unit system in the world, due to its consistency and simplicity. However, in the United States and some other countries, a unit system based on the old British unit system is still widely used. Product specifications and data tables are frequently given in British units. For example, the range of a pressure measurement device might be specified in pounds per square inch (psi). To assist the reader in developing capabilities in both unit systems, both SI and British units systems are used in example problems in this book.

TABLE 1.1   Base SI and British Units

Dimension	SI unit	British unit
Mass	kilogram (kg)	pound mass (lbm)
Length	meter (m)	foot (ft)
Time	second (s)	second (s)
Temperature	Kelvin (K)	Rankine degree (°R)
Electric current	ampere (A)	ampere (A)

The physical world is described with a set of dimensions. Length, mass, time, and temperature are dimensions. When a numerical value is assigned to a dimension, it must be done in a unit system. For example, we can describe the temperature (dimension) of an ice–water mixture in either the SI unit system (0°C) or the British unit system (32°F). International conferences have established a set of SI base units. Table 1.1 lists the base SI units and the corresponding British units. There are two additional base units, the candela for light intensity and the mole for the amount of a substance, but these units are not used in this book. Each of these base units has a corresponding standard such that all users of the unit can compare their results. Standards are discussed in Chapter 2.

Other engineering quantities, such as force and power, are related to the dimensions of the base units through physical laws and definitions. The dimension of force is defined by Newton's second law:

$$F = \frac{m}{g_c}a \qquad (1.1)$$

where F is force, m is mass, and a is acceleration. g_c is a proportionality constant, which depends on the unit system. In the SI unit system, the unit of force is the newton (N) and is defined as the force exerted on a 1-kg mass to generate an acceleration of 1 m/s². In the SI system, g_c has a value of unity. In the British unit system, the unit of force is the pound force (lbf) and is defined as the force exerted on a 1-lb mass at the standard gravitational acceleration of 32.174 ft/sec². In this case, the value of g_c has to be taken as 32.174 lbm-ft/lbf-sec².

In this book, in equations based on Newton's second law, the constant g_c is taken to be unity and does not appear in the equations. All equations will produce a dimensionally correct result if the SI system of units is used properly. Sometimes in the British system, mass is specified using a unit called the slug, defined as 32.174 lbm. When the slug is used to define mass, the constant g_c is also unity. Unfortunately, the slug is not a widely used unit, and most British-unit data tables and specifications use lbm for the mass unit. Consequently, mass numbers supplied in lbm must be converted by dividing by the constant 32.174 when using them in equations in this book. Another characteristic of the British unit system is that two units are used for energy. In mechanical systems, the unit of energy is the ft-lbf, while in thermal systems, it is the Btu. The conversion factor is 1 Btu = 778 ft-lbf.

When using any unit system, great care is required to make sure that the units are consistent, particularly with British units. It is recommended that a units check be performed for all calculations using British units.

1.4   CLOSURE

In engineering in general, one gains a lot of expertise from experience—doing things and finding out what works and what does not. In experimental work this is even more true. It is virtually impossible to document all of the many subtleties of performing an engineering experiment. Quality engineering experimentation takes care, it takes time, and it takes patience.

C H A P T E R   2

General Characteristics of Measurement Systems

A necessary part of planning an experiment is to determine the specifications for the required measurement systems. The characteristics of many specific measuring devices are detailed in Chapters 8 to 10. In this chapter, significant general characteristics of measurement systems are described and definitions are provided for common descriptive terms.

2.1   GENERALIZED MEASUREMENT SYSTEM

In any experiment the experimenter seeks to obtain numerical values for certain physical variables. These unknown variables are known as *measurands*. Examples of measurands include temperature, velocity, and voltage. The measurement system senses the measurand and produces a unique numerical value that describes the measurand. In most cases, the measurement system can be viewed as consisting of three subsystems. As shown in Figure 2.1, these three subsystems are the *sensing element*, the *signal modification subsystem*, and the *recording* or *indicating device*. The sensing element has a significant physical characteristic that changes in response to changes in the measurand. The signal modification subsystem changes the output of the sensing element in some way to make it more suitable for the indicating or recording device. If the user simply reads the output and perhaps copies it to paper, the final device is an indicator. If the output value is saved automatically in some way, the final device is a recorder. A measurement system may have both an indicating device and a recording device. In modern measurement systems, this final stage is often a computer, which can not only display and record the data but can also manipulate the data mathematically.

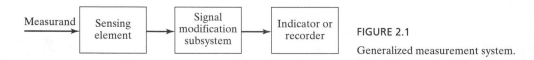

FIGURE 2.1

Generalized measurement system.

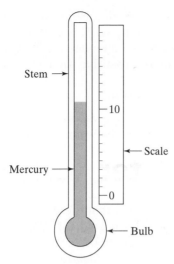

FIGURE 2.2

Mercury-in-glass thermometer.

A simple example of a measurement system is a common mercury-in-glass thermometer (Figure 2.2), which could be used to measure the temperature of water in a container. In this device, the volume of the mercury in the bulb depends on the temperature of the mercury. If the bulb has been in contact with the water for a sufficient time, the mercury will have the same temperature as the surrounding fluid. Hence a measurement of the volume of the mercury can be used to determine the temperature of the water. Unfortunately, it is very difficult to measure the small change in volume of the mercury. If the mercury had the shape of a sphere, the change in diameter would be very small. Therefore, signal modification is required. For the thermometer, signal modification is accomplished by connecting the bulb to the stem. The inside diameter of the stem is very small relative to the diameter of the bulb, and although the change in mercury volume is small, this small change in volume produces a large change in length of the stem mercury column. Actually, the displacement of mercury in the stem is proportional to the differential thermal expansion between the mercury and the glass envelope. Finally, an indicating device is required. In the case of the thermometer, this is accomplished with a scale that is either next to the glass stem or engraved on it directly.

These three subsystems are quite obvious in most measuring devices. This is particularly true for modern measurement systems using electrical output-sensing devices, in which the three subsystems are often physically separate devices. There are, however, some common measuring systems in which all three subsystems are difficult to identify or the components are combined.

2.2   VALIDITY OF MEASUREMENT

It is very important to the experimenter that the output of a measurement system truly states the actual value of the measurand. That is, the experimenter must be convinced that the output of the measurement system is valid. Of course, no measurement system is perfect—there will always be some deviation between the actual value of the measurand and the measurement system output. This deviation must simply be small enough that the output can be used for its intended purpose. Generally speaking, the smaller the allowed deviation, the more expensive will be the measurement system.

2.2.1   Measurement Error and Related Definitions

Several standard terms are used to specify the validity of a measurement. The *error* of a measurement is defined as the difference between the measured value and the true value of the measurand:

$$error = measured\ value - true\ value$$

Error in this technical usage does not imply that there is any mistake in the measurement process, although mistakes can cause errors. Normally, the experimenter can never really know the error of a measurement. If the true value of the measurand were known, there would be no need to make the measurement (except in the process of calibration, where measurements are made of measurands whose values are independently known). What the experimenter can estimate, however, is the *uncertainty interval* (or simply *uncertainty*) of the measurement. The uncertainty is an estimate (with some level of confidence) of the limits of error in the measurement. For example, it might be stated that with 95% confidence, the uncertainty of a voltage measurement is ±1 volt. This means that the error will be greater than 1 V in less than 5% of the cases where we have made such uncertainty predictions. Narrow uncertainty intervals are usually achieved by using calibrated, high-quality measuring systems.

Errors in experiments generally fall into two categories: *systematic errors* (fixed or bias errors) and *random errors* (precision errors). Although both types of error degrade the validity of the data, their causes are different and different actions must be taken to minimize them.

Systematic errors are consistent, repeatable errors. For example, a measuring system might give a consistent 10% high reading. In other cases, the output might be the same absolute amount low for all readings. In general, if the same measuring system is used in the same way more than once to measure the same value of the measurand, the systematic error will be the same each time.

The first major source of systematic error is that resulting from calibration of the measurement system. If the calibration process has some error, that error will be carried into the measurement as a systematic error. Even the most exact calibration will result in a residual systematic error. These are known as *calibration errors*. One very common source of calibration systematic error is nonlinearity. Many modern systems are treated as if they have a linear relationship between the input and the output, and the actual nonlinearity of the system will cause errors.

The second major source of systematic error results from the use of a measuring system in a particular application where the insertion of the measuring device alters the measurand. For example, connecting a temperature-measuring device to a surface may in fact change the local temperature of the surface. Such errors are known as *loading errors*. As another example, consider placing a mercury-in-glass thermometer into a beaker of water. If the beaker and the thermometer are initially at different temperatures, energy will be exchanged between them, and the measured temperature will be neither the initial water temperature nor the initial thermometer temperature (but usually closer to the water temperature). The thermometer is an *intrusive* measurement device and produces a significant loading error. Some measuring devices with negligible loading errors are called *nonintrusive*. For example, devices are available that measure temperature by sensing the infrared radiation emitted. Such a device would have a negligible effect on the measured temperature and is said to be nonintrusive.

A third major systematic error results because the measuring system is affected by variables other than the measurand. For example, a thermometer used to measure the air temperature in a room will read too low, due to thermal radiation effects, if the walls are cooler than the air. A related error is the *spatial error*. If the measurand varies in a spatial region and yet a single measurement or a limited number of measurements are used to determine the average value for the region, there will be a spatial error.

Systematic errors are often not obvious to the experimenter—the measuring system will show clear and consistent changes in output following changes in the measurand, yet it will still have significant error. In setting up an experiment, considerable time may be required to detect and minimize systematic errors. Systematic errors in the measuring system may be detected and reduced by the process of calibration, discussed later. Some systematic errors caused by using the measuring system in a particular application may be reduced by analytical correction of the data for unwanted effects.

Random errors are those caused by a lack of repeatability in the output of the measuring system. The distinction between systematic and random errors is shown graphically in Figure 2.3. The scatter in the data represents random error, and the deviation between the average of the readings and true value demonstrates the systematic error. The random error in a single measurement can be estimated as the difference between the single reading and the average of all readings of the same value of measurand:

$$\text{random error} = \text{reading} - \text{average of readings}$$

which distinguishes the random error from the systematic error. The systematic error can be estimated by using the following equation:

$$\text{systematic error} = \text{average of readings} - \text{true value}$$

For these estimates of systematic and random errors to be reasonable, the number of readings forming the average must be large enough to eliminate the effects of random error in individual measurements on the average.

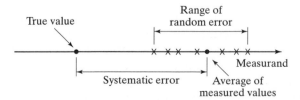

FIGURE 2.3

Distinction between systematic and random errors.

Random errors can originate from the measuring system itself, from the experimental system, or from the environment. Random errors are usually caused by uncontrolled variables in the measurement process. For example, the performance of an amplifier may be slightly sensitive to its temperature. If we do not measure or control the temperature, performance measurements may show a certain variability or scatter. One very important environmental cause of random error is electrical noise. Experiments and measuring systems normally operate in a sea of electric and magnetic fields caused by sources such as building wiring and local radio stations. These electric and magnetic background fields can affect readings by randomly altering voltages in measuring systems and connecting wiring.

Random errors can often be minimized by eliminating uncontrolled variables or properly shielding or grounding the measuring system. Remaining random errors may be amenable to statistical analysis—for example, a large number of readings can be averaged. Statistical analysis of data is discussed in detail in Chapter 6. Example 2.1 shows how to estimate the systematic and maximum random errors for a calibration test of a voltmeter.

Example 2.1

In a calibration test, 10 measurements using a digital voltmeter have been made of the voltage of a battery that is known to have a true voltage of 6.11 V. The readings are: 5.98, 6.05, 6.10, 6.06, 5.99, 5.96, 6.02, 6.09, 6.03, and 5.99 V. Estimate the systematic and maximum random errors caused by the voltmeter.

Solution: First, determine the average of the 10 readings:

$$\text{average } V = 6.03$$

Then the estimate of the systematic error is computed as follows:

$$\text{systematic error} = \text{average value} - \text{true value} = 6.03 - 6.11 = -0.08 \text{ V}$$

To estimate the maximum random error, we need to determine the reading that deviates the most from the average reading. This is the reading of 5.96 V. The maximum precision error is thus

$$\text{maximum random error} = 5.96 - 6.03 = -0.07 \text{ V}$$

Comment: It should be noted that this simple statement of maximum random error may not adequately describe random errors in a measuring system. For example, it may be based on a single bad reading. Statistical methods described in Chapters 6 and 7 provide procedures to determine random errors, which include all of the readings and also provide a basis to eliminate certain bad data.

A measuring system is only designed to operate over a specified range of measurands. The *range* of a measuring system describes the values of the measurand to which that measuring system will respond properly—values of the measurand outside the range are not expected to produce useful output. For example, a voltmeter may have a range of 0 to 10 V and would not give a correct response to measurands of −5 or 13 V. The *span* of a measuring system is the difference between the upper and lower values of the range. For example, a voltmeter with a range of ±3 V has a span of 6 V.

Accuracy, defined as the closeness of agreement between a measured value and the true value, is a common term used to specify uncertainty. Measuring device manufacturers frequently state a value for accuracy as part of the device specifications. Although the term *accuracy* is generally used, it is really the inaccuracy that is specified. As commonly used, manufacturer specifications of accuracy describe residual uncertainty that exists when a device has been properly adjusted and calibrated and is used in a specified manner. Accuracy specifications generally include residual systematic and random errors in the measuring system itself. Accuracy might be given for either a component of a measuring system (e.g., a sensor) or for a complete system and is most often specified as a percentage of full-scale output. For example, if the output of a device can range from 0 to 5 V, and the accuracy is stated as ±5% of full scale, the uncertainty is ±0.25 V, regardless of the reading. The procedure to determine accuracy is described in Section 2.2.2 and is based on information in ANSI/ISA (1979). If more than one component is used in the measurement of a single measurand, a combined uncertainty must be determined. Methods to estimate overall or total uncertainty are described in Chapter 7.

As Figure 2.4 shows for a typical measuring device with an accuracy of ±5% of full scale, at readings below full scale, the percent uncertainty in the reading will be greater than 5%. At readings toward the lower end of the range, the percent uncertainty might be completely unsatisfactory. This problem with high uncertainty at the low end of the range is a major concern in selecting a measuring system. To minimize uncertainty, the experimenter should select measuring systems such that important readings will fall in the middle to upper portions of the range. For example, it would adversely affect uncertainty if a 0-to-200°C thermometer were used to measure a room temperature around 20°C. A 0 to 30°C thermometer would be far more appropriate.

There are other statements of accuracy, such as an accuracy stated as a percent of reading. Manufacturers of high-quality measuring systems and components will normally give enough information about their products so that the experimenter can determine the uncertainty in the measurement that is due to the measuring system itself. The experimenter may have to enlarge the uncertainty interval to account for other error sources that result from the specific application.

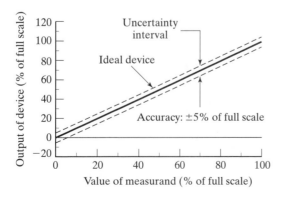

FIGURE 2.4

Accuracy as a percentage of full scale.

Precision is another term frequently used to describe a measuring system or component and characterizes the random error. A highly precise measuring system will give the same value each time it is read, but it may not be very accurate—it may simply give the same inaccurate answer each time the measurement is made. In general, the accuracy of a measuring system cannot be any better than the measurement constraints provided by the instrument precision (although it can be much worse). Accuracy and precision are overall characteristics that describe the validity of measurement. Each characteristic is determined by a number of specific sources of uncertainty (errors).

In measuring devices, accuracy is often degraded by a phenomenon known as *hysteresis*. As shown in Figure 2.5, for the same value of the measurand, different output readings may be obtained if the measurand was increasing prior to taking the reading than if the measurand was decreasing. Hysteresis is caused by such effects as friction, mechanical flexure of internal parts, and electrical capacitance. Errors due to hysteresis are known as *hysteresis errors*. If a measurement is repeated in exactly the same manner, errors due to hysteresis would be repeatable and hence would be considered systematic errors. However, in common measuring processes, the experimenter generally may not know if the measurand was increasing or decreasing when a measurement was made. Hence, the effect of hysteresis will appear random. However, when estimating total uncertainty, it is normally conservative to treat hysteresis errors as a systematic error. Hysteresis error is usually a component of the instrument manufacturers' specification of accuracy.

Another important characteristic of a measuring system is the *resolution*. If a measurand is varied continuously, many measurement devices will show an output that changes in discrete steps. This inability of the measurement system to follow changes in the measurand exactly results in a *resolution error*. Resolution is usually treated as a random error. Internal characteristics of a measuring system may limit resolution. The sensing element itself may not produce a continuous output with a smoothly varying measurand. A wirewound potentiometer (a position-sensing device, discussed in Chapter 8) may have a step type of output. Many digital instruments contain an analog-to-digital converter (discussed in Chapter 4), which places well-defined limits on resolution. Most modern instruments have a digital output display, which will result in a resolution error. If, for example, the digital output device has a reading of 1.372, the reading resolution is simply a value of 1 in the last (least significant) digit.

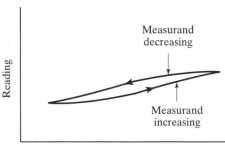

FIGURE 2.5

Effect of hysteresis on instrument reading.

If the device rounds off values, the resolution uncertainty will be ±0.5 in the least significant digit.

In instruments in which the output is read by comparing a pointer to a scale, the ability to resolve a value of the measurand is limited by a characteristic called the scale *readability*. For example, the thermometer shown in Figure 2.2 has a tick mark every degree. One might think that there may be a maximum uncertainty in reading the thermometer of ±0.5 degree (reading to the nearest tick mark). However, the human eye can visually interpolate in the interval between the tick marks—perhaps breaking the interval into five parts. The error due to readability may thus be only ±0.2 degree. Regardless of the spacing of the ticks, the human eye will find it difficult to discriminate differences of less than 0.01 in. (Sweeney, 1953). The manufacturer of the measuring system may well take the output resolution or readability into account in designing the device—the resolution may, in fact, reflect the accuracy of the device. It is pointless to be able to resolve a reading to an interval that is smaller than the uncertainty interval of the measurement.

Repeatability is the ability of a device to produce the same output reading when the same measurand is applied using the same procedure. Inability to repeat a measurement exactly, a random error known as *repeatability error*, is usually a component of the manufacturers' specification of instrument accuracy. It should be noted that hysteresis is not a cause of repeatability error—hysteresis is a separate error. The concept of random error of a measurement is more general than measuring device repeatability and may include variable factors in the measurement process not caused by the measurement device, such as variation of uncontrolled parameters.

Although not a requirement for a measurement system, it is highly desirable that it have a *linear* relationship between input and output, as shown in Figure 2.6. This means that the change in output is proportional to the change in the value of the measurand. A linear response is particularly useful since it simplifies the process of calibration, or checking that the instrument has low error. If it is known that the sensor is basically linear and has good precision, only two points in the span need to be checked. A highly nonlinear device must be calibrated at several points. Deviation from true linearity when linearity is assumed is a systematic error called *linearity error*.

Linearity error is usually a component of the specification of accuracy. There are a number of ways to determine a linearity specification, and these are presented in ANSI/ISA (1979). In the method that determines *terminal-based linearity*, a line is

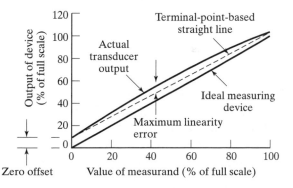

FIGURE 2.6

Example of nonlinearity and zero offset.

drawn connecting the output values at the extreme ends of the span, as shown in Figure 2.6. The linearity error is the maximum deviation between the straight line and the device output. It is normally presented in the form of a percentage of range or a percentage of span.

Measurement systems normally have a point in their range called the *zero* or *null point*. For example, a weight scale should read zero pounds when there is no weight on the platform. Most instruments have some kind of mechanism to adjust the zero, and any error in zero adjustment will affect all measurements made using the device. If the device does not have correct output at the zero point, it is said to have a *zero offset*, as shown in Figure 2.6. Furthermore, if the zero offset is not accounted for in using the device, the offset will result in a systematic error at all readings called a *zero error*. In some cases the null point does not correspond to a zero value for either the measurand or the output, but is simply a measurand value to which the device should initially be adjusted. Manufacturers' specifications of accuracy usually assume that the zero has been adjusted properly. Manufacturers may also specify the maximum expected zero error, often using the term *zero balance*. It is normal to check the zero prior to using a device, and large changes in zero may indicate that the device has been damaged or is malfunctioning.

An important characteristic of a measuring system is the *sensitivity*, defined as the ratio of the change in magnitude of the output to the change in magnitude of the measurand:

$$\text{sensitivity} = \frac{d(\text{output})}{d(\text{input})} \approx \frac{\Delta \text{output}}{\Delta \text{input}} \tag{2.1}$$

For a thermometer, this could be the change in the height of the mercury column in the stem per degree of temperature change. In mechanical measuring devices, sensitivity is an important and limiting characteristic. In systems with electrical output sensors, the sensitivity can normally be increased using simple amplifiers. However, other limits, such as signal-to-noise ratio, may then become important. In linear systems, the sensitivity is constant throughout the range and is given the symbol K. Sensitivity is determined during the process of calibration, and an error in determining the actual sensitivity results in a systematic error called *sensitivity error*, which affects all readings. An error in sensitivity will affect the span, as shown in Figure 2.7. If the span is not as specified, a *span error* will result.

Over a period of time, the output of a measuring system for a fixed measurand may change even though all environmental factors remain constant. This undesirable characteristic is known as *drift*. Many measuring systems are also sensitive to the environmental temperature, a characteristic known as the *thermal stability* of the device. Both drift and thermal stability can affect a number of characteristics of the measuring system and cause additional errors in zero, linearity, hysteresis, and sensitivity. These drift- and thermal-stability–caused errors are not usually included in manufacturers' specifications of accuracy. However, manufacturers often give additional information on drift and thermal stability of instruments, which can be used to estimate random uncertainty.

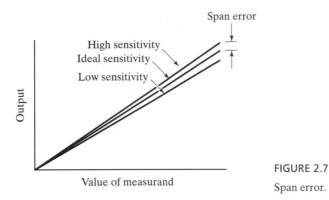

FIGURE 2.7

Span error.

Example 2.2

For a mechanical-shaft, angular-velocity measuring device (tachometer) can measure shaft speed in the range 0 to 5000 rpm. It has an accuracy of ±5% of full scale. You notice that when the shaft speed is zero, the device has a reading of 200 rpm. What is the maximum error that you might estimate in reading a shaft speed of 3500 rpm?

Solution: The accuracy specification indicates an uncertainty of $\pm 0.05 \times 5000 = \pm 250$ rpm. Thus all readings have at least this uncertainty. However, there is a zero offset of 200 rpm. This error is in addition to the accuracy uncertainty. Thus the reading might be as much as $250 + 200 = 450$ rpm high. The error estimate can be reduced if the zero is adjusted or if the data are corrected (by subtracting 200 rpm from the reading). If it has been some time since the instrument was calibrated, there might be an additional error in the sensitivity, but this cannot be determined with the information given.

Comment: Accuracy is an uncertainty specification that usually combines unavoidable systematic and random errors associated with using a measuring device. It typically includes hysteresis, linearity, and repeatability error components. It usually does not include other errors, such as zero, drift, and thermal stability. Errors of these types must be considered separately.

2.2.2   Calibration of Measurement Systems

At some point, all measurement systems should undergo *calibration*, a process in which a set of measurements are made of measurand values that can be determined independently. The readings can then be compared to the known "true" values and the errors determined. The number of different values of measurand required for the calibration process depends on the type of measurement system and the application. In some calibration processes, the value of a measurand is known because it is a *standard*. In other calibration processes, another measurement system of known accuracy can be used to determine the value of the measurand. The use of standards is the more reliable approach, but the latter approach is often more practical.

Standards for Calibration   Standards of measurement have been important to commerce for a very long time since it is important to the purchaser to know that the

TABLE 2.1   SI Base Units

Physical variable	SI unit
Mass	kg
Time	second
Length	meter
Temperature	kelvin
Electric current	ampere
Amount of a substance (mole)	mole
Light intensity	candela

weight or length of a purchase is accurate. The relatively recent expansion in science dramatically increased the need for standards. International conferences have established *primary standards* defining the units for seven physical variables. These units are known as the International System (SI) base units (Table 2.1).

The standard for *mass* is the International Prototype Kilogram, which is a platinum–iridium cylinder kept at the International Bureau of Weights and Measures in France.

The standard for *time*, the second, has been defined as: "the duration of 9,192,631,770 periods of the radiation corresponding to the transition between the two hyperfine levels of the ground state of the cesium-133 atom" (Wildhack and Jennings, 1992). Although this may seem obscure, it is a standard reproducible in any properly equipped laboratory.

The standard for *length*, the standard meter, is defined as "the length of the path traveled by light in a vacuum during a time of 1/299,792,458 of a second" (Wildhack and Jennings, 1992).

The standard for *temperature* is more complicated than for the other base units since it must be specified over a wide range of values. The standard is known as the International Temperature Scale of 1990 (ITS-90) (Preston–Thomas, 1990). The measure of temperature is the thermodynamic temperature, and the unit is the kelvin, defined as 1/273.16 of the thermodynamic temperature of the triple point of water, that temperature where solid, liquid, and vapor phases of pure water exist together in thermal equilibrium. The standard temperature scale extends from 0.65 K to the highest temperatures that can be determined by measuring thermal radiation. While the details of the complete standard are beyond the scope of the present book, in the range of greatest interest to engineer, the standard is fairly straightforward. From the triple point of hydrogen, 13.8033 K, to the freezing point of silver, 961.78 K, the scale is defined by means of a platinum resistance thermometer (see Chapter 9), which is calibrated at a set of fixed points shown in Table 2.2. Above the freezing point of silver, the temperature standard is based on a relationship between the thermal radiation from an object at the measured temperature to the thermal radiation from an object at the temperature of freezing silver, gold, or copper. Guidelines for using the ITS-90 standard are given by Mangum and Furukawa (1990).

The standard for *electric current*, the ampere, is "that constant current which, if maintained in two straight parallel conductors of infinite length and of negligible circular sections, and placed 1 meter apart in a vacuum would produce a force equal to 2×10^{-7} newton per meter of length" (Wildhack and Jennings, 1992).

TABLE 2.2   Fixed Points of International Temperature
Scale of 1990 (ITS-90)

Fixed point[a]	Temperature (K)
Triple point of hydrogen	13.8033
Triple point of neon	24.5561
Triple point of oxygen	54.3584
Triple point of argon	83.8058
Triple point of mercury	234.3156
Triple point of water	273.16
Melting point of gallium	302.9146
Freezing point of lanthanum	429.7485
Freezing point of tin	505.078
Freezing point of zinc	692.677
Freezing point of aluminum	933.473
Freezing point of silver	1234.93
Freezing point of gold	1337.33
Freezing point of copper	1357.77

Source: Preston–Thomas (1990).

[a]Melting and freezing points are at a pressure of 101.325 Pa.

The standard for the *mole* is "the amount of a substance of a system which contains as many elementary entities as there are atoms in 0.012 kg of carbon-12" (Wildhack and Jennings, 1992).

The standard for *light intensity*, the candela, is "the luminous intensity, in a given direction, of a source that emits monochromatic radiation of frequency 540×10^{12} hertz and of which the radiant intensity in that direction is 1/683 watt per steradian" (Wildhack and Jennings, 1992).

All the primary standards except mass can, in theory, be reproduced in any laboratory having the proper equipment. They are called *reproducible standards*. Mass, the only standard that has an exact physical location, is called a *fixed standard*. Although all primary standards except mass can, in principle, be reproduced in any laboratory, this is usually not practical, and the primary standards are difficult to apply for everyday calibration activities. As a result, it is normal to create *secondary standards*. These standards must be traceable to the primary standards—that is, at some point, the secondary standard was compared either to a primary standard or to another secondary standard that is traceable to a primary standard. Laboratories can have masses, accurately sized pieces of metal called gage blocks, quartz crystal clocks, and other practical secondary standards for calibration purposes.

The foregoing list of base primary standards does not include most of the common variables for which calibration is required (force and voltage are obvious examples). This is because standards for all other physical variables are derived from variables having the base primary standards by using physical laws or scientific definitions. The standard for force, for example, is defined by Newton's second law, $F = ma$. Since primary standards have been established for mass, length, and time, the standard for force can be computed. Laboratory secondary standards can also be created for these other variables—a standard battery for voltage is an example. Standard names and definitions are provided for virtually all derived units, and these are documented by the International Organization for Standardization (ISO, 1979).

Some calibration processes are quite difficult, and it is not practical for all laboratories to keep all necessary standards. Laboratories may purchase instruments that have been calibrated by the manufacturers. If it is expected that the uncertainty of a measuring device will change with time, it may have to be returned to the manufacturer (or another laboratory) periodically for calibration. Most laboratories can afford to have some secondary working standards. For example, standards for mass and length can be purchased at moderate cost and, with reasonable care, are very durable.

Static Calibration Process   In the calibration of a measuring system, the system is used to make measurements of known values of the measurand. Various values of the measurand must be used to cover the intended range of use of the measurement system. As mentioned, it is best that the values of the measurand be known because they are standards, while from a practical standpoint, however, it is often easier to determine the values of some measurands using another measuring system of known accuracy. It is preferable that this second measuring system be based on simple physical principles so that the experimenter will have confidence that it maintains its accuracy. For example, a balance scale for measuring force is more likely to maintain its accuracy than a scale that uses springs.

The first step in the calibration process is to take a set of data consisting of measurement system output as a function of the measurand. A graphical presentation of these data is known as the *calibration curve*. Although this curve can be used directly to interpret the instrument output, in most cases the data are correlated using a mathematical function (the process of curve fitting). In some cases, the calibration data will be used to confirm an existing function. The correlating function may be a straight line, a parabola, a higher-order polynomial, or a more complicated function. This function is used to obtain values of the measurand from the measuring device output in the actual experiment. Finally, an estimate should be made of the calibrated system's contribution to the overall experimental uncertainty. An estimate should be made of residual systematic error, and, depending on the expected use of the measurement system, it may also be necessary to estimate the expected random error.

The document ANSI/ISA (1979) describes a standardized procedure for static calibration that is used by many instrument manufacturers. In the first step of this procedure, the calibration data are taken by applying known values of the measurand to the measuring system. The measurand values are incrementally increased from the low end of the range to the top end of the range. The measurand values are then incrementally reduced to the low end of the range. This process is repeated several times (cycles). The procedure then describes how to determine specific values for such errors as linearity, hysteresis, and repeatability as well as overall accuracy. In this procedure, the accuracy statement effectively combines the errors due to nonlinearity, hysteresis, and nonrepeatability. The term *static* indicates that the process is performed in a manner in which time is not a factor—the measurand is changed slowly, and the device is allowed to come to equilibrium prior to taking a reading.

The sequential calibration process just described does not, in general, duplicate the actual measurement process. In a typical measurement situation, the experimenter may not be able to determine if the measurand is increasing or decreasing prior to the measurement. A calibration process in which the values of the measurands are randomly selected is often used. With a random calibration, it may be difficult to separate hysteresis errors from repeatability errors, and hysteresis may have to be treated as part of the random error. This limitation may not be important.

Example 2.3, based on information in ANSI/ISA (1979), is used to demonstrate the static calibration process to determine instrument accuracy and other errors. This process will not determine thermal stability and drift characteristics. Furthermore, it will not account for errors associated with the application, such as spatial errors, nor will it account for dynamic (time-varying) effects. Additional calibration procedures may be required for specific experiments. As discussed in Chapter 7, the error characteristics determined by the method shown in Example 2.3 are in a usable but not ideal form for detailed uncertainty analysis.

Example 2.3   Calibration of a Weighing Scale

A low-cost, nominally 0 to 5-lb spring weighing scale [Figure E2.3(a)] has been calibrated by placing accurate weights on its platform. The values of the applied weights range from 0 to 5 lb in 0.5-lb increments. The weights are applied in a sequential manner, starting at the lowest value, increasing to the largest value (up data) and then decreasing to the lowest value (down data). Five such cycles were performed, and the results of the measurements are presented in Table E2.3(a). As suggested in ANSI/ISA (1979), several cycles were completed before the data recording started. The data recording then started in the middle of the up portion of cycle 1 and ended in the up portion of cycle 6, giving five complete cycles.

Fit a straight line to the data and determine the accuracy, hysteresis, and linearity errors. Also, make estimates of the maximum systematic and random errors.

Solution: The data of Table E2.3(a), which have been plotted in Figure E2.3(b), fall into two bands. The division into two bands is caused by hysteresis in the system—the lower band is for increasing measurand and the upper band for decreasing measurand.

There are a number of methods used to determine a suitable straight-line fit to a set of instrument data. A technique called the method of least squares (linear regression) will be described in Chapter 6. At this point, it is convenient simply to "eyeball" a line through the data to minimize the maximum deviations of the data from the line. This process will approximate a least-squares linear fit. The resulting correlating equation takes the form

$$R = 1.290W - 0.374$$

where R is the reading of the scale and W is the actual weight. Alternatively, we might have given the equation in the form of the weight versus the reading ($W = 0.775R + 0.290$).

FIGURE E2.3(a)

Spring scale.

TABLE E2.3(a)   Scale Calibration Data

True weight (lb)	Scale reading					
	Cycle 1	Cycle 2	Cycle 3	Cycle 4	Cycle 5	Cycle 6
0.5		0.2	0.08	0.17	0.19	0.11
1		0.7	0.78	0.64	0.61	0.7
1.5		1.18	1.26	1.25	1.24	1.23
2		1.81	1.93	1.81	1.93	1.88
2.5	2.62	2.49	2.46	2.46	2.58	2.53
3	3.15	3.18	3.24	3.28	3.13	
3.5	3.9	3.84	3.86	3.97	3.96	
4	4.59	4.71	4.61	4.6	4.6	
4.5	5.41	5.35	5.49	5.46	5.39	
5	6.24	6.27	6.1	6.24	6.16	
4.5	5.71	5.74	5.78	5.87	5.82	
4	4.96	5.11	5.08	5.03	5.03	
3.5	4.22	4.34	4.21	4.22	4.24	
3	3.57	3.64	3.66	3.55	3.67	
2.5	2.98	2.86	2.98	2.98	2.94	
2	2.22	2.23	2.26	2.29	2.26	
1.5	1.57	1.7	1.69	1.63	1.57	
1	1.07	1.07	1.11	1.16	1.11	
0.5	0.52	0.61	0.61	0.61	0.45	
0	0.02	0.08	0.08	−0.03	0.06	

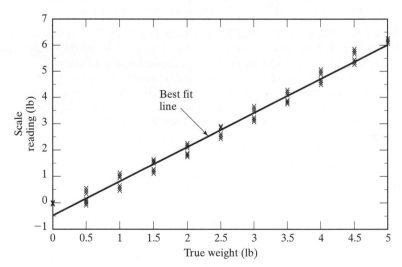

FIGURE E2.3(b)

Plot of scale readings.

Although Figure E2.3(b) is suitable for determining the correlating function, to evaluate the errors it is best to present the data in the form of what is called a *deviation plot*. For each measured value, the difference (deviation) between the measured value and the best-fit equation is evaluated. These values are presented in Table E2.3(b). The last three columns of the table will be discussed later. The deviation data are plotted in Figure E2.3(c). A deviation value of zero indicates that a calibration data point has a value exactly as predicted by the correlating function.

TABLE E2.3(b) Scale Deviation Data

True weight (lb)	Cycle 1	Cycle 2	Cycle 3	Cycle 4	Cycle 5	Cycle 6	Average of cycles	Average up–down	Repeat
							Deviation		
0								0.41	
0.5		−0.07	−0.19	−0.1	−0.08	−0.16	−0.12	0.085	0.12
1		−0.22	−0.14	−0.28	−0.31	−0.22	−0.23	−0.025	0.17
1.5		−0.38	−0.3	−0.31	−0.32	−0.33	−0.33	−0.13	0.08
2		−0.4	−0.28	−0.4	−0.28	−0.33	−0.34	−0.15	0.12
2.5	−0.23	−0.36	−0.39	−0.39	−0.27	−0.32	−0.35	−0.125	0.16
3	−0.35	−0.32	−0.26	−0.22	−0.37		−0.3	−0.09	0.15
3.5	−0.24	−0.3	−0.28	−0.17	−0.18		−0.23	−0.06	0.13
4	−0.2	−0.08	−0.18	−0.19	−0.19		−0.17	0.04	0.12
4.5	−0.02	−0.08	0.06	0.03	−0.04		−0.01	0.17	0.14
5	0.16	0.19	0.02	0.16	0.08		0.12	0.12	0.17
4.5	0.28	0.31	0.35	0.44	0.39		0.35		0.16
4	0.17	0.32	0.29	0.24	0.24		0.25		0.15
3.5	0.08	0.2	0.07	0.08	0.1		0.11		0.13
3	0.07	0.14	0.16	0.05	0.17		0.12		0.12
2.5	0.13	0.01	0.13	0.13	0.09		0.1		0.12
2	0.01	0.02	0.05	0.08	0.05		0.04		0.07
1.5	0.01	0.14	0.13	0.07	0.01		0.07		0.13
1	0.15	0.15	0.19	0.24	0.19		0.18		0.09
0.5	0.25	0.34	0.34	0.34	0.18		0.29		0.16
0	0.39	0.45	0.45	0.34	0.43		0.41		0.11

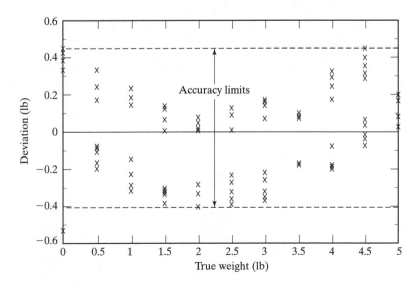

FIGURE E2.3(c)

Plot of deviation data for scale calibration.

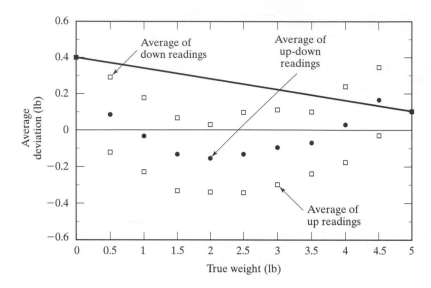

FIGURE E2.3(d)

Average deviation data for scale calibration.

Using Figure E2.3(c), we find that it is a simple matter to estimate the accuracy of this weighting device. Two lines are drawn, parallel to the horizontal axis, such that all the data are contained between the two lines. For this device, the accuracy limits are +0.45 lb and −0.40 lb. Accuracy is frequently presented as a percent of output span. In this case, the output, as predicted by the curve fit, varies from −0.374 to +6.076, giving a span of 6.076 − (−0.374) = 6.45 lb. Accuracy then becomes +7.0% and −6.2% of the output span. It should be noted that accuracy is a bounding error statement—it includes all residual errors that will occur when the instrument is used with the given linear fit to the data. Accuracy does not include errors due to drift or thermal-stability effects, which require additional calibration procedures.

To evaluate the nonlinearity of the data, it is necessary to average all the data taken for each value of the weight. Not only are the values for all the cycles at a given weight averaged, but the "up" and the "down" values are also averaged. These results are presented in the column labeled "Average up–down" in Table E2.3(b). These data are plotted with the solid symbols in Figure E2.3(d). A line has been drawn connecting the terminal points of the data, and the terminal linearity can be evaluated as 0.44 lb, which is about 6.9% of output span. There are other measures of linearity, as described in ANSI/ISA (1979) and Norton (1982).

The repeatability error is the maximum variability of successive measurements of the same value of input approached from the same direction. The column of Table E2.3(b) labeled "Repeat" is this variability for each of the test conditions. The maximum variation, 0.17 lb, occurs at an "up" reading of 1 lb. This corresponds to 2.6% (or ±1.3%) of output span.

ANSI/ISA (1979) indicates that the hysteresis error is evaluated as the maximum difference between the "up" and the corresponding "down" reading for any of the calibration cycles. Evaluating this difference for all the data in Table E2.3(b), the maximum difference is found to be 0.52 lb and occurs in cycles 3 and 4 at 2.5 lb. This becomes 8.1% (or ±4.05%) of output span. Technically, this is not simply the hysteresis error but a combination of the hysteresis error plus another error called *dead band*. These two errors are frequently combined (and often simply called hysteresis error). The interested reader can find a description of the distinction between these two errors in ANSI/ISA (1979).

At this point, it is useful to estimate the total uncertainty due to systematic error and that due to random error, a process not considered in ANSI/ISA (1979). Systematic error is the error that can be expected if a large number of readings are taken of a particular value of the measurand using the same procedure and then averaged. Table E2.3(b) shows these results in the column labeled "Average of cycles," and the data are plotted in Figure E2.3(d). Note that the "up" and "down" values are treated as distinct data. An estimate of the maximum systematic error would be the maximum deviations of these averaged calibration data. The resulting error limits are $+0.41$ lb and -0.35 lb, which become $+6.4\%$ and -5.4% of output span. Systematic errors are discussed further in Chapter 7.

Statistical techniques provide the best approach to evaluating random uncertainty. However, an estimate of the random uncertainty is the repeatability error. The repeatability error limits are $\pm1.3\%$ (see above). Better techniques to evaluate the random error are developed in Chapters 6 and 7. In addition, factors not considered in this calibration procedure (such as uncontrolled changes in ambient temperature) can also contribute to random uncertainty.

Comments: The accuracy numbers determined by the process described above can be used to estimate the maximum uncertainty in measurements made with the scale. The accuracy values $(+0.45/-0.40$ lb) apply to the device output when a given weight is placed on the platform. Using the correlating function, it is a simple matter not only to determine the weight corresponding to a given output reading but also the uncertainty in that weight. It is left for the reader to show that for a given reading, the uncertainty in the applied weight will be $+0.35/-0.31$ lb.

There are techniques by which the uncertainty in weight values determined with this scale might be reduced. First, a nonlinear function such as a parabola might be used to curve-fit the data. Second, the uncertainty interval can be made a function of the reading, resulting in a reduction in the uncertainty interval for some portions of the range. However, hysteresis is a major cause of the poor accuracy, and this cannot be reduced by calibration procedures.

The type of calibration described in Example 2.2 is a *static calibration*. However, many measuring systems are used in situations in which the measurand is changing rapidly. For such situations, a static calibration is inadequate. If possible, the instrument can be subjected to a *dynamic calibration*. Dynamic calibration is usually a more complicated process and the procedure depends on the device being calibrated and often, the specific application. In many cases, dynamic response is determined entirely by analysis.

2.3   DYNAMIC MEASUREMENTS

If a measurand is unchanging in time and if the measurement system instantaneously shows an equilibrium response to the measurand, the measurement process is said to be *static*. In the general case, however, when the measurand is changing in time and the measuring system does not show instantaneous response, the measurement process is said to be *dynamic*. For example, the use of an oral thermometer to measure a person's body temperature is a dynamic measurement since the measurement process must be continued for several minutes for the thermometer to come into equilibrium with body temperature. In making dynamic measurements, the experimenter must account for the dynamic characteristics of the measuring system, the dynamic interaction between the measuring system and the test system, and the dynamic characteristics of the test system.

When a measurement is dynamic, there is usually an error introduced into the measurement, and the experimenter must take actions to minimize this error. For example,

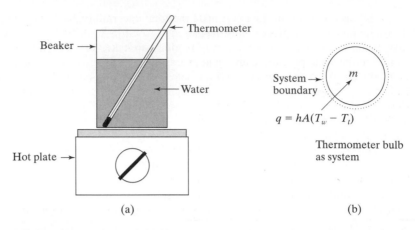

FIGURE 2.8

Dynamic temperature measurement.

consider the measurement of the temperature of the water in a beaker that is being heated on a hot plate as shown in Figure 2.8(a). Considering the bulb of the thermometer as a system [Figure 2.8(b)], we can apply the first law of thermodynamics. Since there is no work term, we obtain

$$q = hA(T_w - T_t) = mc\frac{dT_t}{dt}$$

or

$$(T_w - T_t) = \frac{mc}{hA}\frac{dT_t}{dt} \tag{2.2}$$

where Newton's law of cooling $[q = hA(T_w - T_t)]$ has been used to estimate q, and the symbols are defined as follows:

T_w water temperature
T_t thermometer temperature
h heat transfer coefficient
m bulb mass
c bulb specific heat
dT_t/dt time rate of change of the water temperature

There would be no error in the measurement if $T_w = T_t$. However, if the water is being heated, dT_t/dt is nonzero, so $T_w - T_t$ is nonzero and there exists an inherent dynamic measurement error. The term mc/hA, called the *time constant* of the system (defined below), has dimensions of time and is usually given the symbol τ. To minimize the dynamic error for this situation, τ should be made as small as is practical.

The dynamic response of a measurement system can usually be placed into one of three categories: zero order, first order, and second order. These categories are based on the order of the differential equation needed to describe the dynamic response. Ideal *zero-order* systems respond instantly to measurands, although no instrument is truly zero order. However, the dynamic response of many instruments approximate zero-order behavior when measuring slowly changing measurands. *First-order* measurement systems show capacitance-type energy storage effects. Mechanical analogs to capacitance are springs and devices that store thermal energy. The common thermometer discussed above is an example of a first-order system. *Second-order* systems have inertial effects of inductance or accelerated mass as well as capacitance energy storage. Common spring–mass systems are second order—the mechanical bathroom scale is an example. Second-order systems include a characteristic called damping, which dissipates energy. Second-order systems with low damping are called *underdamped* and can show oscillatory response. Highly damped second-order systems are called *overdamped* and will not show oscillatory behavior. The level of damping that divides these two modes of response is called *critical damping*.

There are specifications for the dynamic response of an instrument that can be used to select an appropriate device. The two most common specifications characterize the response of the measurement system to a step change in input (called *transient response*) and the response to a range of sinusoidal inputs (called *frequency response*). A step change in input to a measuring system is shown in Figure 2.9(a), and some typical system responses are shown in Figure 2.9(b). y is the change in the device output, and y_e is the equilibrium change in the output of the system, the output that will occur after significant time has passed. Response A is typical of first-order and overdamped second-order devices. Response B is typical of underdamped second-order systems.

First-order systems and overdamped second-order systems show an asymptotic response. For first-order systems, this takes an exponential form:

$$\frac{y}{y_e} = 1 - e^{-t/\tau} \tag{2.3}$$

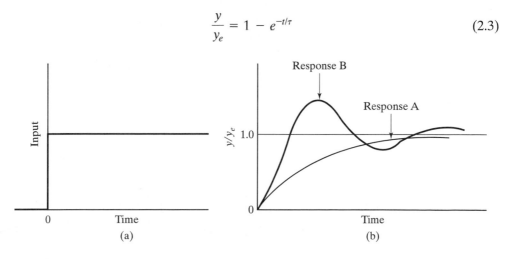

FIGURE 2.9

Application of a step change in measurand; (a) time variation of measurand; (b) typical types of system responses.

The time constant, τ, thus determines the curve and is a useful numerical specification of the transient response of the instrument. It is the time at which the response, y/y_e, has a value of $1 - 1/e = 0.632$. Overdamped second-order systems show a response which is similar to, but somewhat more complicated than, that predicted by Eq. (2.3). As a result, the concept of time constant is not exactly applicable to second-order systems. On the other hand, *response time* is a more general term used to specify transient response and can be used for first- and second-order systems. For example, a 95% response time is the time at which y/y_e has a value of 0.95. Another measure of transient response is the *rise time*. This is usually the time that it takes y/y_e to increase from 0.1 to 0.9.

For oscillatory responses [response B in Figure 2.9(b)], another term, called *settling time*, is more useful. This is the time until the amplitude of the oscillations are less than some fraction (e.g., 10%) of the equilibrium response.

Generally speaking, for minimal dynamic error, the appropriate measure of transient response (e.g., response time or settling time) should be small compared to the expected time for the measurand to change in an experiment. If the measurand changes in a time comparable to the transient response time, there will be a significant dynamic measurement error.

Frequency response is another useful measure of dynamic response. The output of the system is determined for a pure sine-wave input. This process is repeated for a range of frequencies to produce a frequency-response curve such as that shown in Figure 2.10. As shown in Figure 2.10, there is the range of sinusoidal input frequencies over which the measuring system gives a constant ratio of output amplitude to input amplitude. This range of frequencies is called the *bandwidth* of the device. If the device is used for measurements outside the bandwidth, the ratio of output to input amplitudes will change and significant systematic errors are likely to occur. Most input signals are not sinusoidal and are rather complicated functions of time. It is often necessary to decompose complicated waveforms into sinusoidal components, and these components can be used to determine the required frequency response of the measuring system.

In some cases, the frequency response of a second-order system is specified with a single number called the *natural frequency*. In the spring scale example, if a weight is simply dropped on the platform, the device will oscillate for a period of time at the natural frequency. In most cases, dynamic response will be degraded if the measurand varies at a frequency greater than 0.2 to 0.4 times the natural frequency. There are, however, some devices that operate correctly only at frequencies greater than the natural frequency.

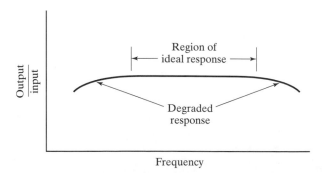

FIGURE 2.10

System frequency response.

Manufacturers' specifications of dynamic response are usually available and can be used for many components of measuring systems. Unfortunately, for some sensing elements and transducers, it is not possible for the manufacturer to supply all the needed dynamic specifications. This is because the dynamic response of the transducer depends not only on the characteristics of the transducer itself but also on the manner in which the transducer is used. For example, the rise time of a thermometer used to measure the temperature of a fluid depends on the properties and motion of the fluid as well as the thermometer itself. Chapter 11 presents appropriate analytical approaches for dynamic analysis of some common measuring systems.

2.4 CLOSURE

In this chapter, we have described the general characteristics of common measuring devices. Common static errors have been described and guidance given on dynamic instrument specifications.

REFERENCES

[1] ANDERSON, N. (1980). *Instrumentation for Process Measurement and Control*, Chilton, Radnor, PA.

[2] ANSI/ISA (1975). *Electrical Transducer Nomenclature and Terminology*, ANSI Standard MC6.1 (ISA S37.1).

[3] ANSI/ISA (1979). *Process Instrumentation Terminology*, ISA Standard S51.1.

[4] BARNEY, G. (1988). *Intelligent Instrumentation*, Prentice Hall, Englewood Cliffs, NJ.

[5] BASS, H. G. (1971). *Introduction to Engineering Measurements*, 5th ed., McGraw-Hill, New York.

[6] BECKWITH, T., MARANGONI, R., AND LIENHARD, V. (1993). *Mechanical Measurements*, Addison-Wesley, Reading, MA.

[7] DALLY, J., RILEY, W., AND MCCONNELL, K. (1993). *Instrumentation for Engineering Measurements*, 2d ed., Wiley, New York.

[8] DOEBELIN, E. O. (1990). *Measurement Systems Application and Design*, 4th ed., McGraw-Hill, New York.

[9] HOLMAN, J. (2001). *Experimental Methods for Engineers*, 6th ed., McGraw-Hill, New York.

[10] ISO (1979). *Units of Measurement*, ISO Standards Handbook 2.

[11] MANGUM, B., AND FURUKAWA, G. (1990). *Guidelines for Realizing the International Temperature Scale of 1990 (ITS-90)*, National Institute of Standards and Technology Technical Note 1265.

[12] NORTON, H. (1982). *Sensor and Analyzer Handbook*, Prentice Hall, Englewood Cliffs, NJ.

[13] PRESTON-THOMAS, H. (1990). The International Temperature Scale of 1990 (ITS-90), *Metrologia*, Vol. 27, pp. 3–10.

[14] SWEENEY, R. (1953). *Measurement Techniques in Mechanical Engineering*, Wiley, New York.

[15] THOMPSON, L. M. (1979). *Basic Electrical Measurements and Calibration*, Instrument Society of America, Research Triangle Park, NC.

[16] TURNER, J. D. (1988). *Instrumentation for Engineers*, Springer-Verlag, New York.

[17] WILDHACK, A., AND JENNINGS, D. (1992). Physical measurement, in *McGraw-Hill Encyclopedia of Science and Technology*, McGraw-Hill, New York.

[18] WOLF, S. (1983). *Guide to Electronic Measurements and Laboratory Practice*, Prentice Hall, Englewood Cliffs, NJ.

PROBLEMS

2.1 Identify the sensing element, signal-modification element, and indicator or recorder element in the following measurement devices. You may need to research the principles of operation to answer the questions.

(a) A mechanical automobile speedometer.

(b) An automobile fuel gage.

(c) A human body thermometer with a liquid crystal display.

2.2 You find a micrometer (a thickness-measuring device) of unknown origin and use it to measure the diameter of a steel rod that is known to have a diameter of 0.5000 in. You use the micrometer to make 10 independent measurements of the rod diameter, and the results are 0.4821, 0.4824, 0.4821, 0.4821, 0.4820, 0.4822, 0.4821, 0.4822, 0.4820, and 0.4822. Estimate the systematic error and the maximum random error in these measurements.

2.3 You attempt to determine the validity of a bathroom scale by repeatedly placing 20 lb of accurate weights on it. Ten readings were obtained with values of 20.2, 20.2, 20.6, 20.0, 20.4, 20.2, 20.0, 20.6, 20.0, and 20.2 lb. Estimate the systematic error and the maximum random error of the measurements.

2.4 Identify which of the following measurements are intrusive and which are nonintrusive. Justify your answers.

(a) Measuring a person's oral temperature with a thermometer.

(b) Measuring the speed of a bullet using high-speed photography.

(c) Determining the temperature of a furnace by an optical thermal-radiation device.

(d) Measuring the speed of an automobile with a radar gun.

2.5 Determine whether the following measurements are intrusive or nonintrusive. You may need to research the details of the measurement process.

(a) Measuring amperage of current in a wire using clamp-on ammeter.

(b) Measuring flow of fluid in a pipe by installing an orifice meter in the pipe.

(c) Measuring composition of gases in an exhaust using a device which optically measures transmitted infrared radiation.

(d) Determining the surface temperature of a pipe using a thermometer that measures the emitted infrared radiation.

(e) Rotational speed of a shaft indicated by a strobotachometer.

2.6 Determine if the following errors are of random or systematic type. Justify your response.

(a) A digital scale, that always shows 0.2 lb when no weight is applied.

(b) Vibration of the needle of an automobile speedometer.

(c) Consistent temperature difference between two sensors reading the air temperature in the same room.

2.7 Determine if the following errors are systematic or random. Justify your response.

(a) Effect of temperature on the circuitry of an electronic measurement device.

(b) Effect of parallax on the reading of a needle-type analog voltmeter.

(c) Effect of using an incorrect value of emissivity in the readings of an infrared thermometer.

2.8 Determine if the following errors are systematic or random. Justify your response.

    (a) Variation in a pressure transducer reading due to variation in ambient temperature.

    (b) Noise induced into the output of an instrument due to magnetic fields in the surroundings.

    (c) Variation in the makeup of a drug due to a malfunction of a mixer in a drug mixing tank.

2.9 A digital device has a decimal indicator showing up to three significant figures. An equivalent analog device has 1000 divisions over the span. Can we determine which instrument is more accurate? Explain.

2.10 A velocity-measuring device can measure velocities in the range 0 to 50 m/s. What is the span of this device?

2.11 You need to measure a pressure, which has a value between 60 kPa and 100 kPa. Four pressure-measuring devices of comparable quality are available:

Device A, range 0–100 kPa

Device B, range 0–150 kPa

Device C, range 50–100 kPa

Device D, range 50–150 kPa

Which device would you choose? Explain your answer.

2.12 A digital output voltmeter has an input range of 0 to 30 V and displays three significant figures XX.X. The manufacturer claims an accuracy of ±2% of full scale. With a voltage reading of 5 V, what are the percent uncertainties of the reading due to accuracy and resolution?

2.13 Digital voltmeters often have a choice of ranges. The ranges indicated on a typical voltmeter are 0–3, 0–30, 0–300, and 0–3000 AC volts. The output is represented with four significant digits. Determine the following:

    (a) Resolution uncertainty for each range (in V).

    (b) If it has an accuracy of ±2% of full range for each range, determine the absolute uncertainty of measurement in each case.

    (c) Determine the relative (percent of reading) uncertainly if, for a measurement of 25 volts, the ranges 30, 300, or 3000 wcrc uscd.

2.14 A bourdon gage (a mechanical device to measure gage pressure, the pressure relative to atmospheric pressure), which has a range of 0 to 50 psi, reads +0.5 psi when measuring atmospheric pressure. It is claimed to have an accuracy of ±0.2% of full-scale reading. What is the expected error in the measurement of 20 psi in psi and in percent of reading? How can you reduce the error produced by this gage?

2.15 A voltmeter with a range of 0 to 100 V reads 2 V when the leads are shorted together. The manufacturer claims an accuracy of ±4% of full scale. Estimate the maximum error when reading a voltage of 80 V in both volts and as a percentage of reading. If the voltmeter is adjusted so that the reading when the leads are shorted together is 0 V, estimate the maximum percent error when reading 80 V.

2.16 A pressure sensing device has an electrical output and is approximately linear. When the input is 100 kPa, the output is 5 mV and when the output is 1000 kPa, the output is 125 mV. What is the sensitivity of this device?

2.17 In a gas thermometer, following the perfect gas law ($PV = mRT$), pressure change is an indication of temperature variation if the volume is constant. Determine if an increase in initial filling pressure will change the sensitivity of the device.

2.18 Figure P2.18 shows the variation of the output of a sensor with respect to its input.

    (a) How does the sensitivity of the sensor vary from A to C?

    (b) Would you recommend using the device in the A–B or B–C range?

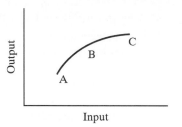

FIGURE P2.18

2.19 In a multirange instrument (such as a multirange voltmeter), how does sensitivity vary with range? Explain.

2.20 In an analog output device, which has a voltage output, how can you increase the sensitivity of the device?

2.21 A common fluid-flow measuring device is called a venturi tube. (See Chapter 10.) For this device, the fluid-flow rate, Q, is proportional to the square root of the measured pressure drop, ΔP, ($Q = C(\Delta P)^{1/2}$ where C is approximately constant).

    (a) Derive a relation for sensitivity of Q with respect to ΔP.

    (b) Determine if the sensitivity increases or decreases with an increase in the pressure drop.

    (c) Would you recommend such a device for larger or smaller values of the pressure drop?

2.22 A linear device converts an input force into an output displacement. The ideal sensitivity of this device is 0.1 cm/N and the input span is 200 N. What will be the ideal output span? What will the output span be if the actual sensitivity is 0.105 cm/N? What will the % error in the output span be?

2.23 A static calibration is performed on a pressure-measuring device called a bourdon gage with a nominal range of 0 to 100 psi. The results of this calibration are shown in Table P2.23.

    (a) Plot the data and fit a straight line through them.

    (b) Prepare a deviation plot and estimate the accuracy and repeatability errors, both as a percentage of the output span.

TABLE P2.23

True pressure (psi)	Measured pressures					
	Cycle 1	Cycle 2	Cycle 3	Cycle 4	Cycle 5	(max.−min.)
20	20	19	20	20	19	1
40	39	40	39	39	39	1
60	59	58	59	58	60	2
80	80	80	79	79	80	1
100	101	100	100	101	102	2
80	84	83	84	84	84	1
60	63	63	63	62	62	1
40	43	42	43	43	44	2
20	24	24	24	24	24	0
0	5	4	4	6	4	2

TABLE P2.24

True pressure (kPa)	Measured pressures					
	Cycle 1	Cycle 2	Cycle 3	Cycle 4	Cycle 5	(max.−min.)
200	192	191	194	193	192	3
400	391	389	391	392	393	4
600	596	597	597	593	596	4
800	807	806	805	803	805	4
1000	1022	1022	1021	1022	1020	3
800	816	814	816	816	817	3
600	606	603	603	603	604	3
400	399	403	402	401	403	4
200	203	201	202	200	205	5
0	10	12	9	11	10	3

2.24 A static calibration is performed on a bourdon gage pressure measuring device with a nominal range of 0 to 1 MPa. The results of this calibration are shown in Table P2.24.

(a) Plot the data and fit a straight line through them.

(b) Using deviation plots, estimate the accuracy and repeatability errors, both as a percentage of the output span.

2.25 A force measuring device called a load cell has an electrical voltage output. A load cell with a range of 0 to 100 N has a nominal full-scale output of 40 mV. By using accurate weights, this load cell has been calibrated and the results are presented in Table P2.25.

(a) Plot the results and fit a straight line to the data. Find the formula for this straight line in both the form force $= f(mV$ output$)$ and the form mV output $= g($force$)$.

(b) Compute the deviations between the data and the correlating straight line. Prepare a deviation plot and determine the accuracy and repeatability of the load cell.

2.26 What is the difference between static and dynamic calibration? What type of calibration would you recommend for

(a) an oral thermometer,

(b) a pressure gage used in a water line, and

(c) a car speedometer? Explain your reasoning.

TABLE P2.25

True force (N)	Measured Output (mV)					
	Cycle 1	Cycle 2	Cycle 3	Cycle 4	Cycle 5	(max.−min.)
20	7.84	7.93	7.96	7.83	7.89	0.13
40	15.96	16.00	15.91	15.85	15.93	0.15
60	24.07	24.04	24.16	24.17	24.04	0.13
80	32.33	32.38	32.42	32.30	32.44	0.14
100	40.79	40.82	40.86	40.72	40.83	0.14
80	32.31	32.29	32.31	32.37	32.40	0.11
60	24.16	24.10	24.03	24.13	24.06	0.13
40	15.97	15.87	15.95	15.89	15.94	0.11
20	7.82	7.87	7.98	7.87	7.84	0.16
0	0.01	−0.02	0.08	0.05	−0.07	0.15

2.27 The time constant of a temperature-measuring device is 0.5 s. This device is used to measure the temperature of the air in a room that is changing at about 5°C/h. Do you consider this to be a static or a dynamic measurement? Why?

2.28 Two thermometers, one larger than the other, are dipped in ice water from ambient temperature. Which one will reach the equilibrium temperature faster? Why?

2.29 Explain the usefulness and appropriateness of the concepts of time constant, response time, rise time, and settling time for

   (a) zero-order,
   (b) first-order,
   (c) overdamped second-order, and
   (d) underdamped second-order systems.

2.30 Answer the following questions:

   (a) High-intensity discharge (HID) lamps come to full brightness in about 5–10 minutes. A light meter, which can detect instant light level, is used to record the light output of an HID lamp. Is this a static or dynamic measurement? Explain.

   (b) The pressure inside a car cylinder cycles at 1500 times per minute at an engine speed of 3000 rpm (revolutions per minute). Can a pressure transducer with a response time of 2 seconds resolve the pressure variation? What value response time would you recommend? Explain your answers.

   (c) A utility meter measures the power draw of a plant every 15 minutes. One of the highest consumers of power in the plant is a 100 kW heater that goes on for 4 minutes every other 10-minute period. Will the meter accurately record the variation in power consumption of this plant? Explain.

2.31 You have just purchased an old sports car. When the engine is idling at 1000 rpm, you press on the throttle and change the rpm to 3000 rpm. You notice that the tachometer overshoots 3000 rpm and oscillates for a couple of seconds about 3000 rpm until it becomes steady. Is this a zero-, first-, or second-order measuring system? What characteristic of the tachometer might you change to eliminate the oscillations?

2.32 A thermometer, initially at a temperature of 20°C, is suddenly immersed in a tank of water with a temperature of 80°C. The time constant of the thermometer is 2 s. What temperature will the thermometer read after 5 s?

2.33 A thermometer, initially at a temperature of 20°C, is suddenly immersed into a tank of water with a temperature of 80°C. The time constant of the thermometer is 4 s. What are the values of the rise time and the 90% response time?

2.34 A nominally linear electrical output force measuring device has the following manufacturer specifications:

Input range	0–100 lb
Full-scale output	30 mV
Linearity	±0.1% full scale
Hysteresis	±0.08% full scale
Repeatability	±0.03% full scale
Zero balance	±2% full scale

Temperature effect on:

Zero	0.002% full scale/°F
Span	0.002% of reading/°F

The device is used in an environment where the ambient temperature is not controlled and can vary between 50 and 95°F. For an input of 20 lb:

(a) What will be the nominal output voltage?

(b) Estimate the uncertainty due to each of the error sources that can be determined from the specifications. Express the uncertainties both in millivolts and as a percentage of reading. Do not attempt to combine the uncertainties.

C H A P T E R   3

Measurement Systems with Electrical Signals

Measuring systems that use electrical signals to transmit information between components have substantial advantages over completely mechanical systems and consequently are very widely used. In this chapter, we describe common aspects of electrical-signal measuring systems.

3.1 ELECTRICAL SIGNAL MEASUREMENT SYSTEMS

Almost all modern engineering measurements can be made using sensing devices that have an electrical output. This means that an electrical property of the device is caused to change by the measurand, either directly or indirectly. Most commonly, the measurand causes a change in a resistance, capacitance, or voltage. In some cases, however, the output of the sensor is a measurand-dependent current, frequency, or electric charge. Electrical output sensing devices have several significant advantages over mechanical devices:

1. Ease of transmitting the signal from measurement point to the data collection point
2. Ease of amplifying, filtering, or otherwise modifying the signal
3. Ease of recording the signal

It should be noted, however, that completely mechanical devices are sometimes still the most appropriate measuring systems.

As discussed in Chapter 2, the typical measuring system can be considered to include three conceptual subsystems, although some thought is required to identify them in many mechanical measuring devices. In electrical signal measuring systems, the subsystems, shown in Figure 3.1, are readily identified and are frequently supplied as separate components: the sensor/transducer stage, the signal conditioning stage, and a

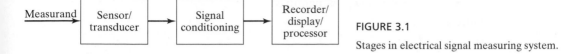

FIGURE 3.1

Stages in electrical signal measuring system.

recording/display/processing stage. These stages correspond directly to the systems shown for a generalized measurement system shown in Figure 2.1.

Electrical output sensing devices are sometimes called *sensors* (or sensing elements) and sometimes called *transducers*. In formal terms, a transducer is defined as a device that changes or converts information in the measurement process (signal modification). A sensor is a device that produces an output in response to a measurand and is thus a transducer. Many electrical output transducers include two stages. In one stage, the measurand causes a physical but nonelectrical change in a sensor. The second stage then converts that physical change into an electrical signal. The term *transducer* may also refer to devices that include certain signal conditioners. Not only are the terms *sensor* and *transducer* often used interchangeably, but there are other words used to name transducers for particular applications—the terms *gage*, *cell*, *pickup*, and *transmitter* being common.

Electrical output transducers are available for almost any measurement. A partial list includes transducers to measure displacement, linear velocity, angular velocity, acceleration, force, pressure, temperature, heat flux, neutron flux, humidity, fluid flow rate, light intensity, chemical characteristics, and chemical composition. If there is a commercial demand for a transducer, it is likely that it is available.

Common sensors and transducers and many associated signal conditioning systems are discussed in Chapters 8 to 10. Certain signal conditioning systems such as amplification and filtering are used with a variety of sensing elements and are discussed in this chapter.

In the simplest systems, the measurement system stage after the signal conditioner may simply record the signal or print or display a numerical value of the measurand. Common recording and display devices are discussed later in this chapter. In the most sophisticated measurement systems, the final stage includes a computer, which can not only record the data but also manage the data taking and perform significant numerical manipulation of the data. These computerized data-acquisition systems are discussed in detail in Chapter 4.

While the three measurement system components shown in Figure 3.1 are by far the most complicated, the characteristics of the interconnecting wiring can have major adverse effects on the signal. Methods for transmitting electrical signals and potential problems associated with signal transmission are discussed later in this chapter.

3.2 SIGNAL CONDITIONERS

There are many possible functions of the signal-conditioning stage. The following are the most common:

Amplification

Attenuation

Filtering (highpass, lowpass, bandpass, or bandstop)

Differentiation

Integration

Linearization

Combining a measured signal with a reference signal

Converting a resistance to a voltage signal

Converting a current signal to a voltage signal

Converting a voltage signal to a current signal

Converting a frequency signal to a voltage signal

More than one signal-conditioning function, such as amplification and filtering, can be performed on a signal. Some signal-conditioning functions may be performed by circuits located in the transducer itself, but in many cases signal conditioning is also performed in separate components that may provide user-controlled adjustments. Signal conditioners are important components of electrical signal measuring systems, and they must be treated carefully since they are capable of introducing major errors into the measuring process. They often must be calibrated, either alone or in combination with a transducer. In the following sections we discuss signal conditioners to amplify, attenuate, and filter signals. Also included are descriptions of circuits to integrate, differentiate, and compare signals.

3.2.1   General Characteristics of Signal Amplification

Many transducers produce signals with low voltages—signals in the millivolt range are common, and in some cases, signals are in the microvolt range. It is difficult to transmit such signals over wires of great length, and many processing systems require input voltages on the order of 1 to 10 V. The amplitude of such signals can be increased using a device called an *amplifier*, shown as a block diagram in Figure 3.2. The low-voltage signal, V_i, appears as a differential voltage on the input side of the amplifier. On the output side of the amplifier appears a higher voltage, V_o.

The degree of amplification is specified by a parameter called the *gain, G*:

$$G = \frac{V_o}{V_i} \tag{3.1}$$

Common instrumentation amplifiers usually have values of gain in the range 1 to 1000; however, higher gains can readily be achieved. The term *gain* is often used even

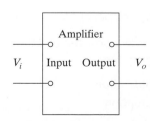

FIGURE 3.2

Generic voltage amplifier.

for devices that attenuate a voltage ($V_o < V_i$), and hence values of G can be less than unity. Gain is more commonly stated using a logarithmic scale, and the result is expressed in decibels (dB). For voltage gain, this takes the form

$$G_{dB} = 20 \log_{10} G = 20 \log_{10} \frac{V_o}{V_i} \qquad (3.2)$$

Using this formula, an amplifier with G of 10 would have a decibel gain, G_{dB}, of 20 dB, and an amplifier with a G of 1000 would have a decibel gain of 60. If a signal is attenuated, that is, V_o is less than V_i, the decibel gain will have a negative value.

Although increase in signal amplitude is the primary purpose of an amplifier, an amplifier can affect the signal in a number of other ways. The most important of these are frequency distortion, phase distortion, common-mode effects, and source loading.

While amplifiers are normally used with signals that include a range of frequencies, most amplifiers do not have the same value of gain for all frequencies. For example, an amplifier might have a gain of 20 dB at 10 kHz and a gain of only 5 dB at 100 kHz. Figure 3.3 shows the *frequency response* of an amplifier plotted as G_{dB} versus the logarithm of the frequency. Typically, the gain will have a relatively constant value over a wide range of frequencies; however, at extreme frequencies, the gain will be reduced (attenuated). The range of frequencies with close to constant gain is known as the *bandwidth*. The upper and lower frequencies defining the bandwidth, called the *corner* or *cutoff frequencies*, are defined as the frequencies where the gain is reduced by 3 dB. Most modern instrumentation amplifiers have constant gain at lower frequencies, even to $f = 0$ (direct current), so f_{c1} in Figure 3.3 is zero. However, all amplifiers have an upper cutoff frequency.

An amplifier with a narrow bandwidth will change the shape of an input time-varying signal by an effect known as *frequency distortion*. Figure 3.4 shows how a square-wave signal is altered by an amplifier that attenuates high frequencies.

Although the gain of an amplifier will be relatively constant over the bandwidth, another characteristic of the output signal, the phase angle, may change significantly. If

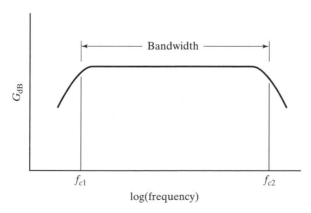

FIGURE 3.3

Amplifier frequency response.

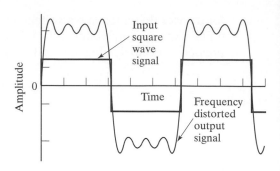

FIGURE 3.4

Frequency distortion of a square wave due to high-frequency attenuation.

the voltage input signal to the amplifier is expressed as

$$V_i(t) = V_{mi} \sin 2\pi ft \qquad (3.3)$$

where f is the frequency and V_{mi} is the amplitude of the input sine wave, the output signal will be

$$V_o(t) = GV_{mi} \sin(2\pi ft + \phi) \qquad (3.4)$$

where ϕ is called the phase angle. In most cases, ϕ is negative and the output waveform will trail the input waveform as shown in Figure 3.5. The phase response of an amplifier is usually presented in a plot of phase angle versus the logarithm of frequency as shown in Figure 3.6. Figure 3.3, which shows the amplitude response, and Figure 3.6, which shows the phase response, are together called the *Bode diagrams* of a dynamic system such as an amplifier.

For pure sinusoidal waveforms, phase shift is usually not a problem. However, for complicated periodic waveforms, it may result in a problem called *phase distortion*. In Chapter 5 it is shown that complicated periodic waveforms can be represented by a (usually infinite) series of sinusoidal terms with different frequencies (Fourier series). If the phase angle varies with frequency, the amplifier can distort the shape of the waveform. It can be shown that if the phase angle varies *linearly* with frequency, the shape of the waveform will not be distorted and the waveform will only be delayed or advanced in time. Phase distortion is demonstrated in Figure 3.7. Figure 3.7(a) shows

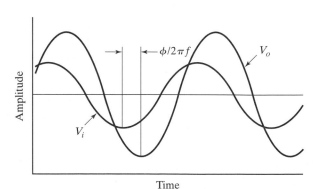

FIGURE 3.5

Effect of phase angle on signal.

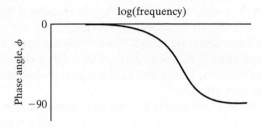

FIGURE 3.6

Typical phase-angle response of amplifier.

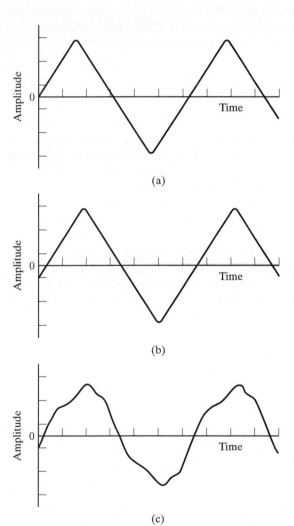

(a)

(b)

(c)

FIGURE 3.7

Effect on signal of linear and nonlinear phase-angle variation with frequency: (a) original signal; (b) phase angle varies linearly with frequency; (c) phase angle varies nonlinearly with frequency.

the original signal. Figure 3.7(b) shows the same signal after it has been passed through a device that varies the phase angle linearly with respect to the frequency. As can be seen, the signal is delayed relative to the original signal but has exactly the same shape. Figure 3.7(c) shows the original signal after it has been passed through a device that varies the signal in a highly nonlinear fashion. The shape of the signal is changed dramatically and shows significant phase distortion.

Another important characteristic of amplifiers is known as *common-mode rejection ratio* (CMRR). When different voltages are applied to the two input terminals (Figure 3.2), the input is known as a *differential-mode voltage*. When the same voltage (relative to ground) is applied to the two input terminals, the input is known as a *common-mode voltage*. An ideal instrumentation amplifier will produce an output in response to differential-mode voltages but will produce no output in response to common-mode voltages. Real amplifiers will produce an output response to both differential- and common-mode voltages, but the response to differential-mode voltages will be much larger. The measure of the relative response to differential- and common-mode voltages is described by common-mode rejection ratio, defined by

$$\text{CMRR} = 20 \log_{10} \frac{G_{\text{diff}}}{G_{\text{cm}}} \tag{3.5}$$

and is expressed in decibels. G_{diff} is the gain for a differential-mode voltage applied between the input terminals, and G_{cm} is the gain for a common-mode voltage applied to both terminals. Since signals of interest usually result in differential-mode input and noise signals often result in common-mode input, high values of CMRR are desirable. High-quality amplifiers often have a CMRR in excess of 100 dB.

Input loading and *output loading* are potential problems that can occur when using an amplifier (and when using many other signal-conditioning devices). The input voltage to an amplifier is generated by an input or source device such as a sensor or another signal conditioning device. If the output voltage of the source device is altered when it is connected to the amplifier, there exists a loading problem. A similar problem occurs when the output of the amplifier is connected to another device—the amplifier output voltage will be changed. To analyze these loading problems, it is possible to use rather simple models of the devices involved (Franco, 2002). As shown in Figure 3.8(a), the source device can be modeled as a voltage generator V_s in series with a resistor R_s. This model does not represent what actually happens in the device, only how the device will behave if someone makes connections to the terminals.

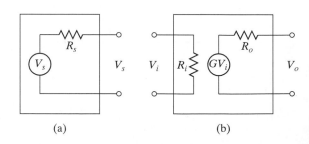

FIGURE 3.8

Models for (a) source and (b) amplifier. (Based on Franco, 2002.)

(a)　　　　　　　　(b)

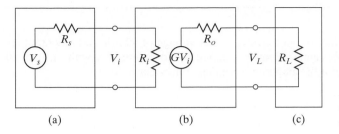

FIGURE 3.9

Combined model of (a) input source, (b) amplifier, and (c) output load. (Based on Franco, 2002.)

Using the same approach, the input of an amplifier can be modeled as an input resistance and the output as a voltage generator GV_i in series with an output resistance R_o, as shown in Figure 3.8(b).

If the source is not connected to the amplifier, the voltage at the source output terminals will be V_s. This is because there will be no current flowing through R_s and consequently, there will be no voltage drop across R_s. When the source is connected to the amplifier, the voltage at the source output terminals will no longer be V_s. As shown in Figure 3.9, V_s, R_s, and R_i form a complete circuit. Consequently, there will be a current flowing through R_s and a resulting voltage drop across R_s. The amplifier has placed a load on the source device. As will be shown, to minimize loading effects at the input and the output, an ideal amplifier (or other signal conditioner) should have a very high value of input resistance (R_i) and a very low value of output resistance (R_o).

When the amplifier is connected to the input source and the output load, the complete system can be modeled as shown in Figure 3.9. To analyze this circuit, we will first solve for the amplifier input voltage V_i in terms of the source voltage V_s. The current through the input loop is $V_s/(R_s + R_i)$, and hence V_i is given by

$$V_i = \frac{R_i V_s}{R_s + R_i} \tag{3.6}$$

Similarly, we can solve the output loop for V_L to obtain

$$V_L = \frac{R_L G V_i}{R_o + R_L} \tag{3.7}$$

Substituting Eq. (3.6) into Eq. (3.7), we obtain

$$V_L = \frac{R_L}{R_o + R_L} G \frac{R_i}{R_i + R_s} V_s \tag{3.8}$$

Ideally, we would like the voltage V_L to be equal to the gain G times V_s:

$$V_L = GV_s \tag{3.9}$$

It can be seen that if $R_L \gg R_o$ and $R_i \gg R_s$, Eq. (3.8) will approximate Eq. (3.9). That is, there will be no loading effects. Thus the ideal amplifier (or signal conditioner) has an infinite input resistance and a zero value of output resistance. For sinusoidal signals, this statement can be made general by requiring an infinite input impedance and a zero output impedance.

Example 3.1

A force-measuring transducer has an open-circuit output voltage of 95 mV and an output impedance of 500 Ω. To amplify the signal voltage, it is connected to an amplifier with a gain of 10. Estimate the input loading error if the amplifier has an input impedance of

  (a) 4 kΩ or
  (b) 1 MΩ.

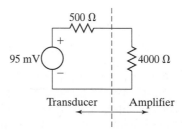

FIGURE E3.1

Solution:

  (a) The transducer can be modeled as a 95-mV voltage generator in series with a 500-Ω resistor. When this is connected to the amplifier, the resulting circuit is as shown in Figure E3.1. Solving for the current yields

$$I = \frac{V}{R} = \frac{0.095}{500 + 4000} = 0.0211 \text{ mA}$$

The voltage across the amplifier input resistor is then

$$V = RI = 4000 \times 0.0211 \times 10^{-3} = 84.4 \text{ mV}$$

The loading error is thus 10.6 mV or 11% of the transducer unloaded output.

  (b) Repeating the analysis replacing the 4-kΩ resistor with a 1-MΩ resistor, the error becomes 0.047 mV or 0.05%.

3.2.2 Amplifiers Using Operational Amplifiers

Practical signal amplifiers can be constructed using a common, low-cost, integrated-circuit component called an *operational amplifier*, or simply an op-amp. An op-amp is represented schematically by a triangular symbol as shown in Figure 3.10(a). The input voltages (V_n and V_p) are applied to two input terminals (labeled + and −), and the output voltage (V_o) appears through a single output terminal. There are two power supply terminals, labeled $V+$ and $V-$. There are also additional terminals (not shown) that may be used for adjustment of certain characteristics.

Figure 3.10(b) shows the model of an op-amp consistent with the modeling used for Figure 3.8(b). The arrangement shown in Figure 3.10(a) is known as the *open-loop configuration* and is distinguished from the closed-loop configuration used in practical amplifier circuits, discussed later. The resistance between the input terminals (r_d) is very high (approaching infinity), the output resistance (r_o) is close to zero, and the op-amp

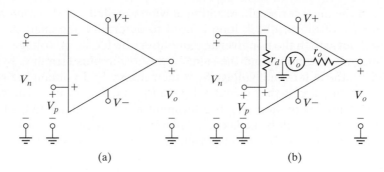

FIGURE 3.10

Operational amplifier symbol and simplified model.

thus approaches an ideal amplifier from a loading standpoint. The amplifier has a very high gain, denoted here by the symbol g, which ideally is infinite. The op-amp gain is given by small g to distinguish it from G, the gain of amplifier circuits using the op-amp as a component. The output of the op-amp in the open-loop configuration shown in Figure 3.10(b) is given by

$$V_o = g(V_p - V_n) \tag{3.10}$$

One op-amp design, identified by the number μA741C (or generically simply a 741), costs less than a cup of coffee and is very widely used. It serves as a model to demonstrate the use of op-amps in a variety of circuits, although the circuits will also show most of the same quantitative characteristics even if op-amps other than the μA741C are substituted. The μA741C has an input impedance r_d on the order of 2 MΩ, an output impedance r_o of about 75 Ω, a gain g of about 200,000, and a CMRR of 70 dB or higher. Other op-amps are available that have one or more superior or at least different characteristics. To describe the characteristics of most circuits, it is not necessary to understand the details of the op-amp internals. A large number of applications of op-amps are discussed in great detail in Franco (2002).

A simple amplifier using an op-amp, called a *noninverting amplifier*, is shown schematically in Figure 3.11. The term *noninverting* means that the sign of the output voltage relative to ground is the same as the input voltage. As is typical of op-amp applications, this circuit shows a *feedback loop* in which the output is connected to

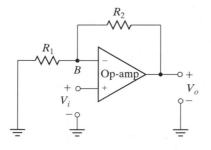

FIGURE 3.11

Simple noninverting amplifier using an op-amp.

one of the input terminals, resulting in what is called a *closed-loop configuration*. The structure of the feedback loop is used to determine the characteristics of op-amp-based devices. In the noninverting amplifier, the feedback voltage is connected to the minus input terminal and the signal input to the plus terminal. Referring to Figure 3.10(b), the + terminal voltage V_p is then simply V_i. To obtain the voltage V_n, we will analyze the circuit that includes V_o, R_2, R_1, and the ground. The current flow from point B into the op-amp negative terminal will be small due to the high op-amp input impedance and will be neglected. Analyzing the circuit, the current through resistor R_1, I_{R1}, is $V_o/(R_1 + R_2)$, and the voltage $V_n = V_B = I_{R1} R_1$ is found to be

$$V_n = V_o \frac{R_1}{R_1 + R_2} \tag{3.11}$$

Substituting these input voltages (V_p and V_n) into Eq. (3.10), we obtain

$$V_o = g(V_p - V_n) = g\left(V_i - V_o \frac{R_1}{R_1 + R_2}\right) \tag{3.12}$$

Solving for V_o, we obtain

$$V_o = \frac{gV_i}{1 + g[R_1/(R_1 + R_2)]} \tag{3.13}$$

For the op-amp, g is very large and the second term in the denominator of Eq. (3.13) is very large compared to 1. With this approximation and noting that the gain for the amplifier is $G = V_o/V_i$, Eq. (3.13) can be solved to give

$$G = \frac{V_o}{V_i} = \frac{R_1 + R_2}{R_1} = 1 + \frac{R_2}{R_1} \tag{3.14}$$

The gain is found to be a function only of the ratio R_2/R_1 and does not depend on the actual resistor values. Values of R_1 and R_2 in the range 1 kΩ to 1 MΩ are typical. Resistances above 1 MΩ complicate circuit design because stray capacitance effects alter the impedance. Low resistance values lead to high power consumption. In some circuits, beyond the scope of this book, high resistance values and specialized op-amps are required.

It should be noted that if V_i is high enough so that $V_o(= GV_i)$ approaches the power supply voltage, further increases in V_i will not increase the output. This situation is called *output saturation*. For the μA741C, saturation will occur when the output is about 2 V less than the power supply voltage. This means that saturation will occur when the output is ±13 V if the power supply is a typical ±15 V.

As discussed by Franco (2002), the feedback loop increases the input impedance and decreases the output impedance relative to the open-loop op-amp. The impedance between the input terminals of the noninverting amplifier is very high. For the circuit shown in Figure 3.11, this input impedance is on the order of hundreds of megaohms, depending on the gain. This means that the amplifier will not appreciably load the input device for most applications. The circuit also shows another major advantage of op-amp amplifiers—a very low output impedance (on the order of 1 Ω), which results in the output voltage being relatively unaffected by connected devices.

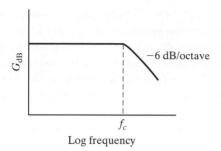

FIGURE 3.12

Frequency response of op-amp amplifier.

The gain computed from Eq. (3.14) is actually the gain that would be expected if the input signal were a dc voltage or an ac voltage at a low frequency. Figure 3.12 shows the frequency response of the amplifier in the form of a Bode plot. It indicates that the gain is essentially constant from low frequency up to a cutoff frequency, f_c. The range of frequencies between $f = 0$ and f_c is the bandwidth of this amplifier. Above f_c, the gain starts to decrease, or roll off, and this roll-off occurs at a rate of 6 dB per octave. Since an *octave* is a doubling of the frequency, this means that each time the frequency doubles in the range above f_c, the gain will decrease by 6 dB from the value computed by Eq. (3.14). Actually, the gain starts to decline at frequencies slightly below f_c since f_c is defined as the frequency at which the gain has declined by 3 dB.

This roll off in gain at high frequencies is an inherent characteristic of op-amps. The cutoff frequency, f_c, depends on the low-frequency gain of the amplifier—the higher the gain, the lower is f_c. This low-frequency gain–cutoff frequency relationship is described by a parameter called the *gain–bandwidth product* (GBP). For most op-amp-based amplifiers, the product of the low-frequency gain and the bandwidth is a constant. Since the lower frequency limit of the bandwidth is zero, the upper cutoff frequency can be evaluated from

$$f_c = \frac{\text{GPB}}{G} \tag{3.15}$$

A value of GPB is a characteristic of the op-amp itself and can be obtained from device handbooks: the μA741C op-amp has a GBP of 1 MHz. The noninverting circuit in Figure 3.11 has the same GBP as the op-amp itself, so, for example, a gain of 10 at low frequencies will result in a bandwidth value of 100 kHz. Some more expensive op-amps have higher values of GBP.

If it is desirable to have a high gain and a larger bandwidth, it is possible to cascade two amplifiers; that is, the output of one amplifier is used as the input for the next. Each amplifier has a lower gain, but the overall gain for the two stages is the same, and the bandwidth will be much higher. Although the gain is constant over the bandwidth, the phase angle between the input and the output, ϕ, shows a strong variation with frequency. For the noninverting amplifier in Figure 3.11, the phase-angle variation with frequency is given by

$$\phi = -\tan^{-1}\frac{f}{f_c} \tag{3.16}$$

At $f = f_c$, ϕ has a value of -0.785 rad ($-45°$). This means that the output trails the input by one-eighth of a cycle. However, the variation of ϕ with f is very close to linear from $f = 0$ to $f = f_c/2$ and is approximately linear from $f = 0$ to $f = f_c$. As a result, signals that are within the bandwidth will be subject to only modest phase distortion. However, if the time relationships of amplified signals from an experiment are to be compared to signals that have not been amplified, the phase angle must be considered.

Example 3.2

Specify values of resistors R_1 and R_2 for a noninverting 741 op-amp amplifier with a gain of 10. Find the cutoff frequency and the phase shift for a sinusoidal input voltage with a frequency of 10,000 Hz.

Solution: The gain is given by Eq. (3.14):

$$G = 10 = 1 + \frac{R_2}{R_1}$$

Selecting $R_1 = 10$ kΩ, R_2 can be evaluated as 90 kΩ. Since the noninverting amplifier using the 741 op-amp has a gain–bandwidth product (GBP) of 1 MHz, the cutoff frequency is given by Eq. (3.15):

$$f_c = \frac{GPB}{G} = \frac{10^6}{10} = 100 \text{ kHz}$$

The phase angle at 10 kHz is given by Eq. (3.16):

$$\phi = -\tan^{-1}\frac{10^4}{10^5} = -5.7°$$

This means that the output trails the input by $5.7°$, or about 1.6% of a cycle.

Another common op-amp circuit, called the *inverting amplifier*, is shown in Figure 3.13. This circuit is called inverting because the output voltage, relative to ground, is of sign opposite to the input voltage. Inverting amplifiers form the basis for many other op-amp circuits, including filters, integrators, and differentiators (discussed later in this chapter). Using analysis similar to that used for the noninverting amplifier,

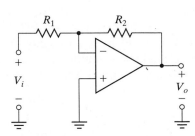

FIGURE 3.13

Inverting op-amp amplifier.

it can be demonstrated that the gain for the inverting amplifier is given by

$$G = -\frac{R_2}{R_1} \tag{3.17}$$

As with the noninverting amplifier, the gain depends on a resistance ratio rather than the absolute resistance values and resistance values are usually in the range 1 kΩ to 1 MΩ.

The inverting amplifier has significantly different characteristics than the noninverting amplifier. Whereas the noninverting amplifier has an input impedance of hundreds of megohms, the inverting amplifier input impedance is about the same as R_1, which frequently is not much larger than 100 kΩ. This may present a loading problem for some input devices. The output impedance of the inverting amplifier shares the low output impedance characteristics of the noninverting amplifier—generally less than 1 Ω. The inverting amplifier also shows a reduction in gain (at 6 dB per octave) above the cutoff frequency. The gain–bandpass product for the inverting amplifier is different from that for the noninverting amplifier. According to Franco (2002), it can be computed from

$$\text{GBP}_{\text{inv}} = \frac{R_2}{R_1 + R_2} \text{GBP}_{\text{noninv}} \tag{3.18}$$

The phase response of the inverting amplifier is the same as that of the noninverting amplifier, as determined by Eq. (3.16).

The amplifiers shown in Figures 3.11 and 3.13 should be satisfactory for some purposes. However, they are not the best for many instrumentation applications. Ambient electric and magnetic fields can induce noise in the connecting input signal wires. Both amplifiers have a single-input signal connection, and electrical noise can be induced in the input wire. Specialized instrumentation amplifiers, often using two or more op-amps (e.g., Figure 3.14) have two ungrounded connections to the source

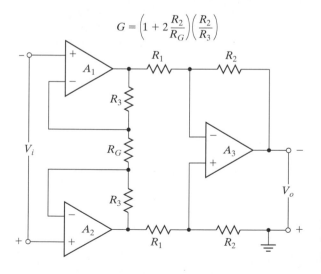

$$G = \left(1 + 2\frac{R_2}{R_G}\right)\left(\frac{R_2}{R_3}\right)$$

FIGURE 3.14

True differential input instrument amplifier. (From Franco, 2002.)

device (true differential input). In this case, electrical noise of the same magnitude and phase is induced in both input wires. This is a common-mode signal. Properly designed and constructed instrumentation amplifiers have a high common-mode rejection ratio and hence can generate output that is largely free of input common-mode noise.

Complete high-quality instrumentation amplifiers without a power supply can be obtained on a single IC chip. Complete instrumentation amplifiers with a power supply and adjustments features for gain and zero offset are readily available commercially. Amplifiers are subject to errors as are other instrument components. They can have nonlinearity errors, hysteresis errors, and thermal-stability errors. If the actual gain is not as predicted, there will be a gain (sensitivity) error.

3.2.3   Signal Attenuation

In some cases a measurement will result in a voltage output with an amplitude higher than the input range of the next component. The voltage must then be reduced to a suitable level, a process known as *attenuation*. The simplest method is to use a voltage-dividing network as shown in Figure 3.15. The resulting output voltage from this network is

$$V_o = V_i \frac{R_2}{R_1 + R_2} \tag{3.19}$$

Dividing networks of this type have potential loading problems. First, it will place a resistive load on the system that generates V_i. The resistive load might draw a significant current, which will change V_i from the value that it would have had if the network had not been installed. This problem can be avoided by having the sum of resistors R_1 and R_2 be very high compared to the output impedance of the system generating V_i. Unfortunately, this means that the output resistance of the network, R_2, will also be high, leading to a problem when the output load is connected. It may be desirable to feed the voltage divider output into a high-input impedance amplifier with unity gain to reduce output loading problems. The effect of loading is evaluated in Example 3.3.

The use of large values of resistance in voltage dividers presents another problem in addition to loading. At high signal frequencies, the impedance due to small amounts of capacitance can be comparable to the divider resistances and can produce attenuation that is frequency dependent.

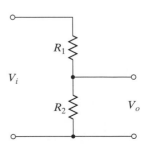

FIGURE 3.15

Attenuation using dividing network.

Example 3.3

The voltage used to power a heater in an experiment is nominally 120 V. To record this voltage, it must first be attenuated using a voltage divider. The attenuator reduces the voltage by a factor of 15 and the sum of the resistors R_1 and R_2 is 1000 Ω.

(a) Find R_1 and R_2 and the ideal voltage output (neglecting loading effects).

(b) If the source resistance R_s is 1 Ω, find the actual divider output V_o and the resulting loading error in V_o.

(c) If the divider output is connected to a recorder that has an input impedance of 5000 Ω, what will be the voltage output (input to the recorder) and the resulting loading error?

Solution:

(a) Using Eq. (3.19), we have

$$\frac{V_o}{V_i} = \frac{1}{15} = \frac{R_2}{R_1 + R_2} = \frac{R_2}{1000}$$

$$R_2 = 66.7$$

$$R_1 = 1000 - R_2 = 933.3$$

The nominal output voltage is $120/15 = 8$ V.

(b) The circuit of the complete system, including the source impedance, is shown in Figure E3.3(a). Solving for the current in the loop gives

$$I = \frac{V}{\sum R} = \frac{120}{1 + 933.3 + 66.7} = 0.1199 \text{ A}$$

We can then evaluate the output voltage:

$$V_o = IR_2 = 0.1199 \times 66.7 = 7.997 \text{ V}$$

The error is thus 0.003 V or 0.04%.

(c) If we connect the divider output to a recorder with an input impedance of 5000 Ω, we will have the circuit shown in Figure E3.3(b). R_2 and R_i are resistors in parallel, which can

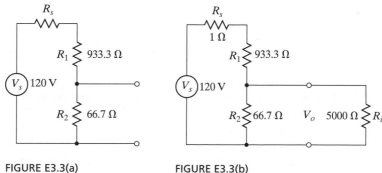

FIGURE E3.3(a)   FIGURE E3.3(b)

be combined to give a value of 65.8. Solving for the loop current, we obtain

$$I = \frac{V}{\sum R} = \frac{120}{1 + 933.3 + 65.8} = 0.1200 \text{ A}$$

$$V_o = IR = 0.1200 \times 65.8 = 7.9 \text{ V}$$

This is a loading error of 0.10 V or 1.3%.

Comments: The primary loading problem was the rather low input impedance of the recorder. In this case R_2 could be adjusted slightly using a calibration process to eliminate the loading error. Alternatively, an amplifier with a high input impedance, such as the noninverting type described in Section 3.2.2 but with unity gain, could be placed between the divider and recorder to eliminate the loading problem. The loading error could be reduced somewhat by reducing the sum $R_1 + R_2$; however, this would raise the power dissipated in these resistors, which is already about 14 W.

3.2.4   General Aspects of Signal Filtering

In many measuring situations, the signal is a complicated, time-varying voltage that can be considered to be the sum of many sine waves of different frequencies and amplitudes. (See Chapter 5.) It is often necessary to remove some of these frequencies by a process called *filtering*. There are two very common situations in which filtering is required. The first is the situation in which there is spurious noise (such as 60-Hz power-line noise) imposed on the signal. The second occurs when a data acquisition system samples the signal at discrete times (rather than continuously). In the latter case, filtering is necessary to avoid a serious problem called *aliasing*. (See Chapter 5.)

A filter is a device by which a time-varying signal is modified intentionally, depending on its frequency. Filters are normally broken into four types: lowpass, highpass, bandpass, and bandstop filters. The characteristics of these categories of filters are shown in Figure 3.16. A *lowpass filter* allows low frequencies to pass without attenuation but, starting at a frequency f_c called the *corner frequency*, attenuates higher-frequency components of the signal. The frequency band with approximately constant gain (V_o/V_i) between $f = 0$ and f_c is known as the *passband*. The significantly attenuated frequency range is known as the *stopband*. The region between f_c and the stopband is known as the transition band. A *highpass filter* allows high frequency to pass but attenuates low frequencies. A *bandpass filter* attenuates signals at both high and low frequencies but allows a range of frequencies to pass without attenuation. Finally, a *bandstop filter* allows both high and low frequencies to pass but attenuates an intermediate band of frequencies. If the band of stopped frequencies is very narrow, the bandstop filter is called a *notch filter*.

Although a very large number of circuits function as filters, four classes of filters are most widely used: Butterworth, Chebyshev, elliptic, and Bessel. Each filter class has unique characteristics that makes it the most suitable for a particular application. Each of these filter classes can have another characteristic, called *order*. For a particular filter class, the higher the order, the greater will be the attenuation of the signal in the

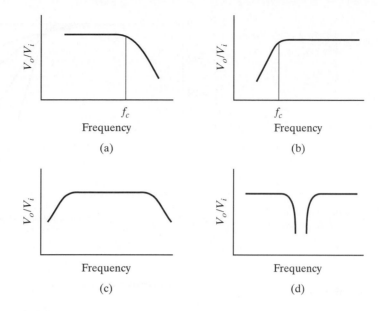

FIGURE 3.16

Categories of electrical filters: (a) lowpass; (b) highpass; (c) bandpass; (d) bandstop.

stopband. To demonstrate some of the general characteristics of filter class and order, filters of the lowpass type are discussed in the following paragraphs.

Butterworth filters have the characteristic that they are maximally flat in the passband. This means that the gain is essentially constant in the passband. For lowpass Butterworth filters with unity dc gain, the gain as a function of frequency f and order n is given by

$$G = \frac{1}{\sqrt{1 + (f/f_c)^{2n}}} \tag{3.20}$$

where n is the order of the filter. This equation is plotted in Figure 3.17. For high frequencies, $f/f_c \gg 1$, the gain G becomes $(f_c/f)^n$. For a first-order filter $(n = 1)$, this means that each time the frequency doubles, the gain will be reduced by a factor of 2. As Eq. (3.2) shows, a halving of the gain corresponds to a 6-dB attenuation. Since an octave is a doubling (or halving) of the frequency, the roll-off for a first-order Butterworth filter is 6 dB/octave. In general, the roll-off in the stopband is $6n$ dB/octave, which, for example, would become 48 dB per octave for $n = 8$.

While high-order Butterworth filters provide large values of roll-off at frequencies in the stopband, in the transition region near the cutoff frequency f_c, the gain does not have a very crisp change in slope. *Chebyshev filters* have a much crisper change in slope but at the price of ripple in the passband gain, as shown in Figure 3.18. The amount of passband ripple is controlled by details of the design, and in the limit,

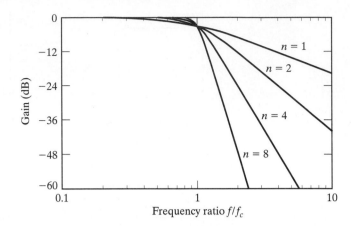

FIGURE 3.17

Gain of lowpass Butterworth filters as a function of order and frequency.

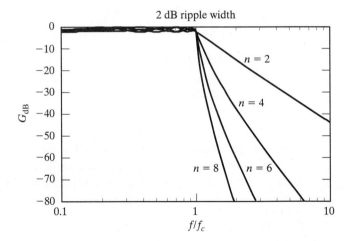

FIGURE 3.18

Gain of lowpass Chebyshev filters as a function of order and frequency.

Chebyshev filters with no passband ripple have the same frequency response as Butterworth filters. High-order Chebyshev filters are more satisfactory than Butterworth designs for notch filters. *Elliptic filters* have a very crisp transition between the passband and the stopband but allow ripples in the stopband as well as the passband.

As with amplifiers, filters alter the phase of components of the signal as a function of frequency. For example, the phase-angle shift for an eighth-order Butterworth filter is 360° at the cutoff frequency. For higher-order filters, this phase response can introduce serious phase distortion. A class of filters called *Bessel filters* are often used because they have a more nearly linear variation of phase angle with frequency in the passband than that of higher-order filters of other classes. Figure 3.19 compares the phase-angle variation for a fourth-order lowpass Bessel filter to a fourth-order Butterworth filter. As can be seen, the Bessel filter shows a nearly linear variation of phase angle within the passband ($f/f_c < 1$). The gain of lowpass Bessel filters is shown in Figure 3.20. Comparing Figure 3.20 with Figure 3.17 shows that for a given filter order,

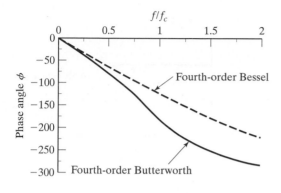

FIGURE 3.19

Comparison of Butterworth and Bessel phase-angle variation with frequency.

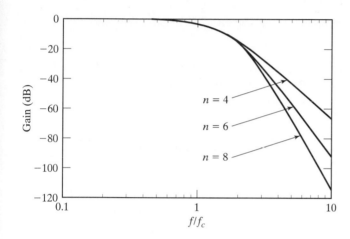

FIGURE 3.20

Gain of lowpass Bessel filters as a function of order and frequency.

Bessel filters have a lower roll-off rate (dB/octave) in the first two octaves of the stopband than do Butterworth filters. After the first couple of octaves above f_c, the attenuation rate of Bessel filters becomes larger and approaches that of Butterworth filters.

In most cases, the instrumentation engineer will need to specify rather than design a filter. Filters are commercially available and can be purchased by specifying appropriate parameters. The engineer must specify the type (e.g., lowpass), class (e.g., Butterworth), the order (e.g., 8), and the corner frequency(ies). For Chebyshev and elliptic filters, specifications must be given that describe the passband and/or stopband ripple.

3.2.5 Butterworth Filters Using Operational Amplifiers

While *passive filters* can be constructed using simple circuits consisting of resistors, capacitors, and inductors, for instrumentation applications, there are significant advantages for *active filters*, the most common of which are based on op-amps. Op-amp filters can be constructed without the use of inductors, which are often bulky and are the least ideal of electronic components. Furthermore, op-amp filters, being based on amplifiers,

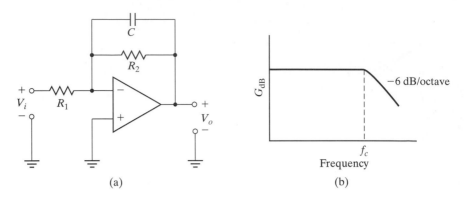

FIGURE 3.21

Lowpass Butterworth filter using op-amp: (a) op-amp circuit; (b) frequency response.

do not cause the signal-amplitude losses associated with passive filters. Figure 3.21(a) shows a simple op-amp circuit that functions as a lowpass filter. This filter circuit, which produces a first-order Butterworth response, is a modification of the inverting amplifier (Figure 3.13) in which a capacitor has been placed in parallel with R_2. The frequency response of this circuit [shown in Figure 3.21(b)] is characterized by a constant gain from low frequencies to the corner frequency f_c and a roll-off in gain of 6 dB per octave above the corner frequency.

To determine the frequency response, we will use ac circuit analysis to examine the behavior of the circuit when a single-frequency sinusoidal signal is input. This input signal takes the form $V_{mi} e^{j2\pi ft}$, where V_{mi} is the sine-wave amplitude and f is the frequency. In deriving Eq. (3.17), the expression for the gain on an inverting amplifier, we used resistances R_1 and R_2 in the feedback loop. If instead we had used complex impedances Z_1 and Z_2, the resulting equation for gain G would have been

$$G = -\frac{Z_2}{Z_1} \qquad (3.21)$$

We can apply this equation to the filter in Figure 3.21(a). Z_2 consists of C and R_2 in parallel. Since the impedance of a capacitor is $1/j2\pi fC$, the value of Z_2 is found to be

$$Z_2 = \frac{1}{\left(\dfrac{1}{R_2}\right) + j2\pi fC} = \frac{R_2}{1 + j2\pi fCR_2} \qquad (3.22)$$

Z_1 is simply R_1, so the expression for G, Eq. (3.21), becomes

$$G = -\frac{1}{R_1}\frac{R_2}{1 + j2\pi fCR_2} = G_o \frac{1}{1 + j2\pi fCR_2} = \frac{V_o}{V_i} \qquad (3.23)$$

where G_o, the low-frequency gain $(-R_2/R_1)$, is the same as that of the simple inverting amplifier (without the presence of capacitor C). It should be noted that the gain G is represented by a complex number.

If we take the absolute value of the terms in Eq. (3.23), we obtain the following expression:

$$\left|\frac{G}{G_o}\right| = \left|\frac{1}{1 + j2\pi fCR_2}\right| = \frac{1}{\sqrt{1 + (2\pi fCR_2)^2}} \qquad (3.24)$$

The corner frequency, f_c, for a Butterworth filter is defined as the frequency where the magnitude of the gain is reduced 3 dB from its low-frequency value, G_o. Equation (3.2) shows that 3 dB corresponds to a reduction in gain, G, by a fraction equal to 0.707. Substituting the value $G = 0.707G_o$ into Eq. (3.24) gives

$$\left|\frac{0.707G_o}{G_o}\right| = 0.707 = \frac{1}{\sqrt{1 + (2\pi fCR_2)^2}} \qquad (3.25)$$

This can be solved for the corner frequency f_c:

$$f_c = \frac{1}{2\pi CR_2} \qquad (3.26)$$

It should be noted that the corner frequency for a lowpass filter computed by Eq. (3.26) cannot be any higher than the cutoff frequency for the inverting amplifier itself.

Equation (3.24) can be used to determine the roll-off rate as frequency becomes high relative to the corner frequency. At high frequencies, $2\pi fCR_2$ is large compared to 1, and Eq. (3.25) reduces to

$$\left|\frac{G}{G_o}\right| = \frac{1}{2\pi fCR_2} \qquad (3.27)$$

This shows that if the frequency is doubled (a change of 1 octave), the gain will be reduced by a factor of 1/2. Based on Eq. (3.2), a reduction in gain G by 1/2 corresponds to -6 dB. Thus the roll-off is 6 dB per octave.

Equations (3.23) and (3.26) can be combined to give an expression for gain in terms of f_c:

$$G = \frac{V_o}{V_i} = G_o \frac{1}{1 + j(f/f_c)} \qquad (3.28)$$

Equation (3.28) can be used to evaluate the phase angle of V_o relative to V_i. The phase of V_o relative to V_i is then the phase of $1/[1 + j(f/f_c)]$:

$$\phi = \tan^{-1}\frac{-f}{f_c} = -\tan^{-1}\frac{f}{f_c} \qquad (3.29)$$

While Eq. (3.16) has the same form as Eq. (3.29), Eq. (3.16) results from characteristics of the op-amp itself, whereas Eq. (3.29) is a consequence of the capacitor in the feedback loop.

As noted earlier, the inverting amplifier has a fairly low input impedance, close to R_1. However, if a higher input impedance is desired, an amplifier of the noninverting

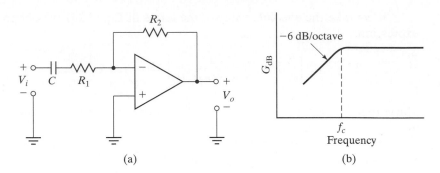

FIGURE 3.22

First-order Butterworth highpass filter using an op-amp: (a) op-amp circuit;
(b) frequency response.

type (or an instrumentation amplifier) can be cascaded with the filter amplifier (i.e.,
the signal is passed through the other amplifier before being input to the filter).

Figure 3.22 shows an op-amp circuit and frequency response for a first-order
high-pass Butterworth filter. The formula for the cutoff frequency, which can be derived in a manner similar to that for the low-pass filter, is given by

$$f_c = \frac{1}{2\pi R_1 C} \tag{3.30}$$

It should also be noted that highpass filters using op-amps are in fact bandpass filters
since the filter amplifier also has a high-frequency cutoff.

Figure 3.23 shows an op-amp circuit and frequency response for a first-order Butterworth bandpass filter. The upper and lower cutoff frequencies for the bandpass filter
are given by

$$f_{c1} = \frac{1}{2\pi R_1 C_1} \qquad f_{c2} = \frac{1}{2\pi R_2 C_2} \tag{3.31}$$

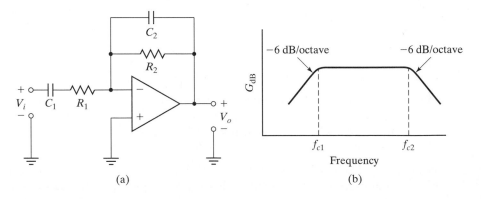

FIGURE 3.23

Bandpass filter using op-amp: (a) op-amp circuit; (b) frequency response.

Although useful for demonstrating the basic theory, the attenuation of first-order filters is often insufficient and higher-order filters are required. Higher-order filters can be constructed by more complex circuitry and by cascading lower-order filters. Higher-order filter circuits have many more parameters that must be determined in the design process, and consequently, the process is commonly performed using computer programs. Johnson (1976) and Franco (2002) discuss methods for designing higher-order filters. As mentioned earlier, filters meeting user specifications are available commercially.

Example 3.4

A transducer measuring pressure needs to respond to oscillations up to 3 Hz but is contaminated by 60-Hz noise. Specify a first-order low-pass Butterworth filter to reduce the 60-Hz noise. How much can the amplitude of the noise be reduced with this filter?

Solution: We are to specify the values of the components in Figure 3.21(a). We pick a corner frequency of 3 Hz. Since no gain is required, we should set $R_2 = R_1$. Pick $R_1 = R_2 = 10\,\text{k}\Omega$. We can now use Eq. (3.30) to obtain the value of C:

$$3 = \frac{1}{2\pi 10{,}000C}$$

Solving for C, we obtain $C = 5.3\ \mu\text{F}$. Since each octave is a doubling of the frequency, we can find the number of octaves between 3 and 60 Hz from

$$3 \times 2^x = 60$$

Solving for x, we obtain $x = 4.3$ octaves. Since the attenuation is 6 dB per octave, the total attenuation is $4.3 \times 6 = 25.9$ dB. Equation (3.2) can be used to evaluate the actual voltage reduction:

$$-25.9 = 20\log_{10}\frac{V_{\text{out}}}{V_{\text{in}}}$$

Solving for $V_{\text{out}}/V_{\text{in}}$, we obtain 0.051. This means that the noise voltage has been reduced to 5.1% of its previous value. If this reduction is insufficient, it will be necessary to use a higher-order filter.

Comment: Using this filter, signals at 3 Hz will be attenuated 3 dB, since 3 Hz is the cutoff frequency.

3.2.6 Circuits for Integration, Differentiation, and Comparison

Circuits for integration, differentiation, and comparison are used in some measurement applications and are presented here for reference. Integrator and comparator circuits are used in analog-to-digital converters (discussed in Chapter 4), and integrating and differentiating circuits are often used for real-time data processing. For the integrator circuit shown in Figure 3.24(a), the output voltage is the time integral of the input voltage:

$$V_o(t) = -\frac{1}{RC}\int_0^t V_i(t)\,dt + V_o(0) \tag{3.32}$$

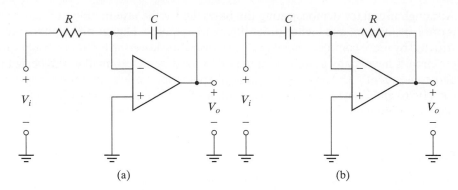

FIGURE 3.24

Op-amp circuits for (a) integration and (b) differentiation.

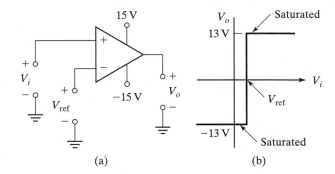

FIGURE 3.25

Op-amp comparator: (a) circuit; (b) output
voltage.

For the differentiator circuit shown in Figure 3.24(b), the output voltage is the time derivative of the input voltage:

$$V_o(t) = -RC \frac{dV_i(t)}{dt} \tag{3.33}$$

Figure 3.25 shows a simple circuit for an open-loop op-amp voltage comparator. In this
circuit the input voltage is compared to a reference voltage V_{ref}. If the input voltage is
greater than the reference voltage $(V_i > V_{ref})$, the output voltage is saturated at about
2 V less than the power supply positive voltage. If the input voltage is less than the reference voltage $(V_i < V_{ref})$, the output voltage will saturate at close to 2 V higher than
the negative power supply voltage.

3.3   INDICATING AND RECORDING DEVICES

3.3.1   Digital Voltmeters and Multimeters

In most cases the output of the signal conditioner is an analog voltage. If the user only
wishes to observe the voltage and the voltage is quasi-steady, a simple digital voltmeter

FIGURE 3.26

Digital multimeter with a digital voltmeter as a mode of operation. (Courtesy of Triplett Corp.)

(DVM) is the best choice of indicating device. Figure 3.26 shows a typical hand-held digital multimeter (DMM), which has a DVM as one of its functions. Modern digital voltmeters have a very high input impedance (in the megohm range), so the voltage measurement process will not usually affect the voltage being measured. An important component of a digital voltmeter is an analog-to-digital (A/D) converter, which converts the input analog voltage signal to a digital code that can be used to operate the display. A/D converters are discussed in greater detail in Chapter 4. Relatively inexpensive digital voltmeters can have an accuracy of better than 1% of reading, and high-quality voltmeters can be significantly better.

Digital multimeters can be used to display other types of input signals, such as current or resistance or frequency. Some can even input and display signals that are already in digital form. It is common in the process control industry to combine a digital voltmeter with a signal conditioner. In this case it is called a *digital panel meter* (DPM). The DPM displays the conditioned signal to assist in adjusting the conditioner while the signal is also transmitted to a central computer data acquisition and control system.

3.3.2 Oscilloscopes

If the output of a sensor is varying rapidly, a digital voltmeter is not a suitable indicating device and an oscilloscope (scope) is more appropriate. In this device, shown in Figure 3.27, the voltage output of the signal conditioner is used to deflect the electron beam in a cathode ray tube (CRT). The CRT, shown schematically in Figure 3.28, consists of a heated cathode that generates free electrons, an anode used to accelerate an electron beam, two sets of deflection plates, and a front face (screen) that is coated with phosphor. These components are contained in an evacuated glass envelope. When voltages of suitable amplitude are applied to the deflection plates, the electron beam will be deflected and cause the phosphors to glow at a particular position on the screen. The deflection plate voltage is proportional to the input voltage, and so the visible deflection is proportional to the input voltage. The CRT in an oscilloscope is simply a rather accurately constructed version of the picture tube in television sets.

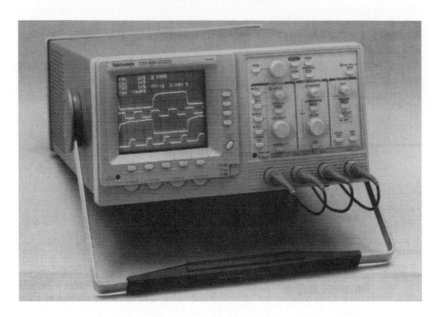

FIGURE 3.27

Cathode ray oscilloscope. (Courtesy of Tektronix, Inc.)

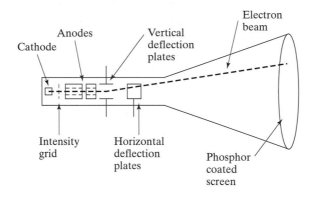

FIGURE 3.28

Schematic view of cathode ray tube.

A block diagram of the basic circuit elements used to control the CRT is shown in Figure 3.29. Normally, the input voltage to be displayed is connected to the amplifier that controls the vertical-plate voltage. A sweep generator is connected to the amplifier controlling the horizontal-plate voltage. This sweep causes the beam to move repeatedly from left to right at a user-controlled constant rate. The result is that the input voltage is displayed as a function of time as the beam sweeps across the screen. For periodic input signals, it is possible to synchronize the horizontal sweep so that it starts at the same point in the input voltage cycle for each sweep. The displays from the

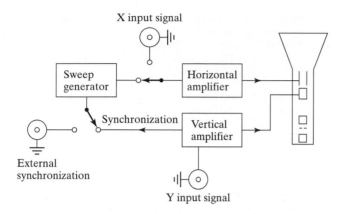

FIGURE 3.29

Block diagram of analog oscilloscope.

repeating sweeps then fall on top of each other, and it is easier for the user to interpret the data on the screen. It is also possible to use a trigger, which causes the horizontal amplifier to sweep the screen only once.

Separate signal voltages can be connected to both the vertical and horizontal amplifiers. In this case, one input voltage is plotted versus the other input voltage. This approach is frequently useful if both input voltages have the same frequency and it is desired to examine the phase relationship between them. Oscilloscopes can display very high frequency signals (up to 100 MHz), but the accuracy of the voltage reading is generally not as good as for digital voltmeters. The thickness of the beam and the resolving power of the human eye limit the accuracy to 1 to 2% in most cases. If only simple information is required, the experimenter can observe the results on the screen and record them on paper. Polaroid-type cameras are also made which attach to the oscilloscope and make it possible to take a photograph of the screen. Videocameras and videotape recorders can also be used to make a permanent record of the display for low frequencies.

An important variation of the simple oscilloscope is the digital-storage oscilloscope. In this device, the input voltage over a period of time is converted to digital form and stored in memory. When the data taking has been completed, circuitry converts the digital data back to analog form and uses the data to drive a conventional CRT. Storage scopes have the advantage that nonrepeating transient input signals can be recorded and then displayed continuously on the screen for analysis. Modern digital oscilloscopes can also record the data onto floppy disks. Hardware is available to convert personal computers into digital storage scopes.

3.3.3   Strip-Chart Recorders

In the past, one of the most common methods of recording time-varying voltages was the strip-chart recorder, an example of which is shown in Figure 3.30. In these devices, the input voltage is used to move a pen that writes on a strip of paper moving at a right angle to the direction of pen motion while unwinding off a continuous roll. Some pens use ink on ordinary paper, others use a heated tip resting on heat-sensitive paper. In either case, a continuous trace of the input voltage is printed on the paper strip. Strip-chart

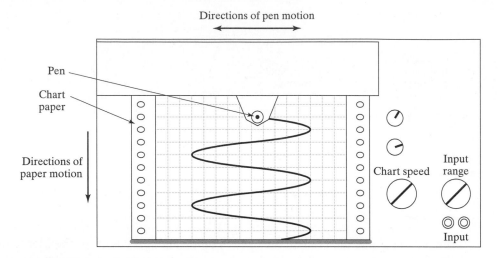

FIGURE 3.30

Strip-chart recorder.

recorders can be accurate to about 0.5% of full scale and can record frequencies up to a few hundred hertz. Although strip-chart recorders are still widely used, particularly in process control, computer data acquisition systems can often record time-varying signals with greater versatility and are often lower in cost.

3.3.4   Data Acquisition Systems

A measurement system could consist of a single sensor, appropriate signal conditioning, and an indicating device such as a digital voltmeter. In most experiments, however, several measurements are required. Data acquisition systems can accept the outputs from several sensors and record them. A block diagram of an experimental setup using a data acquisition system is shown in Figure 3.31. Each of the inputs to the acquisition system is called a *channel*. The acquisition system can be something simple

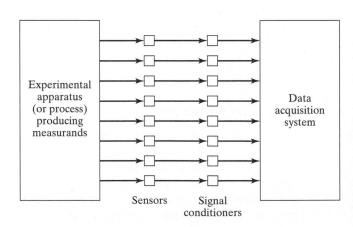

FIGURE 3.31

Block diagram experiment using an eight-channel data acquisition system.

like a multichannel tape recorder. A class of data acquisition system called *data loggers* are often used to keep track of data in processes and either store the results in memory or print the results on a strip of paper.

Personal computers have dramatically revolutionized data acquisition. A personal computer system equipped with data acquisition features can be used, with tremendous advantages, even for small experiments. In Chapter 4 we discuss computer data-acquisition systems in some detail.

3.4   ELECTRICAL TRANSMISSION OF SIGNALS BETWEEN COMPONENTS

Although the three primary components of a measuring system—sensor, signal conditioner, and indicator/recorder—are the most complex, the interconnecting wires can have a significant (in some cases, dominant) effect on the quality of the signals. As a result, close attention has to be given to this wiring. The primary methods of transmitting signals are low-level analog voltage, high-level analog voltage, analog current loop, and digital code.

3.4.1   Low-Level Analog Voltage Signal Transmission

Low-level analog voltage signals (under 100 mV) are common in measuring systems since there are a large number of sensors that have maximum output voltages in this range. Maximum voltages of only 10 mV are frequent. It is difficult to transmit such voltages over long distances because the wires tend to pick up electrical noise (interference). Ambient electric and magnetic fields caused by power conductors and other electrical equipment can induce voltages in the signal wires that distort the signal significantly. While electrical noise is sometimes added to the signal due to direct electrical connections, it is more common for the noise to appear in the signal lines due to capacitance or inductive coupling between the signal lines and noise sources (Figure 3.32).

Another common cause of electrical interference is the *ground loop*. Due to all the electrical equipment around us, there exist significant current flows in the earth itself.

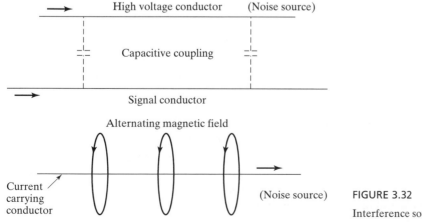

FIGURE 3.32

Interference sources for signal wiring.

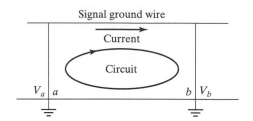

FIGURE 3.33

Source of ground loop.

These currents cause voltage variations (normally time varying) in the earth, which means that two separate connections to the earth ground are not, in general, at the same voltage. Consequently, current is conducted by the doubly grounded conductor shown in Figure 3.33. It is not even necessary for a conductor to be physically connected to the earth ground at more than one point; a second "connection" can be formed from capacitance effects.

In general, these interference effects increase when the length of wiring increases. However, it is difficult to give absolute limits on the length of the connecting wiring since the problems depend on many factors, including the electrical characteristics of the transducer, the strength of electrical noise sources in the vicinity of the wiring, and the nature of the transducer signal. If the signal is quasi-steady, much of the noise can be eliminated by installing filter networks at the receiving end, and wires that are 10 to 20 ft long may be satisfactory. In some cases, wires as short as 2 ft may give unacceptable noise levels.

Although minimizing noise signals is a significant subject in its own right [Morrison (1986) is a complete book on the subject and Wolf (1983) offers a shorter discussion], several guidelines should be followed to minimize the pickup of noise (Figure 3.34):

1. Use shielded wire for signal-wire connections. In a shielded wire, the two signal wires are completely surrounded by a flexible metal shield, the latter normally constructed of braided wire. This shield should be connected to ground at one and only one point. Shielded wire effectively eliminates capacitive coupling to surrounding noise sources. If one of the signal wires is already grounded, the shield may serve as that conductor. Ground does not necessarily mean earth ground. Experimental systems may have a ground point that is not connected to earth ground. Ground is simply a common connection point that keeps grounded components at the same potential. It should be noted that the ground lead available in power plugs is primarily for safety purposes and is not a good ground for instrumentation. The power-plug ground may connect to the actual earth ground a great distance from the experimental setup. In addition, power grounds usually run in conduits parallel to power conductors, which induce noise voltages.

2. Keep the two signal wires very close to each other. It is preferable that the two signal wires actually be twisted around each other (twisted-pair wires). This creates a balanced condition in which both conductors are affected in the same way (common-mode instead of differential-mode noise signals). The two conductors of power leads should also be close together and twisted if possible to minimize the radiated magnetic fields. Avoid getting power and signal wires too close to each other.

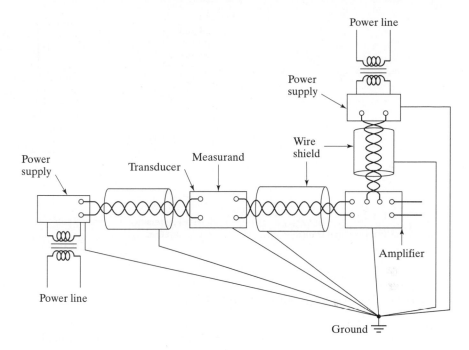

FIGURE 3.34

Grounding and shielding of system components.

3. If possible, use a single ground for the entire experimental setup. This is to mini-mize problems with ground loops. Each shield should be connected to ground with a single connection.
4. Amplify all signals with amplifiers with true differential inputs and a high com-mon-mode rejection ratio (CMRR).
5. Use high-quality power supplies. In some cases it may be better to use batteries as power supplies for instruments instead of ac line-powered power supplies to minimize power-line interference noise.

The experimenter should always check for noise signals in the instrumentation prior to taking the data. One way to do this is to observe the results on the indicating or recording device with a static measurand applied to the transducer. Any variation in output with time represents noise. With improperly wired millivolt systems, it is not un-usual to find noise with a larger amplitude than the desired signal.

3.4.2 High-Level Analog Voltage Signal Transmission

Some transducers' output voltages are higher than 100 mV, usually in the range 0 to 10 V. These transducers are normally more expensive than low-level-output transducers, but the signals are much less susceptible to interference. The reason they are more expensive is that an instrumentation amplifier is normally included directly in the transducer. These high-level signals can normally be transmitted for a distance of 30 to 100 ft without major problems.

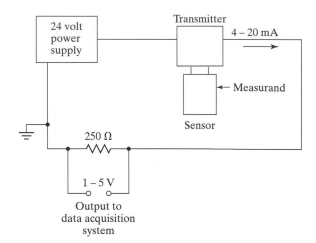

FIGURE 3.35

Current-loop signal transmission.

3.4.3 Current-Loop Analog Signal Transmission

In current-loop systems the output of the sensor is converted to a *current* rather than a voltage at the location of the sensor. Over the range of the transducer, this current will vary between 4 and 20 mA. A typical current-loop system shown in Figure 3.35 has significant advantages and is widely used in process applications such as power plants and refineries. At the transducer, a device called a *transmitter* converts the sensor output into a current. This current is then converted to a voltage at the receiving end. The minimum current of 4 mA means that open circuits in the wiring are easy to detect—a reading of 0 mA is possible only under open-circuit conditions. The current signals are also much less susceptible than analog voltage signals to environmental noise since the power associated with current signals is very large compared to most analog voltage systems. In the latter, the impedance of the recorder/indicator is usually extremely high (1 MΩ or more), and the current, and hence the power, are close to zero. Current-loop signals can be transmitted for up to 2 miles without major degradation due to noise. Long cable lengths will, however, constrain the upper-frequency limit of the signals.

3.4.4 Digital Signal Transmission

Digital signal transmission is by far the most reliable method but also the most difficult and expensive to implement. In digital signal transmission, the information in the transducer signal is converted to a series of voltage pulses, called *bits*, which transmit the information in digital code. If the voltage of the pulse exceeds a certain level, the pulse is "on"; if the voltage is below another level, the pulse is "off." There is a band between the two levels to avoid ambiguity. When such systems are designed properly, the signals are almost immune to problems from environmental noise. Signals from satellites visiting planets millions of miles away are transmitted successfully in digitally coded radio waves. In these cases the background noise is very strong relative to the radio signal, yet it is still possible to extract high-quality data.

There are some standard methods of transmitting digital data along cables. The most common are known as RS232, RS422, and IEEE488. (See Beckwith et al., 1993.) RS232 and RS422 are known as serial communication standards. The information is transmitted one bit at a time using two conductors. RS232 is normally limited to about 50 ft, whereas RS422 is suitable for hundreds of feet. Another serial standard that has become popular is USB—universal serial bus. IEEE488 (sometimes called GPIB) is a parallel standard—8 bits are transmitted simultaneously along eight conductors (plus eight more conductors for other purposes) and transmission is normally limited to cable lengths of 60 ft. Many components of measuring systems are designed to use one of these digital standards. Digital communication between measurement system components is common except for the transducer.

Unfortunately, most transducers do not output digital signals. The limited available digital-output transducers (or conversion devices connected to the transducer) are for the most part expensive and are only suitable for low-frequency variations in the measurand. If digital data transmission is used, the recording or indicating device must also have digital input capability. However, it is likely that in future years, improved high-speed digital output transducers will become commonplace, due to the inherent low noise of the systems.

REFERENCES

[1]   ANDERSON, N. (1980). *Instrumentation for Process Measurement and Control*, Chilton, Radnor, PA.

[2]   ANSI/ISA (1975). *Electrical Transducer Nomenclature and Terminology*, ANSI Standard MC6.1 (ISA S37.1).

[3]   ANSI/ISA (1979). *Process Instrumentation Terminology*, ISA Standard S51.1.

[4]   BARNEY, G. (1988). *Intelligent Instrumentation*, Prentice Hall, Englewood Cliffs, NJ.

[5]   BASS, H. G. (1971). *Introduction to Engineering Measurements*, 5th ed., McGraw-Hill, New York.

[6]   BECKWITH, T., MARANGONI, R., AND LIENHARD, V. (1993). *Mechanical Measurements*, Addison-Wesley, Reading, MA.

[7]   DALLY, J., RILEY, W., AND McCONNELL, K. (1993). *Instrumentation for Engineering Measurements*, 2d ed., Wiley, New York.

[8]   FRANCO, S. (2002). *Design with Operational Amplifiers and Analog Integrated Circuits*, McGraw-Hill, New York.

[9]   HOLMAN, J. (2001). *Experimental Methods for Engineers*, 6th ed., McGraw-Hill, New York.

[10]   JOHNSON, D. (1976). *Introduction to Filter Theory*, Prentice Hall, Englewood Cliffs, NJ.

[11]   MORRISON, R. (1986). *Grounding and Shielding Techniques in Instrumentation*, 3d ed., Wiley, New York.

[12]   NORTON, H. (1982). *Sensor and Analyzer Handbook*, Prentice Hall, Englewood Cliffs, NJ.

[13]   THOMPSON, L. M. (1979). *Basic Electrical Measurements and Calibration*, Instrument Society of America, Research Triangle Park, NC.

[14]   TURNER, J. D. (1988). *Instrumentation for Engineers*, Springer-Verlag, New York.

[15]   TURNER, J. D. and HILL, M. (1998). *Instrumentation for Engineers and Scientists*, Oxford University Press, New York.

[16]   WOLF, S. (1983). *Guide to Electronic Measurements and Laboratory Practice*, 2d ed., Prentice Hall, Englewood Cliffs, NJ.

PROBLEMS

3.1 An amplifier produces an output of 5 V when the input is 5 μV. What are the gain G and the decibel gain G_{dB}?

3.2 An amplifier has a gain of 60 dB. If the input voltage is 3 mV, what is the output voltage?

3.3 An amplifier has a programmable (user selectable) gain (V_{out}/V_{in}) of 10, 100, or 500. What is the gain in dB for each of these selections?

3.4 A pressure transducer has an output impedance of 120 Ω and is to be connected to an amplifier. What must be the minimum input impedance of the amplifier to keep the loading error less than 0.1%?

3.5 The output of a force transducer is 5 mV when measured with a DMM that has a 5-M Ω input impedance. The output (for the same force) is 4.8 mV when connected to a filter with a 10-k Ω input impedance. What is the output impedance of the force transducer?

3.6 A noninverting amplifier like that shown in Figure 3.11 is to be constructed with a μA741C op-amp. It is to have a gain of 100.

    (a) Specify values for the two resistors.

    (b) What will be the gain and phase angle at 10 kHz?

3.7 A noninverting amplifier like Figure 3.11 is to be constructed with a μA741C op-amp. It is to have a gain of 100.

    (a) Specify values for the two resistors.

    (b) Sketch the Bode plots for this amplifier using specific numerical values.

3.8 A noninverting amplifier like Figure 3.11 is to be constructed with a μA741C op-amp. It is to have a gain of 1000. Specify values for the two resistors. What will be the gain and phase angle at 10 kHz?

3.9 An inverting μA741C-based amplifier like Figure 3.13 is to have a gain of 100 and an input impedance of 1000 Ω. What will be the cutoff frequency and the values of R_1 and R_2?

3.10 An inverting μA741C-based amplifier like Figure 3.13 is to have a gain of 10 and an input impedance of 10 kΩ. What will be the cutoff frequency and the values of R_1 and R_2?

3.11 Change in frequency is often measured in *decades* rather than *octaves* where a decade is an increase in frequency by a factor of 10.

    (a) How many octaves correspond to a decade?

    (b) The roll-off for a first order Butterworth filter is 6 dB per octave. What will be the roll-off in dB per decade?

3.12 Derive Eq. (3.17).

3.13 The output of a dc generator that produces a maximum voltage of 90 V is to be attenuated to 10 V for input to a filter. Specify values of the resistors in an attenuation network such that the loading error of the voltage at the output terminals of the generator is 0.1%. The output impedance of the generator is 10 Ω and the filter has an input impedance of 100 kΩ.

3.14 A data acquisition system is to be used to measure the value of a maximum 120-V line voltage. The maximum input voltage to the data acquisition system is 8 V and its input impedance is 1 MΩ. The output impedance of the power line circuit is 0.5 Ω.

    (a) Determine the value for resistor R_2 in Figure 3.15 if R_1 has a value of 100 kΩ.

(b) Determine the power dissipated by the two resistors.

(c) Determine the loading error of the line voltage at the attenuator input.

3.15 How many octaves higher is a 7600-Hz sine wave than a 2100-Hz sine wave?

3.16 A lowpass Butterworth filter has a corner frequency of 1 kHz and a roll-off of 24 dB per octave in the stopband. If the output amplitude of a 3-kHz sine wave is 0.10 V, what will be the output amplitude of a 20-kHz sine wave if the input amplitudes are the same?

3.17 A recording device has a frequency response which shows that the output is down 2 dB at 200 Hz. If the actual input is 5.6 V at 200 Hz, what will be the expected error of the voltage reading (in volts)?

3.18 A transducer measures a sinusoidal signal with an amplitude of ±5 V and a frequency up to 10 Hz. Superimposed on this signal is 60-Hz noise with an amplitude of 0.1 V. It is desired to attenuate the 60-Hz signal to less than 10% of its value using a Butterworth filter. Select a filter order to perform this task if the corner frequency is 10 Hz.

3.19 A vibration transducer measures a sinusoidal signal with up to a 100-Hz frequency and an amplitude of ±5 V. Superimposed on this signal is an additional signal with a frequency of 1000 Hz and an amplitude of 0.2 V. It is desired to attenuate the 1000-Hz signal to 1% of its value using a Butterworth filter. Using a corner frequency of 100 Hz, select a filter order to perform this task.

3.20 If the corner frequency of a lowpass filter is 1500 Hz, calculate the attenuation from the bandpass gain at 3000 Hz for the following filters:

(a) A fourth-order Butterworth filter.

(b) A fourth-order Chebyshev filter with 2-dB ripple width.

(c) A fourth-order Bessel filter.

3.21 Specify the values for the resistors and capacitor for a first-order Butterworth lowpass op-amp filter with a dc gain of unity, a corner frequency of 12 kHz, and a dc input impedance of 1000 Ω.

3.22 How would you solve Problem 3.21 if the required input impedance were 10 MΩ?

3.23 If a fourth-order Butterworth lowpass filter has a unity dc gain and a corner frequency of 1500 Hz, what will be the gain at 25 kHz?

3.24 An oscilloscope shows a vertical deflection of 4.3 divisions. The oscilloscope is set on a vertical sensitivity of 2 V/div. What is the input voltage?

3.25 An oscilloscope has eight divisions in the vertical direction, and the vertical sensitivity is set to 100 mV/div. What is the range of the oscilloscope on that scale?

3.26 An oscilloscope has vertical divisions that are 1 cm high, and the beam trace is 0.5 mm in thickness. The vertical scale is set to 5 mV per division. What would you consider to be a reasonable estimate of resolution of this device in mV?

C H A P T E R   4

Computerized Data-Acquisition Systems

Computers so facilitate the process of data acquisition that they are the appropriate tool for a very broad spectrum of experiments. In this chapter we describe the basic components and operation of computerized data-acquisition systems. Since some readers might not be familiar with computer systems, a brief background is provided. An important related subject, sampling of time-varying signals, is covered in Chapter 5.

4.1   INTRODUCTION

As in other aspects of technology, computers have had a major impact on the field of data acquisition. Since the late 1950s, computers have been used to monitor, and in many cases to control, the performance of large process plants such as refineries and chemical plants and to acquire data in major testing programs such as the space program. These systems were extremely expensive and required highly skilled personnel to set them up. In the late 1960s, lower-cost computer data-acquisition systems became available for smaller tests, but since they were still difficult to set up and program, their use was limited. The appearance in the late 1970s of the simple and reliable Apple II personal computer led to inexpensive and simple data-acquisition systems that could readily be used for small experiments. Personal computer systems (mostly using the Windows® operating system) are now so capable that they can be used for a significant fraction of all engineering testing.

Computers significantly enhance the process of data acquisition, largely because of their versatility and speed. Computers manage the acquisition of data from multiple sensors (at high sampling rates if desired), save the data, manipulate and display the data, and, if required, make use of the results to perform control functions. Figure 4.1 shows a block diagram of a typical computerized data-acquisition system (DAS or DAQ) with eight channels for the input of analog voltage signals. The computer plus two other components, the multiplexer and the analog-to-digital converter, replace the

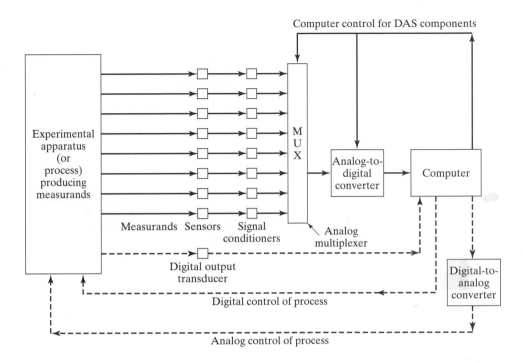

FIGURE 4.1

Computerized data-acquisition system.

single data-acquisition system box shown for a general data-acquisition system in Figure 3.31. Although most transducers produce an analog voltage (at least after signal conditioning), some transducers produce digital output. These signals do not pass through the analog multiplexer and analog-to-digital converter but take a more direct path to the computer (possibly through a digital multiplexer). The computer may be used to control the process, either directly with digital signals or with analog signals from digital-to-analog converters. The functions of these components will be discussed later in the chapter.

4.2   COMPUTER SYSTEMS

4.2.1   Computer Systems for Data Acquisition

The computer systems most commonly used for data acquisition in experiments are personal computers using the Windows® operating system. Sampling rates of over 10 million samples per second are possible, and more than 3000 separate sensors can be sampled (although not in the same system at the same time). Several major companies supply software and hardware to make a personal computer into a data-acquisition system, and some of the hardware is available in low-cost generic form. Figure 4.2 shows a board designed to plug into the interior bus of a personal computer and a

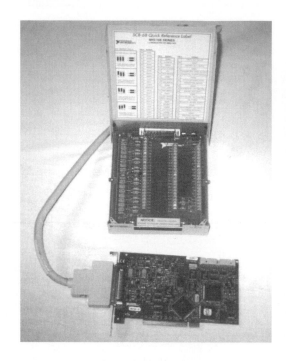

FIGURE 4.2

Data-acquisition board with eight analog input channels and two analog output channels also showing an analog signal connection box. Manufactured by National Instruments.

shielded box for connecting the transducers used for the experiment. This board not only permits reading the output of various transducers, but also has outputs for control of the experiment if required. By carefully selecting suppliers, integrating the systems and software can be seamless, making implementation straightforward. Hardware and software are also available for the Macintosh® personal computer systems.

In some cases, the performance of personal computers is not sufficient, and higher performance computers called workstations and servers are more suitable. Data-acquisition hardware and software are readily available for these computers. These higher performance systems often use operating systems such as Unix instead of Windows®.

For monitoring and controlling many production systems, specialized embedded computers are used. Probably the most common embedded computers are those used for fuel control in modern automobiles. However, embedded computer systems are used in a wide variety of devices from medical imaging equipment to assembly-line robots.

4.2.2   Components of Computer Systems

Although some computers used in data acquisition are highly specialized (the engine-control computer in an automobile, for example), the computers normally used are quite standard and have the following components:

Central processing unit (CPU)
Program (software)
Random access memory (RAM)
Mass storage systems

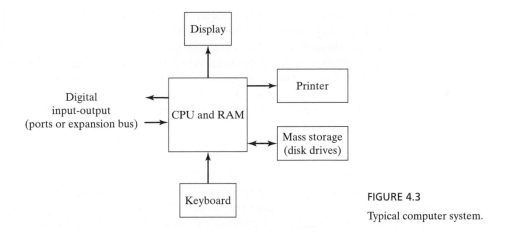

FIGURE 4.3

Typical computer system.

Display
User input device (keyboard, mouse, etc.)
Printers and plotters

Figure 4.3 is a block diagram of these components.

The *central processing unit* (CPU) controls all aspects of computer system operation and performs all of the arithmetic operations (multiplication, addition, etc.). The CPU operations follow instructions contained in the user-provided program. The CPU also follows instructions from the computer operating system programs and from built-in programs.

Most personal computers have a single central processor, but may have similar processors controlling certain peripheral devices, such as the display or keyboard. However, since there is only one central processor, for the most part, computers execute instructions in a sequential manner. That is, only one instruction can be executed at a time.

The *program* provides a set of instructions that cause a computer to perform a specific function. The program is often known as *software* to distinguish it from the electronic and mechanical components of the computer, known as *hardware*. Software may either be written by the user completely, or it may be commercial software in which the user only specifies certain inputs. In some cases (and commonly in the data-acquisition field), the user may combine his or her own programming and commercial software.

The *random access memory* is a subsystem of the computer that can store information temporarily. In particular, it stores the program instructions and numerical data when the computer is being operated. RAM is made up of purely electronic components with no moving mechanical parts, and as a result, it can supply information to the central processor at a very high rate (and store information at the same rate). The information in RAM can readily be changed by the CPU. RAM is volatile, and all stored information is lost when the computer power is interrupted. Consequently, programs and data are permanently stored in mass storage devices. (See next paragraph.) There is normally another kind of memory in a computer, *read-only memory* (ROM), used for permanent storage of information required by the computer to operate. The user cannot normally modify information stored in ROM.

Mass storage devices are used to store large volumes of information, such as numerical data and programs, in a permanent or semipermanent form. Mass storage can take a number of forms. The simplest form of mass storage is the magnetic tape recorder. These units are similar in operation to audiocassette recorders and can store substantial quantities of data. These data can be transferred to the CPU at quite high rates (but slower than RAM) but the data are arranged sequentially. For example, to read data in the middle of a tape, it is necessary to wind through the first part of the tape prior to reading the desired data.

At the present time, the most common mass storage system is the *disk drive*. In this system, information is stored in concentric bands on rotating disks coated with a magnetic material. An electromagnetic read/write head moves radially over the disk surface and alters the magnetic properties of the surface in a retrievable manner. Disk drives store large amounts of data, which can be transferred to the CPU rapidly (but again much more slowly than RAM). Disk drives have the advantage over tape drives that data files can be retrieved randomly—only the needed portion of information need be read. The disk drive has the advantage over tape drives that a compact disk has over a cassette tape—that any segment can be accessed almost immediately.

Another technology is the optical disk storage device. The primary advantage of these devices (which are based on the same technology as that used in audio compact disks) is that they store very large volumes of data. Optical disks can be used to store large databases such as library indices and encyclopedias. They can also serve as an everyday medium for file storage.

The display (or monitor or screen) system usually uses a cathode-ray tube (CRT). Displays in portable computers use liquid crystal display (LCD) technology, which is much more compact and uses less power. The display shows the user a limited amount of information and is also used to prompt the user for required input or actions.

The most common user input device is the keyboard. The user simply types required information. However, a "mouse" or a similar device is often used to point to regions on the screen to communicate with the computer. Some computers have what are known as *touch screens*, in which the user can touch regions of the screen to communicate.

Usually, computers have some kind of printer in order to produce printed output (*hard copy*). The method of printing depends on the quantity and quality of printing required. Optical devices such as laser printers are common for all types of computers. Moderately priced ink jet printers usually have the capability to print in color, which is often desirable in experimental work. Plotters are another type of output device. Mechanical plotters using conventional ink pens are the most common type, largely due to the high-quality print and the ability to use multiple colors.

Computers also have features that enable the user, optionally, to connect to other devices. Inside the computer, it is possible to connect to the *bus*, a series of conductors connecting the internal components of the computer. Outside the computer, there are generally plug connections known as *ports*, which are connected to the bus internally. The components that convert a computer to a computerized data-acquisition system connect either directly to the bus or to one of the ports.

4.2.3   Representing Numbers in Computer Systems

While numbers used in the everyday world are normally represented in base 10 (decimal), it is far more practical in computers to represent numbers in base 2 (binary). Information in computers is stored in bistable devices called *flip-flops*, which can have two possible states. One state is defined as "on" and is assigned a numerical value of 1, and the other state is defined as "off" and is assigned a numerical value of 0. A series of flip-flops are required to represent a number. For example, the binary number 1001, which corresponds to the decimal number 9, can be represented in a computer using four flip-flops. Each of these flip-flops represents a *"bit"* of the number. The leftmost "1" in the binary number 1001 is the *most significant bit* (MSB). The rightmost "1" is the *least significant bit* (LSB). It is common in computers to break long binary numbers up into segments of 8 bits, which are known as *bytes*.

There is a one-to-one correspondence between binary numbers and decimal numbers. For example, the 4-bit binary number above can be used to represent the positive decimal integers from 0 (represented by 0000) to 15 (represented by 1111). Example 4.1 shows how to convert a binary number to decimal, and Example 4.2 shows how to convert a decimal number to binary.

Example 4.1

Convert the 8-bit binary number 01011100 to decimal.

Solution: In moving from left to right, the bits represent $2^7, 2^6, 2^5, 2^4, 2^3, 2^2, 2^1$, and 2^0. Hence to get the decimal equivalent, we simply have to add the contributions of each bit.

$$N_{10} = 0(2^7) + 1(2^6) + 0(2^5) + 1(2^4) + 1(2^3) + 1(2^2) + 0(2^1) + 0(2^0)$$
$$= 0 + 64 + 0 + 16 + 8 + 4 + 0 + 0$$
$$= 92$$

Example 4.2

Find the 8-bit binary number with the same value as that of the decimal number 92.

Solution: This problem can be solved by a series of divisions by 2:

	Remainder
2/92	
2/46	0
2/23	0
2/11	1
2/5	1
2/2	1
2/1	0
0	1

The answer is contained in the remainders. The top number is the least significant bit, and the bottom number is the most significant bit. The answer is, thus, 1011100. However, we are asked for an 8-bit number, so the most significant bit is 0 and the answer is

$$N_2 = 01011100$$

In Examples 4.1 and 4.2, binary numbers are used to represent positive decimal integers. However, it is also necessary to represent negative numbers and floating-point numbers (such as 3.56×10^3). Special techniques are used to represent these numbers.

Negative numbers are most commonly represented inside computers by a technique known as *2's complement*. As noted above, 4 bits can be used to represent the decimal integers 0 to 15. These same 4 bits can alternatively be used to represent numbers from −8 to +7. The positive numbers from 0 to 7 are represented by the three least significant bits, ranging from 0000 to 0111. The negative numbers from −8 to −1 are represented by the binary numbers 1000 to 1111, respectively. For positive numbers, the most significant bit is always 0, while it is always 1 for negative numbers. To convert a negative decimal integer to 2's-complement binary, the following procedure can be followed:

1. Convert the integer to binary as if it were positive.
2. Invert all of the bits—change 0's to 1's and 1's to 0's.
3. Add 1 LSB to the final result.

This process is demonstrated in Example 4.3. While 2's complement may appear to be an awkward method to represent negative numbers, computers can perform arithmetic very efficiently using this numeric representation.

Floating-point numbers are handled by keeping separate track of the two parts of the number, the mantissa and the exponent. Separate arithmetic operations are performed on the two parts, similar to hand calculations.

Example 4.3

Convert the decimal integer −92 to an 8-bit 2's-complement binary number.

Solution: In Example 4.2 we converted +92 to binary and the result was

01011100

Next we invert all the bits, to obtain

10100011

and then add 1 to obtain the final result, 10100100.

In most computerized data-acquisition-system applications, the user will not actually need to use binary numbers. This is because the interface between the user and the computer normally makes binary-to-decimal or decimal-to-binary conversions automatically. When a user requests that a number stored in the computer be printed out, the computer converts the internal binary number to decimal prior to printing. Similarly,

TABLE 4.1   ASCII Characters

Decimal code	Character	Decimal code	Character	Decimal code	Character	Decimal code	Character	Decimal code	Character
33	!	52	4	71	G	90	Z	109	m
34	"	53	5	72	H	91			110	n
35	#	54	6	73	I	92		111	o
36	$	55	7	74	J	93			112	p
37	%	56	8	75	K	94	∧	113	q
38	&	57	9	76	L	95	-	114	r
39	'	58	:	77	M	96		115	s
40	(	59	;	78	N	97	a	116	t
41	)	60	<	79	O	98	b	117	u
42	*	61	=	80	P	99	c	118	v
43	+	62	.	81	Q	100	d	119	w
44	,	63	?	82	R	101	e	120	x
45	−	64	(a)	83	S	102	f	121	y
46	.	65	A	84	T	103	g	122	z
47	/	66	B	85	U	104	h	123	{
48	0	67	C	86	V	105	i	124		
49	1	68	D	87	W	106	j	125	}
50	2	69	E	88	X	107	k	126	~
51	3	70	F	89	Y	108	l	127	

when a decimal number is typed on the keyboard, it is converted to binary before it is stored in the computer. Although the DAS user will not frequently convert bases between binary and decimal, it is necessary to be familiar with the binary number system to understand the operation of some system components.

When computers communicate with the outside world through various input/output ports, other binary codes are normally used that differ from the internal code. The most common code used in computers is known as the *American Code for Information Interchange* (ASCII). In this code, 8 bits are used to represent up to 256 characters. The first 128 of these (0 to 127) are standardized, and ASCII codes 33 to 127 are shown in Table 4.1. For example, the character lowercase k is represented by the binary number 01101011, which has the decimal value 107. Codes 0 to 32 are special control characters and are not shown. While ASCII-coded data are used to transfer information, they are not useful for arithmetic operations.

4.3   DATA-ACQUISITION COMPONENTS

To make a computer into a data-acquisition system, several additional components are required, as shown in Figure 4.1. Multiplexers and analog-to-digital converters are almost always present and other components are added as required. If it is necessary to take readings of several measurands at precisely the same time, simultaneous sample-and-hold devices (or related systems) are required. If it is necessary to control the experiment (or process), the computer must supply outputs in digital or analog form. If the required control signal is analog (to control such components as valves and heaters), the computer-system digital output must be processed through a digital-to-analog converter.

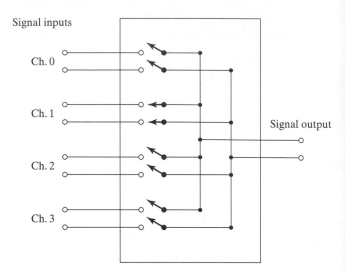

Signal inputs

FIGURE 4.4

Four-channel analog multiplexer (channel 1 connected).

4.3.1   Multiplexers

As mentioned above most computers perform instructions in a sequential manner. Thus, the simplest method to take data is to read the outputs of the sensors sequentially. In most cases the computer reads information from the various channels one at a time using a device called a *multiplexer* (MUX). The MUX, as used in this application, is essentially an electronic switch. The computer instructs the MUX to select a particular channel, and the data are read and processed. The computer then causes the MUX to select another channel, and so forth. A mechanical analog to the DAS multiplexer is shown in Figure 4.4. The switches in the multiplexers used in data-acquisition systems are semiconductor devices.

Although multiplexers are rather simple devices, they are subject to some errors. One of these is *crosstalk*. Adjacent channels may interfere with the channel being read. This error is normally a precision error. An additional error might occur if the output signal voltage is not exactly the same as the input—that is, the multiplexer alters the signal. The measure of this characteristic is known as *transfer accuracy*. In the most common situation, an analog multiplexer is used to select analog signals from the connected transducers and directly or indirectly connect these signals to the analog-to-digital converter. (See Section 4.4.2.) However, some transducers have a digital output that can be input to the computer directly, bypassing the multiplexer and the analog-to-digital converter.

4.3.2   Basics of Analog-to-Digital Converters

To explain the function of the A/D converter it is necessary to describe two distinct methods by which electronic systems process numerical information: analog and digital. Many everyday electronic devices, such as television sets and audio amplifiers, are basically analog devices (although they may have some digital components). Modern computers, on the other hand, are digital devices. If we are trying to represent a value of 5 V in an analog device, we could, for example, charge a capacitor to 5 V. In a digital device, 5 V will be represented by a digital code (a digital binary number such as 0101), which is stored on bistable flip-flops. The actual digital coding scheme used to represent an analog voltage is well defined in a particular situation, but is not universal.

Most transducers generate an analog output signal. For example, a temperature transducer might generate an output voltage that is proportional to the sensed temperature. If the temperature were to be varied in a continuous manner, the transducer output will show a continuous variation. There are possibly an infinite number of possible output values. The analog-to-digital converter (A/D converter or ADC) is the device required in data-acquisition systems to convert the analog transducer signals into the digital code used by the computer.

The digital representation of the value of the transducer output is a code that is related to the analog transducer output but does not describe the output exactly. For example, we could connect the output of the transducer to a relay, which in turn operates a light bulb. If the voltage is less than or equal to 5 V, the bulb is off. If the transducer output voltage is greater than 5 V, the bulb is on. The on–off status of the light bulb is a digital code that crudely represents the transducer output. Similarly, the output of the transducer could be represented by two light bulbs. There are four possible states for the pair of light bulbs:

	Bulb 1	Bulb 2
$V < 2.5$	off	off
$2.5 \leq V < 5.0$	off	on
$5.0 \leq V < 7.5$	on	off
$7.5 \leq V$	on	on

With two light bulbs, the transducer output is now represented more accurately in digital form than with a single light bulb. These light bulb devices represent primitive forms of analog-to-digital converters. The single-light-bulb device is a 1-bit A/D converter, and the two-light-bulb device is a 2-bit A/D converter. A bit has two possible states, on or off. Inside a computer, flip-flops instead of light bulbs are used to represent these states.

In general, the output of an analog-to-digital converter has 2^N possible values, where N is the number of bits used to represent the digital output. The 1-bit device has two possible output states, 0 and 1; and the 2-bit device has $2^2 = 4$ possible output states (00, 01, 10, and 11 in binary representation). Computerized data-acquisition systems usually use A/D converters with at least 8 bits, where the number of possible states is 2^8 (which equals 256). The possible states are then represented by binary numbers with values between 00000000 and 11111111. An example of the output is 10000001. Since there is a direct conversion between binary (base 2) and decimal (base 10) numbers, this output of 10000001 could be stated as 129 in decimal. The actual physical representation at the output of the A/D converter is, however, binary.

Although the circuitry of analog-to-digital converters can vary widely, from an external performance viewpoint, they can normally be described by three primary characteristics. The first of these is the number of bits used to represent the output. The greater the number of bits, the greater the number of possible output states and the more accurately the digital output will represent the analog input. The second characteristic is the input range. The third characteristic is the conversion speed, which is the time it takes to create a digital output after the device is instructed to make a conversion.

The input range of an A/D converter is the range of analog input voltages over which the converter will produce a representative digital output. Input voltages outside the range will not produce a meaningful digital representation of the input. The input range of A/D converters can be classified as unipolar or bipolar. A *unipolar* converter can only respond to analog inputs with the same sign. Examples of the input range are 0 to 5 V or 0 to −10 V. *Bipolar* converters can convert both positive and negative analog inputs, with ±5 V or ±10 V being typical input ranges. Many computer data-acquisition systems offer the user the option of selecting a unipolar or bipolar mode.

A conceptually simple type of A/D converter, called a unipolar single-slope integrating converter, serves to demonstrate the A/D conversion process. In this device a fixed reference voltage is used to charge an integrator at a constant rate. (See Section 3.2.6.) The integrator output voltage will then increase linearly with time. As shown in Figure 4.5, a digital clock (counter) is started at the same time that the charging is begun. The integrator output voltage is compared continuously with the analog input voltage, using a comparator. (See Section 3.2.6.) When the integrator voltage exceeds the analog input voltage, the digital clock is stopped. The count of the digital clock is the digital output of the A/D converter. This process is simulated numerically in Example 4.4.

A block diagram of a single-slope integrating converter is shown in Figure 4.6. The FET (field-effect transistor) is used to zero the integrator at the start of the conversion. Although they demonstrate the principle of A/D conversion in a simple manner, ramp-type converters are not commonly used. Ramp converters are not as accurate as other types and are relatively slow. The most common practical A/D converters are discussed in the next section.

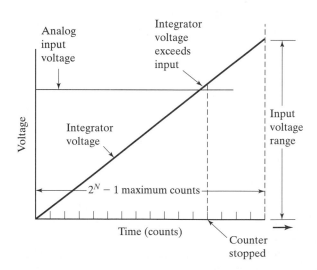

FIGURE 4.5

Ramp A/D converter process.

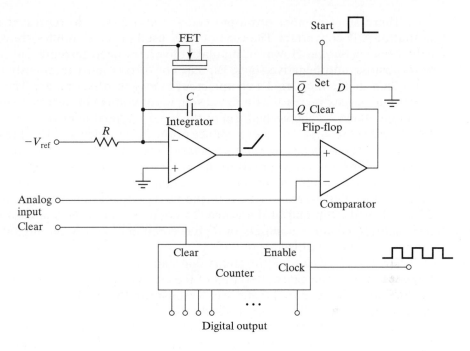

FIGURE 4.6

Single-slope integrating A/D converter circuit. (Based on Turner, 1988.)

Example 4.4

A 4-bit single-slope integrating A/D converter has an input range of 0 to 10 V. Compute the digital output for an analog input of 6.115.

Solution: The incremental voltage increase in each clock tick is

$$V_{ref} = \frac{10 - 0}{2^4} = 0.625$$

We must find the integrator voltage at the end of each clock tick and compare it to the input voltage:

$0.625 \times 1 = 0.625$	$0.625 \times 6\ \ = 3.750$
$0.625 \times 2 = 1.250$	$0.625 \times 7\ \ = 4.375$
$0.625 \times 3 = 1.875$	$0.625 \times 8\ \ = 5.000$
$0.625 \times 4 = 2.500$	$0.625 \times 9\ \ = 5.625$
$0.625 \times 5 = 3.125$	$0.625 \times 10 = 6.250 > 6.115$ stop counter

When the clock has counted to 10, the slope voltage exceeds the input voltage and the conversion is complete. The output of the converter is the decimal number 10, which has the value 1010 when expressed in binary.

There are a number of output codes that are used to represent the digital output of A/D converters. The type of code used in the examples above is simply called *binary output*. However, bipolar converters need to represent inputs with both positive and negative signs. Bipolar converters most frequently use one of two other common codes, 2's-complement binary or offset binary. The output binary number may increase with increasing input voltage or decrease with increasing input. Table 4.2 shows the form of the binary numbers for these codes. V_{ru} and V_{rl} represent the upper and lower values of the input range and V_i is the input voltage. Offset binary is just like simple binary except that for bipolar converters, the output code of zero corresponds to the lower end of the input range instead of an input of zero. 2's complement starts with a binary number on $2^N/2$ at the lower end of the range, has a value of zero in the center of the input range, and rises to $2^N/2 - 1$ at the top end of the range. 2's complement is the standard method of representing negative numbers in digital computers and is the digital converter output most easily utilized.

Figure 4.7 gives some formulas that can be used to estimate the output of A/D converters. The exact operation of the converter may cause the output to differ by one bit. Examples 4.5 and 4.6 demonstrate the use of these formulas. If an input voltage is outside the input range, the A/D converter is said to be *saturated*. Although this is an error, it is really more of an experimental mistake and cannot be estimated through error analysis. The experimenter must ensure that the device is used within its range. When the experimenter evaluates the data, output values corresponding to the bottom or top of the range should be viewed with suspicion. Worse yet, if the input voltage has a very large magnitude, it might damage the converter.

Analog-to-digital converters are subject to a number of errors. There are three systematic errors to be considered: linearity, zero, and sensitivity (or gain) errors. These errors are the same as for a general device, as described in Chapter 2 where the analog voltage is the input and the digital code is the output. A/D converters are also subject to thermal-stability effects that can alter zero, sensitivity, and linearity.

TABLE 4.2 Common Binary Output Codes of A/D Converters

Input voltage V_i	Form of binary output code			
	Offset binary	Inverted offset binary	2's complement	Inverted 2's complement
V_{ru}	$111\ldots11$	$000\ldots00$	$011\ldots11$	$100\ldots00$
⋮	⋮	⋮	⋮	⋮
.	$100\ldots00$	$011\ldots11$	$000\ldots00$	$111\ldots11$
.	$011\ldots11$	$100\ldots00$	$111\ldots11$	$000\ldots00$
⋮	⋮	⋮	⋮	⋮
V_{rl}	$000\ldots00$	$111\ldots11$	$100\ldots00$	$011\ldots11$

Definitions of variables:

V_i analog input voltage
V_{ru} upper value of input range
V_{rl} lower end of input range
N number of bits
D_o digital output

To find the output of a 2's-complement A/D converter, given the input analog voltage, use

$$D_o = \mathrm{int}\left\{\frac{V_i - V_{rl}}{V_{ru} - V_{rl}}\,2^N\right\} - \frac{2^N}{2} \tag{A}$$

"int" indicates that the calculation should be rounded to the nearest integer. The maximum allowable positive output is $(2^N/2 - 1)$, and the maximum negative output is $(-2^N/2)$. If the computed output is outside these limits, the converter is saturated and the actual output will have the value of the nearest limit.

For an offset binary or simple binary converter, the output will be

$$D_o = \mathrm{int}\left\{\frac{V_i - V_{rl}}{V_{ru} - V_{rl}}\,2^N\right\} \tag{B}$$

and the output will range from 0 at the lower limit to 2^N-1 at the upper limit.

To find the analog input, given the digital output, the following formula can be used for a 2's-complement converter:

$$V_i = \left(D_o + \frac{2^N}{2}\right)\left(\frac{V_{ru} - V_{rl}}{2^N}\right) + V_{rl} \tag{C}$$

For binary or offset binary, the formula is

$$V_i = D_o\left(\frac{V_{ru} - V_{rl}}{2^N}\right) + V_{rl} \tag{D}$$

FIGURE 4.7

Formulas to estimate A/D converter digital output.

Example 4.5

For the problem of Example 4.4, use Eq. (B) of Figure 4.7 to compute the digital output.

Solution: Since this is a simple binary device, Eq. (B) is applicable:

$$D_o = \mathrm{int}\left(\frac{6.115 - 0}{10 - 0} \times 2^4\right) = \mathrm{int}(9.78) = 10$$

The output of 10 is the same as that obtained in Example 4.4.

Example 4.6

A 12-bit A/D converter with 2's-complement output has an input range of -10 to $+10$ V. Find the output codes when the input is $-11, -5, 0, 6.115,$ and 12 V.

Solution: Substituting into Eq. (A) in Figure 4.7, the results are as follows:

Input	Computed output	Actual output
-11	-2253 (below range)	-2048
-5	-1024	-1024
0	0	0
6.115	1252	1252
12	2458 (above range)	2047

The value -11 is below the input range, so the output is simply the lowest possible output state, which is -2048. The value 12 is above the input range, so the output is the largest output state, which is 2047 for a 12-bit converter. The experimenter should be suspicious of any result at the limits of the range. For example, an analog input of -13 V would produce the same output as the -11-V input in the table.

Since the output of an A/D converter changes in discrete steps (one LSB), there is a resolution error (uncertainty), known as a *quantizing error*, which is treated as a (random) error. This quantizing error is ± 0.5 LSB. In input units, this translates to

$$\text{input resolution error} = \pm 0.5 \frac{V_{ru} - V_{rl}}{2^N} \text{ volts} \tag{4.1}$$

Thus, changes in the input as large as this resolution may not show up as changes in the digital output. For an 8-bit converter, the error is about $\pm 0.2\%$ of the input range, and for a 12-bit converter it is about $\pm 0.01\%$. The user must select a converter with suitable resolution to obtain acceptable accuracy. Example 4.7 demonstrates the use of Eq. (4.1).

Example 4.7

A 12-bit A/D converter has an input range of -10 to $+10$ V. Find the resolution error of the converter for the analog input.

Solution: Using Eq. (4.1) gives us

$$\text{input resolution error} = \pm 0.5 \left[\frac{10 - (-10)}{2^{12}} \right] = \pm 0.00244$$

The resolution uncertainty of ± 0.00244 V is the best that can be achieved.

Comment: If the input voltage were 0.1 V (which has a magnitude at the low end of the input range), the quantization error would then represent $\pm 2.5\%$ of the reading, which is probably not acceptable. The solution to this problem is to amplify the input (with some well-defined gain such as 10) before the signal enters the converter.

If the input signal is varying rapidly, A/D converters will cause another error, due to the converter aperture time, which is the time required to convert the input signal. Since aperture time is always greater than zero, the input voltage will, in general, change during the conversion process. The value of the input at the start of the conversion can only be determined with some error, depending on the aperture time and the rate at which the input signal is changing. To minimize this problem in computerized data-acquisition systems, an additional component, called a *sample-and-hold device*, is inserted between the multiplexer and the A/D converter in Figure 4.1. This device reads the input value very rapidly and then holds a constant value for the input to the A/D converter during the conversion time. For example, the conversion time is typically from 10 to 25 μs for a successive-approximations converter (see Section 4.3.3), but a sample-and-hold device can read the data in 1.5 μs or less.

4.3.3   Practical Analog-to-Digital Converters

The most common type of A/D converter used in computerized data-acquisition systems, the *successive-approximations converter*, employs an interval-halving technique. In this device a series of known analog voltages are created and compared to the analog input voltage. In the first trial, a voltage interval of one-half the input span is compared with the input voltage. This comparison determines whether the input voltage is in the upper or lower half of the input range. If the input voltage is in the upper half of the range, the most significant bit is set to 1; otherwise it is set to zero. This process is repeated with an interval half the width of the interval used in the first trial to determine the second most significant bit, and so forth until the least-significant bit is determined. This process is shown graphically in Figure 4.8, and Example 4.8 shows a numerical simulation. For an 8-bit converter, only eight comparisons are needed, and in general, for an N-bit converter, only N comparisons have to be made. Hence the device is very fast even for converters with large values of N (such as 16 or 18). A typical 12-bit successive-approximations converter can complete a conversion in 10 to 25 μs. A block diagram of a successive-approximations converter is shown in Figure 4.9. (The digital-to-analog converter component, which creates an analog signal to compare with the input signal, is described in the next section.)

Example 4.8

An 8-bit A/D unipolar successive-approximations converter has an input range of 0 to 5 V. An analog voltage of 3.15 V is applied to the input. Find the analog output.

Solution: First, a reference voltage increment is created:

$$\Delta V = \frac{\text{input span}}{2^N} = \frac{5}{2^8} = 0.01953$$

Trial digital output (D_o)	$D_o \times \Delta V$	Pass/Fail	Actual digital output
10000000 (128)	2.5	Pass	10000000 (128)
11000000 (192)	3.75	Fail	10000000 (128)
10100000 (160)	3.125	Pass	10100000 (160)
10110000 (176)	3.438	Fail	10100000 (160)
10101000 (168)	3.281	Fail	10100000 (160)
10100100 (164)	3.203	Fail	10100000 (160)
10100010 (162)	3.164	Fail	10100000 (160)
10100001 (161)	3.144	Pass	10100001 (161)

The numbers in parentheses are the decimal equivalents of the binary numbers. At the end of the process, the output is 10100001, which is the same as 161 in decimal.

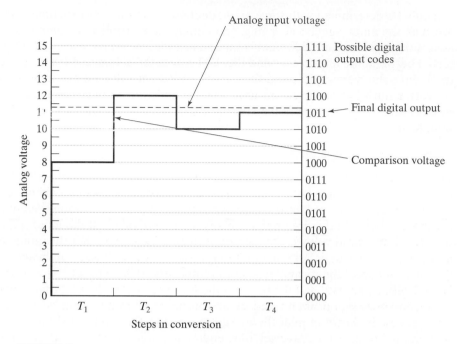

FIGURE 4.8

Graphical description of the method of successive approximations for a 4-bit A/D converter.
Based on Franco (2002).

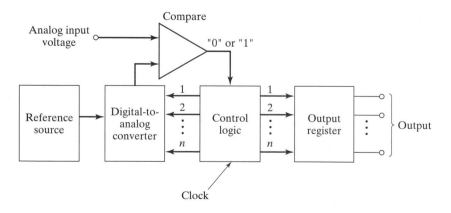

FIGURE 4.9

Block diagram of a successive-approximations A/D converter. Copyright 1990 Instrument
Society of America. From *Computer-based Data Acquisition Systems* by J. Taylor.

The fastest type of A/D converter is the *parallel* or *flash A/D converter*. With this device the conversion process is completed in a single step. Conversion times can be as fast as 10 ns. These converters essentially make all possible comparisons at once and select the digital output that corresponds to the correct answer. A block diagram of a flash converter is shown in Figure 4.10. Parallel converters require a large number of components ($2^N - 1$ comparators, for example), but modern high-component-density chip technology has made them practical. Parallel converters are commonly used for applications where 8-bit resolution is adequate. They are also available with 12-bit resolution but at a higher cost. A variation of this parallel concept, the *half-flash converter*, is common in data-acquisition systems where high sampling rates and more than 8-bit resolution are required. The half-flash converters use two flash converters and take two steps to complete a conversion. Half-flash converters can complete a 12-bit conversion in about 100 ns.

A common A/D converter design used in digital voltmeters is the *dual-slope integrating converter*. A block diagram is shown in Figure 4.11(a), and the operation is shown graphically in Figure 4.11(b). In this device, the analog input voltage signal V_i is connected to an integrator for a fixed period of time t_0. This process results in a voltage on the integrator. The analog input voltage is then disconnected from the integrator, and a reference voltage V_r of a polarity opposite to V_i is connected to the integrator. A counter is then used to measure the time t_1 it takes to reduce the integrator voltage to zero. Mathematically, the time t_1 can be expressed as

$$t_1 = t_0 \frac{V_i}{V_r} \qquad (4.2)$$

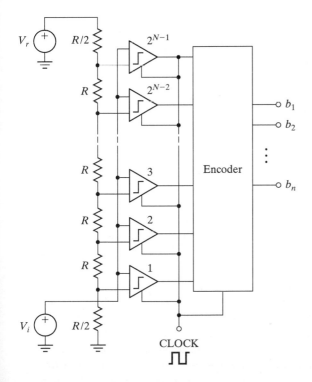

FIGURE 4.10

Block diagram of a parallel A/D converter. Based on S. Franco, *Design with Operational Amplifiers and Analog Integrated Circuits*, McGraw-Hill, 1988. Used with permission of McGraw-Hill Inc.

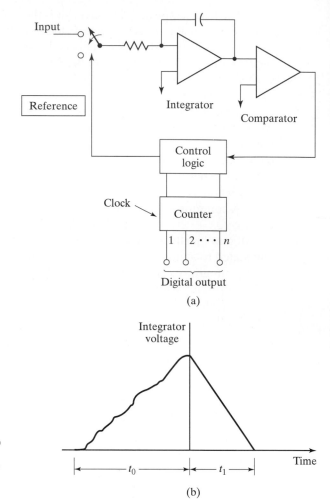

FIGURE 4.11

Block diagram of a dual-slope integrating A/D
converter: (a) block diagram; (b) integration
process. Copyright 1990 Instrument Society of
America. From *Computer-based Data
Acquisition Systems* by J. Taylor.

Since t_0 and V_r are fixed, the time t_1 is proportional to $V_i t_1$, which is the counter output and hence is in digital form, is a digital representation of the analog input voltage. These devices are probably the most accurate type of A/D converter and have the advantage that the input voltage is integrated over time, thus averaging out noise (such as 60-Hz power-line interference). The conversion time is relatively slow, on the order of 4 to 8 ms (sometimes slower), and the devices are only suitable for the slowest data-acquisition systems.

4.3.4   Digital-to-Analog Converters

While in some situations a data-acquisition system is used only to collect data, it is common for the computer to use the acquired results to change some aspect of the measured system. In process plants, this will be a control function. In experiments, it may be a control or it may be a variation of some independent variable of the experimental

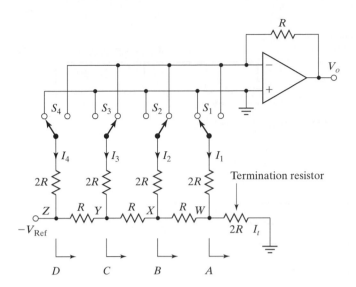

FIGURE 4.12

Digital-to-analog converter. (Based on Turner, 1988.)

apparatus. For example, if we are determining the performance of a semiconductor device as a function of temperature, the computer may acquire all relevant data, then change the power of a heater to alter the temperature, monitor the temperature until it has stabilized again, and then acquire more data.

Most devices used to control a system, such as solenoids, heaters, and valves, are analog devices. These devices operate on the basis of an analog input voltage. To operate these devices under computer control, the computer digital signals must be converted to analog signals. If the analog device is simply an on–off component, digitally controlled relays can be used. If the analog device requires proportional control, then a *digital-to-analog converter* (D/A converter or DAC) is required. As shown schematically in Figure 4.12, the various bits of the digital output signal are used to operate a set of electronic switches. In Figure 4.12 there are four bits and four switches. The resulting analog output voltage will be proportional to the digital input number. D/A converters are specified like the A/D converter—the number of input bits, the analog output range, and the conversion speed. As for A/D converters, a number of digital codes are also used.

Generally, the output electrical power of a D/A converter is insufficient to directly operate most control devices, such as proportional valves. Consequently, the output of the converter will be used to operate an amplifier, which will in turn supply the power for the activated control device.

4.3.5   Simultaneous Sample-and-Hold Subsystems

When a DAS uses a multiplexer, the individual channels are not read at precisely the same time. If the input signals are changing slowly relative to the time it takes to complete all the readings, this is not a problem. In some cases it is important that all channels (or a subset of the channels) be recorded at precisely the same time. To solve this problem, the channels of interest can each be connected to a simultaneous sample-and-hold subsystem. Figure 4.13 is a block diagram showing how a simulta-

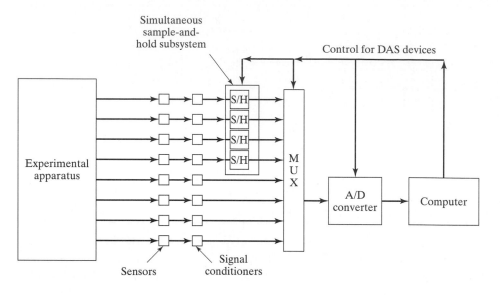

FIGURE 4.13

Simultaneous sample and hold subsystem.

neous sample-and-hold module is added to the DAS of Figure 4.1. Each of the top four channels connect to a sample-and-hold device. These sample-and-hold devices are instructed to take a reading simultaneously. However, the data from these channels are not immediately processed. The computer subsequently reads each of the channels connected to the sample-and-hold subsystem (using the multiplexer) and processes the data. Some high-speed data-acquisition systems actually have an A/D converter for each channel that must be read simultaneously. These systems are known as simultaneous sample-and-convert systems.

4.4   CONFIGURATIONS OF DATA-ACQUISITION SYSTEMS

While a generic data-acquisition system generally has the components shown in Figure 4.1, there are many ways in which these components are packaged. There are differences in system size, performance, durability, and features. A large number of companies market data-acquisition systems and components, and in most cases, selecting equipment involves significant research and discussions with supplier representatives. Furthermore, since the field is advancing rapidly, new features will frequently appear. The following general discussion is intended only as an overview.

The simplest and least-expensive data-acquisition systems are those created by plugging one or more data-acquisition circuit boards into the bus of a personal computer. At a minimum, these circuit boards include a multiplexer and an A/D converter and will be capable of handling eight analog input channels. A common additional feature is an amplifier, which is located between the multiplexer and the A/D converter. This amplifier usually has programmable gain, in the range 1 to 1000, which can be set to a different value for each channel. An amplifier shared by several channels in this way results in considerable cost savings compared with individual amplifiers for each channel. Additional features that may be included are provisions for digital inputs, D/A converters (for process/experiment control), simultaneous sample-and-hold systems, counters for transducers with frequency output, and clock-timers to take samples automatically.

These plug-in board systems can have maximum specifications of 100 to 200 channels, sampling rates up to 1 million samples per second, and resolutions up to 18 bits. Special boards are also available that will convert a personal computer into a digital oscilloscope with sampling rates up to 20 million samples per second. However, not all of these maximum figures can be obtained at the same time. The digital oscilloscope boards, while giving high sampling rates, are normally restricted to one or two channels and 8-bit resolution. The specified sampling rate is generally the total number of samples that can be taken in a second and that must be distributed among the sampled channels. Using programmable amplifiers also reduces the maximum sampling rate. Since the amplifier requires a short settling time whenever it is connected to a new signal, the sampling rate is reduced. For the maximum sampling rate, each channel should have its own amplifier.

More expensive types of personal computer–based data-acquisition systems locate data-acquisition components in separate modules that connect to the computer through ports or adapters connected to the internal bus. These systems can normally handle much larger numbers of channels—numbers in the thousands are within reason. They also generally have more convenient features for connecting signal conditioners. One particularly important class of these modular data-acquisition systems is the IEEE488, or GPIB, system. IEEE488 systems use various types of instruments which arc connected together with a standard IEEE488 digital bus. The bus, in turn, connects to a controlling computer. These systems are relatively simple to program compared with early non-IEEE488 systems, and they are fairly straightforward and reliable. Digital oscilloscopes frequently connect to a host computer using the IEEE488 bus.

Process control computers, used in such places as automated production lines, chemical and petroleum process plants, and electric power plants, are normally based on computers with higher performance than personal computers. These systems can acquire data from hundreds of sensors, perform real-time computations, and control the process. Since process-control applications often have harsh environments and demand durable and reliable systems, rugged construction and packaging are paramount.

A more recent development in process control is the distributed data-acquisition system. In systems of this type, modular components provide signal conditioning and analog-to-digital conversion in close proximity to the sensors. The resulting digital data are then transmitted to the central computer using simple standard network connections (as in an office), wireless connections, or standard computer serial connections. Each modular package can handle several sensors, and there can be multiple modular packages. The modular packages can also include components for control such as D/A converters and relays. The primary advantage of these systems is the elimination of a large number of wires connecting individual sensors to the central computer. This reduces cost, improves reliability, and minimizes electrical noise. For the most part, these systems are not high speed, being limited to about 1000 samples per second.

4.5   SOFTWARE FOR DATA-ACQUISITION SYSTEMS

For a computerized data-acquisition system (with possible control functions) to per-form satisfactorily, the system must be operated using suitable software. To take a data sample, for example, the following instructions must be executed:

1. Instruct the multiplexer to select a channel.
2. Instruct the A/D converter to make a conversion.
3. Retrieve the result and store it in memory.

In most applications, other instructions are also required, such as setting amplifier gain or causing a simultaneous sample-and-hold system to take data. The software required depends on the application.

4.5.1   Commercial Software Packages

In the process-control industry, sophisticated computer programs have been available for some time. Using selections from various menus, the operator can configure the program for the particular application. These programs can be configured to take data from transducers at the times requested, display the data on the screen, and use the data to perform required control functions. These systems are often configured by technicians rather than engineers or programmers, so it is important that the software setup be straightforward. For complicated processing or control functions, it is possible to include instructions programmed in a higher-level language such as C.

There are a number of very sophisticated software packages now available for per-sonal computer–based data-acquisition systems. These packages are very capable—they can take data, display it in real time, write the data to files for subsequent processing by another program, and perform some control functions. The programs are configured for a particular application using menus or icons. They may allow for the incorporation of C program modules. These software packages are the best choice for the majority of ex-perimental situations.

REFERENCES

[1] ANDERSON, N. (1980). *Instrumentation for Process Measurement and Control*, Chilton, Rad-nor, PA.

[2] BARNEY, G. (1988). *Intelligent Instrumentation*, Prentice Hall, Englewood Cliffs, NJ.

[3] DALLY, J., RILEY, W. AND MCCONNELL, K. (1993). *Instrumentation for Engineering Mea-surements*, 2d ed., Wiley, New York.

[4] FRANCO, S. (2002). *Design with Operational Amplifiers and Analog Integrated Circuits*, Mc-Graw-Hill, New York.

[5] INTELLIGENT INSTRUMENTATION (1994). *The Handbook of Personal Computer Instrumenta-tion*, Intelligent Instrumentation, Tucson, AZ.

[6] SHAPIRO, S. F. (1987). Board-level systems set the trend in data acquisition, *Computer De-sign*, Apr. 1.

[7] SHEINGOLD, D. H., Ed. (1986). *Analog Digital Conversion Handbook*, Prentice Hall, Engle-wood Cliffs, NJ.

[8] SONY. (1989). *Semiconductor IC Data Book 1990, A/D, D/A Converters*, Sony Corp.

[9] TAYLOR, J. L. (1990). *Computer-Based Data Acquisition*, Instrument Society of America, Research Triangle Park, NC.

[10] TURNER, J. D. (1988). *Instrumentation for Engineers*, Springer-Verlag, New York.

[11] WOLF, S. (1983). *Guide to Electronic Measurements and Laboratory Practice*, 2d ed., Prentice Hall, Englewood Cliffs, NJ.

PROBLEMS

4.1 Convert the decimal number 147 to 8-bit simple binary.

4.2 Convert the decimal number 1149 to 12-bit simple binary.

4.3 Convert the numbers +121 and −121 to 2's-complement 8-bit binary numbers.

4.4 Find the 12-bit 2's-complement binary equivalent of the decimal number 891.

4.5 The number 10010001 is an 8-bit 2's-complement number. What is its decimal value?

4.6 How many bits are required for a digital device to represent the decimal number 27,541 in simple binary? How many bits for 2's-complement binary?

4.7 How many bits are required to represent the number −756 in 2's-complement binary?

4.8 A 12-bit A/D converter has an input range of ±8 V, and the output code is offset binary. Find the output (in decimal) if the input is

(a) 4.2 V.

(b) −5.7 V.

(c) 10.9 V.

(d) −8.5 V.

4.9 An 8-bit A/D converter has an input range of 0 to 10 V and an output in simple binary. Find the output (in decimal) if the input is

(a) 5.75 V.

(b) −5.75 V.

(c) 11.5 V.

(d) 0 V.

4.10 A 12-bit A/D converter has an input range of ±10 V and an amplifier at the input with a gain of 10. The output of the A/D converter is in 2's-complement format. Find the output of the A/D converter if the input to the amplifier is

(a) 1.5 V.

(b) 0.8 V.

(c) −1.5 V.

(d) −0.8 V.

4.11 A 16-bit A/D converter has an input range of 0 to 5 V. Estimate the quantization error (as a percent of reading) for an input of 1.36 V.

4.12 An A/D converter has an input range of ±8 V. If the input is 7.5 V, what is the quantization error in volts and as a percent of input voltage if the converter has 8 bits, 12 bits, and 16 bits.

4.13 A 12-bit A/D converter has an input range of −8 to +8 V. Estimate the quantization error (as a percentage of reading) for an input −4.16 V.

4.14 An A/D converter uses 12 bits and has an input range of ±10 V. An amplifier is connected to the input and has selectable gains of 10, 100, and 1000. A time-varying signal from a transducer varies between $+15$ and -15 mV and is input to the amplifier. Select the best value for the gain to minimize the quantizing error. What will be the quantizing error (as a percentage of the reading) when the transducer voltage is 3.75 mV? Could you attenuate the signal before amplification to reduce the quantizing error?

4.15 A 12-bit A/D converter has an input range of ±10 V and is connected to an input amplifier with programmable gain of 1, 10, 100, or 500. The connected transducer has a maximum output of 7.5 mV. Select the appropriate gain to minimize the quantization error, and compute the quantization error as a percent of the maximum input voltage.

4.16 An 8-bit digital-to-analog converter has an output range of 0 to 5 V. Estimate the analog voltage output if the input is simple binary and has the decimal value of 32.

4.17 A 3.29-V signal is input to a 12-bit successive approximations converter with an input range of 0 to 10 V and simple binary output. Simulate the successive-approximations process to determine the simple binary output.

4.18 Repeat Problem 4.8(a) using the successive-approximations simulation used in Example 4.8.

4.19 What are the errors that a digital data-acquisition system may introduce into a measurement? Specify whether these errors are of bias or precision type.

4.20 Imagine that you want to purchase a digital data-acquisition system. List the questions that you want to discuss with an application engineer working for a supplier.

C H A P T E R 5

Discrete Sampling and Analysis of Time-Varying Signals

Unlike analog recording systems, which can record signals continuously in time, digital data-acquisition systems record signals at discrete times and record no information about the signal in between these times. Unless proper precautions are taken, this discrete sampling can cause the experimenter to reach incorrect conclusions about the original analog signal. In this chapter we introduce restrictions that must be placed on the signal and the discrete sampling rate. In addition, techniques are introduced to determine the frequency components of time-varying signals (spectral analysis), which can be used to specify and evaluate instruments and also determine the required sampling rate and filtering.

5.1   SAMPLING-RATE THEOREM

When measurements are made of a time-varying signal (measurand) using a computerized data-acquisition system, measurements are only made at a discrete set of times, not continuously. For example, a reading (sample) may be taken every 0.1 s or every second, and no information is taken for the time periods between the samples. The experimenter is then left with the problem of deducing the actual measurand behavior from selected samples. The rate at which measurements are made is known as the *sampling rate*, and incorrect selection of the sampling rate can lead to misleading results.

   Figure 5.1 shows a sine wave with a frequency, f_m, of 10 Hz. We are going to explore the output data of a discrete sampling system for which this continuous time-dependent signal is an input. The important characteristic of the sampling system here is its sampling rate (normally expressed in hertz). Figures 5.2 to 5.5 show the

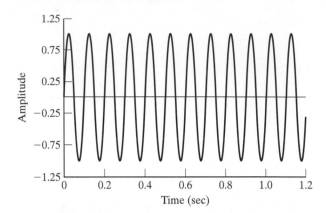

FIGURE 5.1

10-Hz sine wave to be sampled.

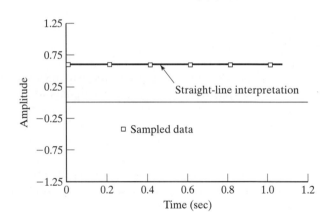

FIGURE 5.2

Results of sampling a 10-Hz sine wave at a
rate of 5 Hz.

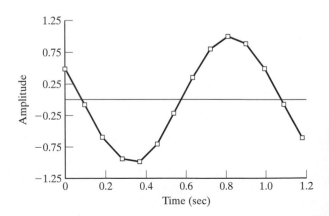

FIGURE 5.3

Results of sampling a 10-Hz sine wave at 11 Hz.

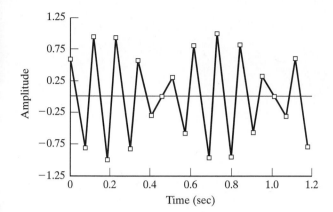

FIGURE 5.4

Results for sampling a 10-Hz sine wave at a rate of 18 Hz.

sampled values for sampling rates of 5, 11, 18, and 20.1 samples per second. To infer the form of the original signal, the sample data points have been connected with straight-line segments.

In examining the data in Figure 5.2, with the sampling rate of 5 Hz, it is reasonable to conclude that the sampled signal has a constant (dc) value. However, we know that the sampled signal is, in fact, a sine wave. The amplitude of the sampled data is also misleading—it depends on when the first sample was taken. This behavior (a constant value of the output) occurs if the wave is sampled at any rate that is an integer fraction of the base frequency f_m (e.g., f_m, $f_m/2$, $f_m/3$, etc.).

The data in Figure 5.3 appear to be a sine wave, as are the sampled data, but only one cycle appears in the time that 10 cycles occurred for the sampled data. The frequency, 1 Hz, is the difference between the sampled-data frequency, 10 Hz, and the sampling rate, 11 Hz. The data in Figure 5.4, sampled at 18 Hz, also represent a periodic wave. The apparent frequency is 8 Hz, the difference between the sampling rate and the signal frequency, and is again incorrect relative to the input frequency. These incorrect frequencies that appear in the output data are known as *aliases*. Aliases are false frequencies that appear in the output data, that are simply artifacts of the sampling process, and that do not in any manner occur in the original data.

Figure 5.5, with a sampling rate of 20.1 Hz, can be interpreted as showing a frequency of 10 Hz, the same as the original data. It turns out that for any sampling rate greater than twice f_m, the lowest apparent frequency will be the same as the actual frequency. This restriction on the sampling rate is known as the *sampling-rate theorem*. This theorem simply states that the sampling rate must be greater than twice the highest-frequency component of the original signal in order to reconstruct the original waveform correctly. In equation form this is expressed as

$$f_s > 2f_m \qquad (5.1)$$

where f_m is the signal frequency (or the maximum signal frequency if there is more than one frequency in the signal) and f_s is the sampling rate. The theorem also specifies

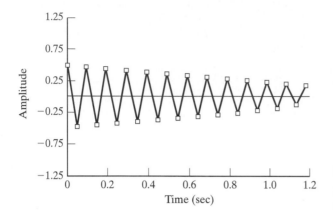

FIGURE 5.5

Results for sampling a 10-Hz sine wave at a rate of 20.1 Hz.

methods that can be used to reconstruct the original signal. The amplitude in Figure 5.5 is not correct, but this is not a major problem, as discussed in Section 5.4.

The sampling-rate theorem has a well-established theoretical basis. There is some evidence that the concept dates back to the nineteenth-century mathematician Augustin Cauchy (Marks, 1991). The theorem was formally introduced into modern technology by Nyquist (1928) and Shannon (1948) and is fundamental to communication theory. The theorem is often known by the names of the latter two scientists. A comprehensive but advanced discussion of the subject is given by Marks (1991). In the design of an experiment, to eliminate alias frequencies in the data sampled, it is necessary to determine a sampling rate and appropriate signal filtering. This process will be discussed in some detail later in the chapter.

Even if the signal is correctly sampled (i.e., at a frequency greater than twice the signal frequency), the data can be interpreted to be consistent with specific frequencies that are higher than the signal frequency. For example, Figure 5.6 shows the same data as in Figure 5.5: a 10-Hz signal sampled at 20.1 samples per second. The sampled data are shown as the small squares. However, these data are not only consistent with a 10-Hz sine wave but in this case, the data are also consistent with 30.1 Hz. Actually, there are an infinite number of higher frequencies that are consistent with the data. If, however, the requirements of the sampling-rate theorem have been met (perhaps with suitable filtering), there will be no frequencies less than half the sampling rate that are consistent with the data except the correct signal frequency. The higher frequencies can be eliminated from consideration since it is known that they don't exist.

In some cases, the requirements of the sampling-rate theorem may not have been met, and it is desired to estimate the lowest alias frequency. The lowest is usually the most obvious in the sampled data. A simple method to estimate alias frequencies involves the folding diagram as shown in Figure 5.7 [Taylor (1990)]. This diagram enables one to predict the alias frequencies based on a knowledge of the signal frequency and the sampling rate. To use this diagram, it is necessary to compute a frequency f_N called the folding frequency. f_N is half the sampling rate, f_s. The use of this diagram is demonstrated in Example 5.1.

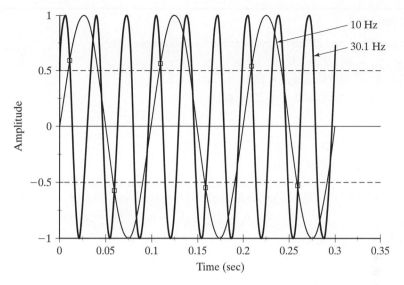

FIGURE 5.6
Higher frequency aliases.

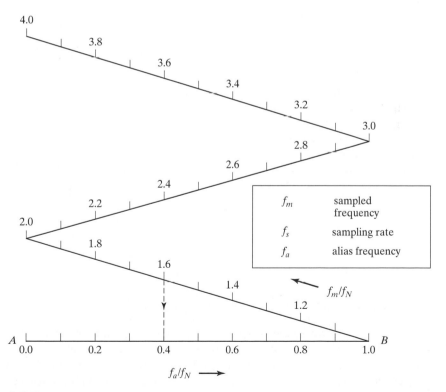

FIGURE 5.7
Folding diagram.

Example 5.1

Compute the lowest alias frequencies for the following cases:

(a) $f_m = 80$ Hz and $f_s = 100$ Hz.
(b) $f_m = 100$ Hz and $f_s = 60$ Hz.
(c) $f_m = 100$ Hz and $f_s = 250$ Hz.

Solution:

(a) $f_N = 100/2 = 50$ Hz
$f_m/f_N = 80/50 = 1.6$

Find f_m/f_N on the folding diagram, draw a vertical line down to the intersection with line AB, and read 0.4 on line AB. The alias frequency can then be determined from

$$f_a = (f_a/f_N)f_N = 0.4 \times 50 = 20 \text{ Hz}$$

(b) $f_N = 60/2 = 30$
$f_m/f_N = 100/30 = 3.333$

Finding 3.333 on the folding diagram and drawing the vertical line down to AB, we find $f_a/f_N = 0.667$. The alias frequency is then

$$f_a = 0.667 \times 30 = 20 \text{ Hz}$$

(c) $f_N = 250/2 = 125$ Hz
$f_m/f_N = 100/125 = 0.8$

This falls on the line AB, so $f_a/f_N = 0.8$ and $f_a = 125$ Hz, which is the same as the sampled frequency.

Comment: In part (a), the sampling frequency is between the signal frequency and the minimum frequency for correct sampling. The lowest alias frequency is the difference between the sampling frequency and the signal frequency.

In part (b), the sampling frequency is less than the signal frequency. The folding diagram is the simplest method to determine the lowest alias frequency.

In part (c), the requirement of the sampling-rate theorem has been met, and the alias frequency is in fact the signal frequency.

We will always find a lowest frequency using the folding diagram, whether it is a correct frequency or a false alias. To know that the frequency is correct, we must insure that the sampling rate is at least twice the actual frequency, usually by using a filter to remove any frequency higher than half the sampling rate.

5.2 SPECTRAL ANALYSIS OF TIME-VARYING SIGNALS

When a signal is a pure sine wave, determining the frequency is a simple process. However, the general time-varying signal does not have the form of a simple sine wave; Figure 5.8 shows a typical example. As discussed below, complicated waveforms can be considered to be constructed of the sum of a set of sine or cosine waves of different frequencies. The process of determining these component frequencies is called *spectral analysis*.

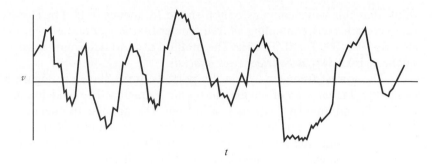

FIGURE 5.8

Typical measured time-varying waveform.

There are two times in an experimental program when it may be necessary to perform spectral analysis on a waveform. The first time is in the planning stage and the second is in the final analysis of the measured data. In planning experiments in which the data vary with time, it is necessary to know, at least approximately, the frequency characteristics of the measurand in order to specify the required frequency response of the transducers and other instruments and to determine the sampling rate required. While the actual signal from a planned experiment will not be known, data from similar experiments may be used to determine frequency specifications.

In many time-varying experiments, the frequency spectrum of a signal is one of the primary results. In structural vibration experiments, for example, acceleration of the vibrating body may be a complicated function resulting from various resonant frequencies of the system. The measurement system is thus designed to respond properly to the expected range of frequencies, and the resulting data are analyzed for the specific frequencies of interest.

To examine the methods of spectral analysis, we first look at a relatively simple waveform, a simple 1000-Hz sawtooth wave as shown in Figure 5.9. At first, one might

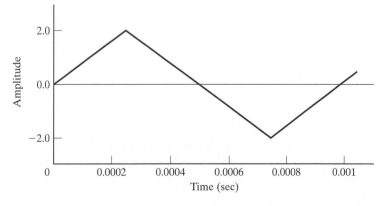

FIGURE 5.9

1000-Hz sawtooth waveform.

think that this wave contains only a single frequency, 1000 Hz. However, it is much more complicated, containing all frequencies that are an odd-integer multiple of 1000, such as 1000, 3000, and 5000 Hz. The method used to determine these component frequencies is known as *Fourier-series analysis*.

The lowest frequency, f_0, in the periodic wave shown in Figure 5.9, 1000 Hz, is called the *fundamental* or the *first harmonic frequency*. The fundamental frequency has period T_0 and angular frequency ω_0. (*Note:* The angular frequency, $\omega = 2\pi f$, where $f = 1/T$.) As discussed by Den Hartog (1956), Churchill (1987), and Kamen (1990), any periodic function $f(t)$ can be represented by the sum of a constant and a series of sine and cosine waves. In symbolic form, this is written

$$f(t) = a_0 + a_1 \cos \omega_0 t + a_2 \cos 2\omega_0 t + \cdots + a_n \cos n\omega_0 t$$
$$+ b_1 \sin \omega_0 t + b_2 \sin 2\omega_0 t + \cdots + b_n \sin n\omega_0 t \qquad (5.2)$$

The constant a_0 is simply the time average of the function over the period T. This can be evaluated from

$$a_0 = \frac{1}{T} \int_0^T f(t)\, dt \qquad (5.3)$$

The constants a_n can be evaluated from

$$a_n = \frac{2}{T} \int_0^T f(t) \cos n\omega_0 t\, dt \qquad (5.4)$$

and the constants b_n can be evaluated from

$$b_n = \frac{2}{T} \int_0^T f(t) \sin n\omega_0 t\, dt \qquad (5.5)$$

Although it can be tedious, the constants a_n and b_n can be computed in a straightforward manner for any periodic function. Of course, Eq. (5.2) is an infinite series, so the constants a and b can only be determined for a limited number of terms. Since $f(t)$ cannot, in general, be expressed in equation form, it is normal to evaluate Eqs. (5.3), (5.4), and (5.5) by means of numerical methods.

The function $f(t)$ is considered to be an even function if it has the property that $f(t) = f(-t)$. $f(t)$ is considered to be an odd function if $f(t) = -f(-t)$. If $f(t)$ is even, it can be represented entirely with a series of cosine terms, which is known as a *Fourier cosine series*. If $f(t)$ is odd, it can be represented entirely with a series of sine terms, which is known as a *Fourier sine series*. Many functions are neither even nor odd and require both sine and cosine terms. If Eqs. (5.3) through (5.5) are applied to the sawtooth wave in Figure 5.9 (either using direct integration or a numerical method), it will be found that all the a's are zero (it is an odd function) and that the first seven b's are

$$b_1 = 1.6211 \qquad b_5 = 0.0648$$
$$b_2 = 0.0000 \qquad b_6 = 0.0000$$
$$b_3 = -0.1801 \quad b_7 = -0.0331$$
$$b_4 = 0.0000$$

It is not surprising that the a's are zero since the wave in Figure 5.9 looks much more like a sine wave than a cosine wave. b_1, b_3, b_5, and b_7 are the amplitudes of the first, third, fifth, and seventh harmonics of the function $f(t)$. These have frequencies of 1000, 3000, 5000, and 7000 Hz, respectively. It is useful to present the amplitudes of the harmonics on a plot of amplitude versus frequency as shown in Figure 5.10. As can be seen, harmonics beyond the fifth have a very low amplitude. Often, it is the energy content of a signal that is important, and since the energy is proportional to the amplitude squared, the higher harmonics contribute very little energy.

Figure 5.11 shows the first and third harmonics and their sum compared with the function $f(t)$. As can be seen, the sum of the first and third harmonics does a fairly good job of representing the sawtooth wave. The main problem is apparent as a rounding near the peak—a problem that would be reduced if the higher harmonics (e.g., fifth, seventh, etc.) were included. Fourier analysis of this type can be very useful in specifying the frequency response of instruments. If, for example, the experimenter considers the first-plus-third harmonics to be a satisfactory approximation to the sawtooth wave, the sensing instrument need only have an upper frequency limit of 3000 Hz. If a better representation is required, the experimenter can examine the effects of higher harmonics on the representation of the wave and then select a suitable transducer.

Example 5.2 demonstrates the process of determining Fourier components for another function. The process of determining Fourier coefficients using numerical methods is demonstrated in Section A.1, Appendix A.

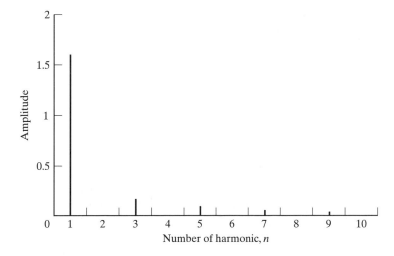

FIGURE 5.10

Amplitudes of harmonics for a sawtooth wave.

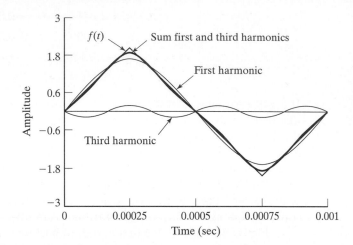

FIGURE 5.11

Harmonics of sawtooth wave.

Example 5.2

For the periodic function shown in Figure E5.1, find the amplitude of the first, second, and third harmonic components.

Solution: The fundamental frequency for this wave is 10 Hz and the angular frequency, $\omega = 2\pi f$, is 62.83 rad/s. The first complete cycle of this function can be expressed in equation form as

$$f(t) = 60t \qquad 0 \le t < 0.05$$
$$f(t) = 60t - 6 \qquad 0.05 \le t < 0.10$$

Since the average value for the function is zero over one cycle, the coefficient a_0 will have a value of zero. Also, by examination, we can conclude that it is an odd function and that the cosine terms will be zero and only the sine terms will be required.

Using Eq. (5.5), the first harmonic coefficient can be computed from

$$b_1 = \frac{2}{0.1}\left[\int_0^{0.05} 60t \sin(62.83t)\, dt + \int_{0.05}^{0.1} (60t - 6) \sin(62.83t)\, dt \right]$$

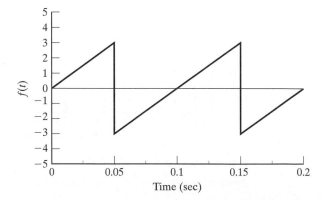

FIGURE E5.1

Sawtooth wave.

This can be evaluated using standard methods to give a value of 1.9098. Similarly, the second and third harmonics can be evaluated from

$$b_2 = \frac{2}{0.1}\left[\int_0^{0.05} 60t \sin(2 \times 62.83t)\,dt + \int_{0.05}^{0.1}(60t - 6)\sin(2 \times 62.83t)\,dt\right]$$

$$b_3 = \frac{2}{0.1}\left[\int_0^{0.05} 60t \sin(3 \times 62.83t)\,dt + \int_{0.05}^{0.1}(60t - 6)\sin(3 \times 62.83t)\,dt\right]$$

to give values of -0.9549 and 0.6366, respectively.

One problem associated with Fourier-series analysis is that it appears to only be useful for periodic signals. In fact, this is not the case and there is no requirement that $f(t)$ be periodic to determine the Fourier coefficients for data sampled over a finite time. We could force a general function of time to be periodic simply by duplicating the function in time as shown in Figure 5.12 for the function in Figure 5.8. If we directly apply Eqs. (5.3), (5.4), and (5.5) to a function taken over a time period T, the resulting Fourier series will have an implicit fundamental angular frequency ω_0 equal to $2\pi/T$. However, if the resulting Fourier series were used to compute values of $f(t)$ outside the time interval 0–T, it would result in values that would not necessarily (and probably would not) resemble the original signal. The analyst must be careful to select a large enough value of T so that all wanted effects can be represented by the resulting Fourier series. An alternative method of finding the spectral content of signals is that of the Fourier transform, discussed next.

5.3 SPECTRAL ANALYSIS USING THE FOURIER TRANSFORM

Although a discussion of Fourier series is useful to introduce the concept that most functions can be represented by the sum of a set of sine and cosine functions, the technique most commonly used to spectrally decompose functions is the *Fourier transform*. The Fourier transform is a generalization of Fourier series. The Fourier transform can be applied to any practical function, does not require that the function be periodic, and for discrete data can be evaluated quickly using a modern computer technique called the *Fast Fourier Transform*.

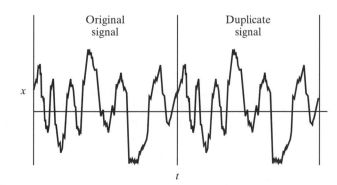

FIGURE 5.12

Duplicating a signal to make it harmonic.

In presenting Fourier transform, it is common to start with Fourier series, but in a different form than Eq. (5.2). This form is called the *complex exponential form*. It can be shown that sine and cosine functions can be represented in terms of complex exponentials:

$$\cos x = \frac{e^{jx} + e^{-jx}}{2}$$

$$\sin x = \frac{e^{jx} - e^{-jx}}{2j} \tag{5.6}$$

where $j = \sqrt{-1}$. These relationships can be used to transform Eq. (5.2) into a complex exponential form of the Fourier series. [See basic references on signals, such as Kamen (1990)] The resulting exponential form for the Fourier series can be stated as

$$f(t) = \sum_{n-\infty}^{\infty} c_n e^{jn\omega_0 t} \tag{5.7}$$

where

$$c_n = \frac{1}{T} \int_{-T/2}^{T/2} f(t) e^{-jn\omega_0 t} \, dt \tag{5.8}$$

Each coefficient, c_n, is, in general, a complex number with a real and an imaginary part.

In Section 5.2 we showed how a portion of a nonperiodic function can be represented by a Fourier series by assuming that the portion of duration T is repeated periodically. The fundamental angular frequency, ω_0, is determined by this selected portion of the signal ($\omega_0 = 2\pi/T$). If a longer value of T is selected, the lowest frequency will be reduced. This concept can be extended to make T approach infinity and the lowest frequency approach zero. In this case, frequency becomes a continuous function. It is this approach that leads to the concept of the Fourier transform. The Fourier transform of a function $f(t)$ is defined as

$$F(\omega) = \int_{-\infty}^{\infty} f(t) e^{-j\omega t} \, dt \tag{5.9}$$

$F(\omega)$ is a continuous, complex-valued function. Once a Fourier transform has been determined, the original function $f(t)$ can be recovered from the inverse Fourier transform:

$$f(t) = \frac{1}{2\pi} \int_{-\infty}^{\infty} F(\omega) e^{j\omega t} \, d\omega \tag{5.10}$$

In experiments, a signal is measured only over a finite time period, and with computerized data-acquisition systems, it is measured only at discrete times. Such a signal is not well suited to analysis by the continuous Fourier transform. For data taken at discrete times over a finite time interval, the discrete Fourier transform (DFT) has been defined as

$$F(k\Delta f) = \sum_{n=0}^{N-1} f(n\,\Delta t) e^{-j(2\pi k\Delta f)(n\,\Delta t)} \qquad k = 0, 1, 2, \ldots, N - 1 \tag{5.11}$$

where N is the number of samples taken during a time period T. The increment of f, Δf, is equal to $1/T$, and the increment of time (the sampling period Δt) is equal to T/N. The F's are complex coefficients of a series of sinusoids with frequencies of $0, \Delta f, 2\Delta f, 3\Delta f, \ldots, (N-1)\Delta f$. The amplitude of F for a given frequency represents the relative contribution of that frequency to the original signal.

Only the coefficients for the sinusoids with frequencies between 0 and $(N/2 - 1)\Delta f$ are used in the analysis of signal. The coefficients of the remaining frequencies provide redundant information and have a special meaning, as discussed by Bracewell (2000). The requirements of the Shannon sampling-rate theorem also prevent the use of any frequencies above $(N/2)\Delta f$. The sampling rate is N/T, so the maximum allowable frequency in the sampled signal will be less than one-half this value, or $N/2T = N\Delta f/2$.

The original signal can also be recovered from the DFT using the inverse discrete Fourier transform, given by

$$f(n\,\Delta t) = \frac{1}{N}\sum_{k=0}^{N-1} F(k\,\Delta f)e^{j(2\pi k\Delta f)(n\Delta t)} \qquad n = 0, 1, 2, \ldots, N-1 \qquad (5.12)$$

The F values from Eq. (5.11) can be evaluated by direct numerical integration. The amount of computer time required is roughly proportional to N^2. For large values of N, this can be prohibitive. A sophisticated algorithm called the *Fast Fourier Transform* (FFT) has been developed to compute discrete Fourier transforms much more rapidly. This algorithm requires a time proportional to $N \log_2 N$ to complete the computations, much less than the time for direct integration. The only restriction is that the value of N be a power of 2: for example, 128, 256, 512, and so on. Programs to perform fast Fourier transforms are widely available and are included in major spreadsheet programs. The fast Fourier transform algorithm is also built into devices called *spectral analyzers*, which can discretize an analog signal and use the FFT to determine the frequencies.

It is useful to examine some of the characteristics of the discrete Fourier transform. To do this, we will use as an example a function that has 10- and 15-Hz components:

$$f(t) = 2 \sin 2\pi 10t + \sin 2\pi 15t \qquad (5.13)$$

This function is plotted in Figure 5.13. Since this function is composed of two sine waves with frequencies of 10 and 15 Hz, we would expect to see large values of F at these frequencies in the DFT analysis of the signal. If we discretize one second of the signal into 128 samples and perform an FFT (we used a spreadsheet program as demonstrated in Section A.2), the result will be as shown in Figure 5.14, which is a plot of the magnitude of the DFT component, $|F(k\,\Delta f)|$, versus the frequency, $k\,\Delta f$. As expected, the magnitudes of F at $f = 10$ and $f = 15$ are dominant. However, there are some adjacent frequencies showing appreciable magnitudes. In this case, these significant magnitudes of F at frequencies not in the signal are due to the relatively small number of points used to discretize the signal. If we use $N = 512$, the situation improves significantly, as shown in Figure 5.15.

It should be noted that the requirements of the Shannon sampling-rate theorem are satisfied for the results shown in both Figures 5.14 and 5.15. For Figure 5.14, the

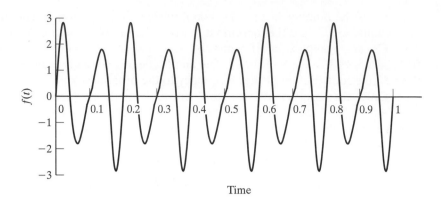

FIGURE 5.13

The function $2 \sin 2\pi 10t + \sin 2\pi 15t$.

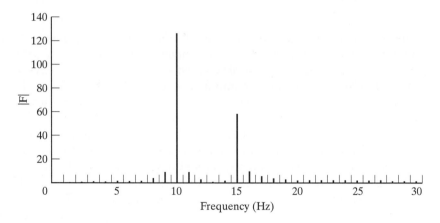

FIGURE 5.14

FFT of Eq. (5.13), $N = 128$, $T = 1$ s.

sampling rate is 128 samples per second, so the maximum frequency in the sample should be less than 64 Hz. The actual maximum frequency is 15 Hz.

For the FFTs shown in Figures 5.14 and 5.15, the data sample contained integral numbers of complete cycles of both component sinusoids. In general, the experimenter will not know the spectral composition of the signal and will not be able to select a sampling time T such that there will be an integral number of cycles of any frequency in the signal. This complicates the process of Fourier decomposition. To demonstrate this point, we will modify Eq. (5.13) by changing the lowest frequency from 10 Hz to 10.333 Hz.

$$f(t) = 2\sin 2\pi 10.3333t + \sin 2\pi 15t \tag{5.14}$$

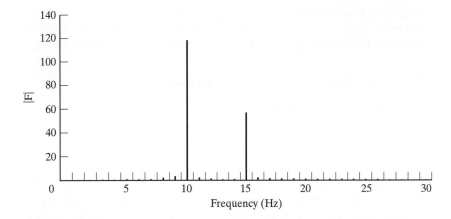

FIGURE 5.15

FFT of Eq. (5.13), $N = 512, T = 1$ s.

An FFT is then performed in a 1-s period with 128 samples. Although 15 complete cycles of the 15-Hz component are sampled, 10.3333 cycles of the 10.3333-Hz component are sampled. The results of the DFT are shown in Figure 5.16.

   Although we would expect the Fourier coefficient for 10.333 Hz to be distributed between 10 and 11 Hz in the FFT result, in fact the entire spectrum has been altered. These significant coefficients at frequencies not in the original signal are caused by an effect called *leakage*. They are caused by the fact that there are a nonintegral number of cycles of the 10.333-Hz sinusoid. The common method to reduce leakage is known as *windowing*. Since leakage is caused by incomplete cycles at the ends of the sampling period, windowing seeks to minimize the end effects by multiplying the signal by a

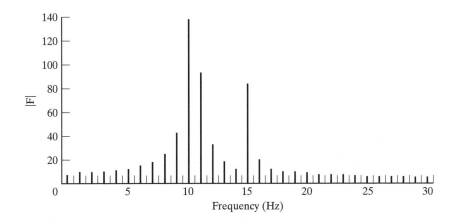

FIGURE 5.16

FFT of Eq. (5.14), $N = 128, T = 1$ s.

weighting function that is larger at the center of the signal sampling period than it is at the ends. Turner (1988) and VanDoren (1982) introduce the windowing concept, and Brigham (1974) discusses the process of windowing in greater detail.

5.4   SELECTING THE SAMPLING RATE AND FILTERING

5.4.1   Selecting the Sampling Rate

As shown in previous sections, a typical signal can be considered to be composed of a set of sinusoids of different frequencies. In most cases the experimenter can determine the maximum signal frequency of interest, which we shall call f_c. However, the signal frequently contains significant energy at frequencies higher than f_c. If the signal is to be recorded with an analog device, such as an analog tape recorder, these higher frequencies are usually of no concern. They will either be recorded accurately or attenuated by the recording device. If, however, the signal is to be recorded only at discrete values of time, the potential exists for the generation of false, alias signals in the recording.

The sampling-rate theorem does not state that to avoid aliasing, the sampling rate must be twice the maximum frequency of interest but that the *sampling rate must be greater than twice the maximum frequency in the signal*, here denoted by f_m. As an example, consider a signal that has Fourier sine components of 90, 180, 270, and 360 Hz. If we are only interested in frequencies below 200 Hz, we might set the sampling rate at 400 Hz. In our sampled output, however, we will see frequencies of 130 Hz and 40 Hz, which are aliases caused by the 270- and 360-Hz components of the signal. Section 5.1 provided a simple method to determine alias frequencies for a given signal frequency and sampling rate.

If the sampling rate is denoted by f_s, the sampling rate theorem is formally stated as

$$f_s > 2f_m \qquad (5.15)$$

If we set the sampling rate, f_s, to a value that is greater than twice f_m, we should not only avoid aliasing but also be able to recover (at least theoretically) the original waveform. In the foregoing example, in which f_m is 360 Hz, we would select a sampling rate, f_s, greater than 720 Hz.[†]

[†]Although it is normally desirable to select such a sampling rate, it is possible to relax this in some cases. The requirement that $f_s > 2f_m$ is more stringent than is necessary to eliminate alias frequencies in the $0-f_c$ range. As we noted in Section 5.1, for sampling rates between a sampled frequency f and twice that frequency, the lowest alias frequency is the difference between f and the sampling rate. Assume that we have set the sampling rate such that the minimum alias frequency f_a has a value just equal to f_c. The highest frequency in the signal that could cause aliasing is f_m. (All frequencies above f_m have zero amplitude.) Then this alias frequency, f_a, is the difference between f_s and f_m:

$$f_a = f_s - f_m = f_c$$

Solving, we obtain $f_s = f_c + f_m$. If this basis is used to select the sampling rate, there will be aliases in the sampled signal with frequencies in the $f_c - f_s$ range but not in the $0-f_c$ range. Digital filtering techniques using software can be used to eliminate these alias frequencies. If f_c has a value of 200 Hz and f_m has a value of 400 Hz, we would select a sampling rate of 600 Hz, significantly lower than twice f_m (800 Hz). For further discussion, see Taylor (1990).

If this sampling-rate restriction is met, the original waveform can be recovered using the following formula called the *cardinal series* (Marks, 1991):

$$f(t) = \frac{1}{\pi} \sum_{n=-\infty}^{\infty} f(n\Delta T) \frac{\sin[\pi(t/\Delta T - n)]}{t/\Delta T - n} \tag{5.16}$$

In this formula, $f(t)$ is the reconstructed function. The first term after the summation, $f(n\,\Delta T)$, represents the discretely sampled values of the function, n is an integer corresponding to each sample, and ΔT is the sampling period, $1/f_s$. One important characteristic of this equation is that it assumes an infinite set of sampled data and is hence an infinite series. Real sets of sampled data are finite. However, the series converges and discrete samples in the vicinity of the time t contribute more than terms not in the vicinity. Hence, the use of a finite number of samples can lead to an excellent reconstruction of the original signal.

As an example, consider a function $\sin(2\pi 0.45t)$, that is, a simple sine wave with a frequency of 0.45 Hz and a peak amplitude of 1.0. It is sampled at a rate of 1 sample per second and 700 samples are collected. Note that f_s exceeds $2f_m$, so this requirement of the sampling-rate theorem is satisfied. A portion of the sampled data is shown in Figure 5.17. The sampled data have been connected with straight-line segments and do not appear to closely resemble the original sine wave. Using Eq. (5.16) and the 700 data samples, the original curve has been reconstructed for the interval of time between 497 and 503 s, as shown by the heavy curve on Figure 5.15. The original sine wave has been recovered from the data samples with a high degree of accuracy. Reconstructions with small sample sizes or at the ends of the data samples will, in general, not be as good as this example. Although Eq. (5.16) can be used to reconstruct data, other methods, beyond the scope of the present text, are often used (Marks, 1991).

5.4.2   Use of Filtering to Limit Sampling Rate

Many measurement systems effectively eliminate unwanted high frequencies at the sensing stage. As discussed in Chapter 11, many temperature-measuring devices (first-order systems) show a sharp drop in response as the frequency of the measurand increases. Most other sensing devices will show an attenuated response if the frequency

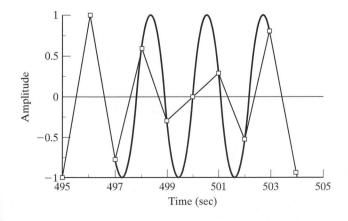

FIGURE 5.17

0.45-Hz sine wave sampled at 1 sample per second.

is high enough. In many of these systems, the sampling rate can be made sufficiently high to avoid aliasing effects.

As noted by Taylor (1990), a caution is in order here concerning the effect of system noise on the measurement of low-frequency measurands. As an example, consider a measuring device being used to measure a quasi-steady temperature. Since the data are being taken with a computerized data-acquisition system, this temperature sensor is sampled at 30 Hz, the same as for the other sensors. If 60-Hz power-line noise is imposed on the signal, there will be an alias signal with a frequency of 0 Hz (dc) appearing in the discrete data. (See Section A.3.) This alias can produce a substantial error in the quasi-steady temperature measurement, even if the discrete data are time averaged.

If the maximum signal frequency, f_m, is substantially higher than the maximum frequency of interest, f_c, it may be impractical to set the sampling frequency at $2f_m$. It is then better to pass the input analog signal through a lowpass filter (called an *antialiasing filter*). This filter can reduce the amplitude of the frequencies above f_c to values that will cause negligible aliasing. We will now have a new value of f_m, which is still higher than f_c, but much lower than before the filter was inserted.

When we specify f_m, we need to define what is meant by an effectively zero amplitude since filters can attenuate signals but will not reduce them to zero amplitude. The question then arises as to what level of attenuation is adequate for practical purposes. For systems using an analog-to-digital converter, this decision can be based on the characteristics of the converter. For a given converter, there is an input voltage below which the converter will produce an output that is the same as would be produced by a zero input voltage.

The required signal attenuation can be determined by computing a parameter of the A/D converter called the *dynamic range*. For a converter with N bits, the dynamic range is

$$\text{dynamic range} = 20 \log_{10}(2^N) \quad \text{dB} \tag{5.17}$$

For monopolar 8- and 12-bit converters, the dynamic ranges are 48 and 72 dB, respectively. For bipolar converters N is reduced by one since one bit is effectively used to represent the sign of the input. For 8- and 12-bit bipolar converters, the values of dynamic range would be respectively 42 and 66 dB.

If we know that the amplitude of an input frequency component does not exceed the input range of the A/D converter, and we then attenuate that signal component an amount equal to the converter dynamic range, we can be sure that it will produce no digital output. To select a filter, we need to define a corner frequency and an attenuation rate. In most cases, we can select the corner frequency as the maximum frequency of interest, f_c. We can then determine f_m based on the filter attenuation rate (dB/octave). Following Taylor (1990), we use the dynamic range to determine the required sampling rate (or conversely, the required filtering for a given sampling rate). The number of octaves required to attenuate a signal the number of decibels corresponding to the dynamic range is

$$N_{\text{oct}} = \frac{\text{dynamic range}}{\text{filter attenuation rate (dB/octave)}} \tag{5.18}$$

f_m can then be evaluated from

$$f_m = f_c 2^{N_{\text{oct}}} \tag{5.19}$$

Equation (5.15) is used to set the sampling rate at twice f_m. If the sampling rate so determined is too high, a higher filter attenuation rate must be specified.

This approach is bounding and assumes only that the voltage of the composite signal (sum of all Fourier components) is scaled to be just within the A/D converter input range. In most cases, the higher-frequency components of the signal that require attenuation have amplitudes much lower than the composite signal and hence require less attenuation. Using a more detailed analysis that takes account of the actual amplitudes of the frequency components above f_c, it is usually possible to use a lower sampling rate or a lesser filter attenuation rate.

Antialiasing filters often require substantial rates of attenuation. These relatively high levels of attenuation present practical problems in actual systems. As discussed in Chapter 3, a first-order Butterworth filter attenuates at only 6 dB per octave (an octave is a doubling of frequency) and eight octaves are required to attenuate a signal 48 dB. As a result, higher-order filters are usually required—an eighth-order Butterworth filter can attenuate a signal 48 dB in one octave. High-order filters have other effects, however, such as large phase shifts in the passband. Since there is a trade-off between filter order and sampling rate, a design compromise is typically required. The process of specifying filtering and sampling rate is demonstrated in Example 5.3.

Example 5.3

A computerized data-acquisition system is to be used to record the data from a pressure transducer connected to the combustion chamber of a spark ignition internal combustion engine running at 4000 rpm. The data-acquisition system has an 8-bit bipolar A/D converter and a maximum sampling rate of 10,000 samples per second. Specify a suitable filter and the required sampling rate.

Solution: Since the actual form of the pressure versus time is not known in advance, this problem statement is, in fact, rather vague. To estimate the pressure–time characteristics, data from another engine are used, as shown in Figure E5.3(a) (Obert, 1973). The time scale in this figure has been scaled to correspond to 4000 rpm. The first step is to perform a Fourier spectral analysis on

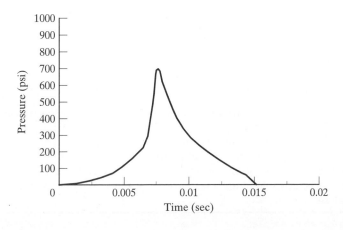

FIGURE E5.3(a)

Typical pressure–time plot.

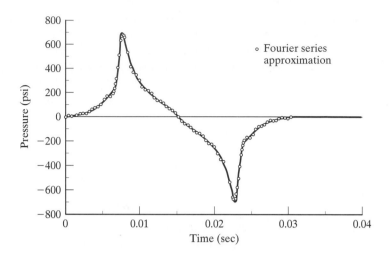

FIGURE E5.3(b)

Fourier analysis of pressure–volume diagram.

these data. Since the data are not periodic, the data are mirrored and inverted about the maximum time to create a periodic function as shown in Figure E5.3(b). The frequency of the first harmonic (the lowest-frequency component) is 32.9 Hz and the angular frequency, ω, is 206.68 rad/s. Equation (5.5) was used to determine the first 20 terms of a Fourier sine series using numerical methods. (See Appendix A.) Since the curve in Figure E5.3(b) is an odd function, only sine terms are required. The sum of these 20 Fourier components is presented on Figure E5.3(b) with circular plot symbols. The agreement between the data and the Fourier series is considered adequate. [Alternatively, the Fourier transform of the curve in Figure E5.3(a) could have been computed and used to determine the maximum significant frequency.] Since the twentieth harmonic has a frequency of 20 times the first harmonic, the highest frequency of interest, f_c, in this problem is 658 Hz. We will select this frequency as the corner frequency of the filter.

As the A/D converter is 8 bit and bipolar, it has a 42-dB dynamic range. We can use this value as the required attenuation between $f_c = 658$ Hz and the frequency f_m where the voltage has been attenuated to effectively zero. We will use a fourth-order Butterworth filter, which attenuates the signal $24 (= 4 \times 6)$ dB per octave. Since we require 42 dB, we can use Eq. (5.18) to determine that this filter will attenuate the signal sufficiently in $1.75 (= 42/24)$ octaves. As a result, f_m will have a value of 2213 Hz $(= 2^{1.75} \times 658)$ [Eq. (5.19)]. The minimum sampling rate can then be determined from Eq. (5.1):

$$f_s = 2f_m = 2 \times 2213 = 4426 \text{ Hz}$$

4426 Hz is then the minimum sampling rate to avoid aliases and is within the capability of the DAS.

Comments: It might be desirable to increase the sampling rate and use a lower-order filter. Filters have significant phase distortion within the filter bandpass, and this distortion would be less if the filter order were lower. This selected filtering is quite conservative. The higher-frequency Fourier components have amplitudes that are substantially smaller than the low-frequency components. Their amplitude can be reduced to less than the A/D converter threshold with lesser attenuation. It would thus be possible to use a lesser order of filter or a lower sampling rate using a more detailed analysis.

Actually, the sampling constraints discussed above may be too stringent for some applications. As discussed in Taylor (1990), if we allow the sampled signal to contain a certain amount of distortion due to aliases, the sampling frequency can be reduced further. This approach, which allows some distortion, should be undertaken only if the experimenter has a good understanding of the analog signal and the experimental requirements.

REFERENCES

[1]   BRACEWELL, R. (2000). *The Fourier Transform and Its Applications*, McGraw-Hill, New York.

[2]   BRACEWELL, R. (1989). The Fourier transform, *Scientific American*, June.

[3]   BRIGHAM, E. (1974). *The Fast Fourier Transform*, Prentice Hall, Englewood Cliffs, NJ.

[4]   CHURCHILL, R. (1987). *Fourier Series and Boundary Value Problems*, McGraw-Hill, New York.

[5]   DEN HARTOG, J. P. (1956). *Mechanical Vibrations*, McGraw-Hill, New York. Republished by Dover, New York, 1985.

[6]   KAMEN, E. (1990). *Introduction to Signals and Systems*, Macmillan, New York.

[7]   LIGHTHILL, M. (1958). *Introduction to Fourier Analysis and Generalized Functions*, Cambridge, MA.

[8]   MARKS, R. (1991). *Introduction to Shannon Sampling and Interpolation Theory*, Springer–Verlag, New York.

[9]   NYQUIST, H. (1928). Certain topics in telegraph transmission theory, *AIAA Transactions*, Vol. 47, pp. 617–644.

[10]   OBERT, E. (1973). *Internal Combustion Engines and Air Pollution*, Harper & Row, New York.

[11]   OPPENHEIM, A., AND SHAFER, R. (1989). *Discrete-Time Signal Processing*, Prentice Hall, Englewood Cliffs, NJ.

[12]   SHANNON, C. (1948). A mathematical theory of communications, *Bell Systems Technical Journal*, Vol. 27, pp. 379, 623.

[13]   TAYLOR, J. (1990). *Computer Based Data Acquisition Systems*, 2d ed., Instrument Society of America, Research Triangle Park, NC.

[14]   TURNER, J. (1988). *Instrumentation for Engineers*, Springer–Verlag, New York.

[15]   VANDOREN, A. (1982). *Data Acquisition Systems*, Reston Publishing, Reston, VA.

PROBLEMS

Note: An asterisk denotes a spreadsheet problem.

5.1 A single cycle of a ramp function of voltage versus time has the form $v(t) = 100t$ from -0.01 to 0.01 s. Using direct integration, evaluate the Fourier coefficients a_0, a_1, a_2, b_1, and b_2. Could you have deduced the values of a_0, a_1, and a_2 without performing the integrations?

5.2 A single cycle of a ramp function has the form $v(t) = 25t$ from $t = 0$ to $t = 1$ s. Evaluate the Fourier constants a_0, a_1, a_2, b_1, and b_2 by direct integration.

*5.3 For the ramp function described in Problem 5.1, evaluate the coefficients a_0, a_1, a_2, b_1, and b_2 using a numerical method such as described in Section A.1. Use 100 equally spaced time intervals.

*5.4 For the ramp function described in Problem 5.2, evaluate the coefficients a_0, a_1, a_2, b_1, and b_2 using a numerical method such as described in Section A.1. Use 100 equally spaced time intervals.

5.5 A transient voltage has the form $v(t) = 20t - 200t^2$ in the time interval 0–0.1 s. Specify a periodic form of this function so that only sine terms will be required for a Fourier-series approximation.

5.6 A transient voltage has a value of 3.5 V from $t = 0$ to $t = 1.5$ s and has a value of zero at other times. Specify a periodic form of this function so that it can be represented entirely with a Fourier cosine series.

*5.7 The function shown in Figure P5.7 can be represented by the Fourier cosine series

$$f(t) = 0.2 + \frac{2}{\pi} \sum_{n=1}^{\infty} \frac{(-1)^n}{n} \sin 0.2n\pi \cos 4n\pi t$$

Using a spreadsheet program, compute the sum of a_0, a_1, a_2, and a_3 terms at subintervals of 0.01 s over the interval 0 to 0.5 s. Plot this sum and compare it to the original function.

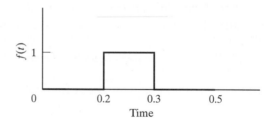

FIGURE P5.7

*5.8 Using a spreadsheet program as shown in Section A.2, generate 128 values of the function $f(t) = 5 \sin 10\pi t + 3 \cos 40\pi t$ between $t = 0$ and $t = 1$ s. Perform an FFT on the results and find the magnitude of the coefficients (as produced by the FFT; they will be complex). Plot the results versus frequency in a bar chart such as Figure 5.12. Interpret the peaks observed.

*5.9 Using a spreadsheet program as shown in Section A.2, generate 128 values of the function $f(t) = 3 \cos 500\pi t + 5 \cos 800\pi t$ between $t = 0$ and $t = 0.1$ s. Perform an FFT on the results and find the magnitude of the coefficients (as produced by the FFT, they will be complex). Plot the results versus frequency in a bar chart such as Figure 5.14. Interpret the peaks observed.

*5.10 Consider the function $f(t) = 7.5 \sin^2 10\pi t$. Using a spreadsheet program as shown in Section A.2, generate 128 values between $t = 0$ and $t = 2$ s. Perform a FFT on the results, evaluate the magnitude of the coefficients, and present the result in a bar graph such as Figure 5.14. What is the meaning of the coefficient at $f = 0$?

5.11 A function has the form $f(t) = 5 \sin 10\pi t + 3 \cos 40\pi t$ and is sampled at a rate of 30 samples per second. What alias frequency(ies) would you expect in the discrete data?

5.12 The function in Problem 5.11 is sampled at 5 samples per second. What alias frequencies would you expect in the discrete data?

5.13 What would be the minimum sampling rate to avoid aliasing for the function in Problem 5.11?

5.14 The function $f(t) = 3 \cos 500\pi t + 5 \cos 800\pi t$ is sampled at 400 samples per second starting at $t = 0.00025$ s. What alias frequencies would you expect in the output?

5.15 What would be the minimum sampling rate to avoid aliasing for the function in Problem 5.14?

*__5.16__ Using a spreadsheet program, plot the function of Problem 5.14 from $t = 0.00025$ to 0.1 s at 0.0025-s intervals. Connect the points with straight lines. Explain the shape of the resulting plot.

5.17 A 3.5-kHz sine wave signal is sampled at 2 kHz. What would be the expected alias frequency?

5.18 A 1-kHz sine wave is sampled at 1.5 kHz. What is the expected alias frequency?

5.19 What is the dynamic range of a 16-bit bipolar A/D converter?

5.20 What is the dynamic range of a 14-bit unipolar A/D converter?

5.21 For an experiment, f_c is 10 kHz and the A/D converter is bipolar with 12 bits. If a second-order Butterworth lowpass filter is used, what will f_m be if the signal is attenuated to the converter threshold?

5.22 In a vibration experiment, an acceleration-measuring device can measure frequencies up to 3 kHz, but the maximum frequency of interest is 500 Hz. A data acquisition system is available with a 12-bit bipolar A/D converter and a maximum sampling rate of 10,000 samples per second. Specify the corner frequency and order of a Butterworth antialiasing filter for this application (attenuating the signal to the converter threshold) and the actual minimum sampling rate.

5.23 In a time-varying pressure measurement, the maximum frequency of interest is 100 Hz, but there are frequencies in the signal with significant energy up to 1000 Hz. A first-order Butterworth filter is to be constructed to filter out the higher frequencies. If the A/D converter has 8 bits and is bipolar, determine the minimum sampling rate for an attached data-acquisition system.

C H A P T E R   6

Statistical Analysis of Experimental Data

Measurement processes usually introduce a certain amount of variability or randomness into the results, and this randomness can affect the conclusions drawn from experiments. In this chapter, we discuss important statistical methods that can be used to plan experiments and interpret experimental data.

6.1   INTRODUCTION

Random features are observed in virtually all measurements. Even if the same measuring system is used to measure a fixed measurand repeatedly, the results will not all have the same value. This randomness can be caused by uncontrolled (or uncontrollable) variables affecting the measurand or lack of precision in the measurement process. In some cases, the randomness of the data so dominates the data that it is difficult to distinguish the sought-after trend. This is common in experiments in the social sciences and sometimes occurs in engineering. In such cases, statistics may offer tools that can identify trends from what appears to be a set of confused data. In engineering, the general trends of the data are usually evident; however, statistical tools are often needed to identify and generalize the characteristics of test data or determine bounds on the uncertainty of the data.

The types of errors in measurement were discussed in Chapter 2 and are generally broken into two categories: systematic and random. Systematic errors are consistent, repeatable errors that can often be minimized by calibration of the measurement system. It is the random errors that are most amenable to the methods of statistical analysis. Statistical concepts are not only useful for interpreting experimental data but also for planning experiments, particularly those with a large number of independent variables or parameters.

As an example of random behavior of experimental results, consider the measurement of the temperature of hot gas flowing through a duct. Due to factors outside the experimenter's control, the temperature shows variability when measured over a period of an hour. This is despite the fact that all instruments have been calibrated and are operated

TABLE 6.1   Results of 60
Temperature Measurements in a
Duct

Number of readings	Temperature (°C)
1	1089
1	1092
2	1094
4	1095
8	1098
9	1100
12	1104
4	1105
5	1107
5	1108
4	1110
3	1112
2	1115

properly. The temperature of the gas is then considered to be, in the nomenclature of statistics, a *random variable*. Table 6.1 shows the results of a set of 60 measurements of air
temperature in the duct. These temperature data are observed values of a random variable. A typical problem associated with data such as these would be to determine
whether it is likely that the temperature might exceed certain limits. Although these data
show no temperatures less than 1089°C or greater than 1115°C, we might, for example,
ask if there is a significant chance that the temperature will ever exceed 1117°C or be less
than 1085°C (either of which might affect the manufacturing process in some applications). The methods described later in the chapter will enable us to answer this question.

To help us visualize the data, it is helpful to plot them in the form of a bar graph
as shown in Figure 6.1. A bar graph of this type used for statistical analysis is called a
histogram. To create this graph, we first arrange the data into groups, called *bins*, as
shown in Table 6.2. Each bin has the same width (range of temperature values). Figure 6.1

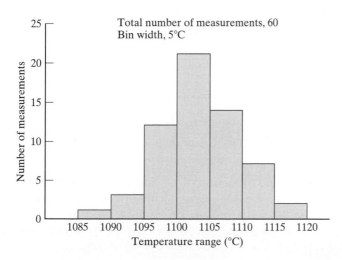

FIGURE 6.1

Histogram of temperature data.

TABLE 6.2 Temperature
Measurements Arranged into Bins

Bin (°C)	Number of measurements
$1085 \leq T < 1090$	1
$1090 \leq T < 1095$	3
$1095 \leq T < 1100$	12
$1100 \leq T < 1105$	21
$1105 \leq T < 1110$	14
$1110 \leq T < 1115$	7
$1115 \leq T < 1120$	2

is then plotted such that the height of each bar is proportional to the number of read-ings occurring in each bin (frequency of occurrence). Some comments should be noted about the figure. First, there is a peak in the number of readings near the center of the temperature range. Second, the number of readings at temperatures less than or greater than the central value drops off rapidly. Finally, the "curve" is bell shaped, not parabolic— the number of samples for bins away from the center, while small, is not zero. These characteristics of the data in Table 6.1 are common, though not necessary, properties of experimental results. Figure 6.2 shows some other forms of distributions that may be found in engineering applications.

Some general guidelines apply to the construction of histograms. (See Rees, 1987.) It is customary to have from 5 to 15 bins. It is simplest if each of the bins has the same width (difference between the smallest and largest values in the bin). There are special rules if bins of unequal width are used. The bins should cover the entire range of the data, with no gaps, but the bins should not overlap. In Figure 6.1, there are seven bins and each bin has a width of 5°C. This example illustrates a random variable that can vary continuously and can take any real value in a certain domain. Such a variable is called a *continuous random variable*. Some experiments produce discrete (noncon-tinuous) results, which are considered to be values of a *discrete random variable*. Exam-ples of discrete random variables are the outcome of tossing a die (which has the only possible values of 1, 2, 3, 4, 5, or 6) and fail/no-fail products in a quality control process.

To apply statistical analysis to experimental data, the data are usually character-ized by determining parameters that specify the central tendency and the dispersion of the data. The next step is to select a theoretical distribution function that is most suit-able for explaining the behavior of the data. The theoretical function can then be used to make predictions about various properties of the data.

6.2 GENERAL CONCEPTS AND DEFINITIONS

6.2.1 Definitions

Population. The *population* comprises the entire collection of objects, mea-surements, observations, and so on whose properties are under consideration and about which some generalizations are to be made. Examples of population are the en-tire set of 60-W electric bulbs that have been produced in a production batch and val-ues of wind speed at a certain point over a defined period of time.

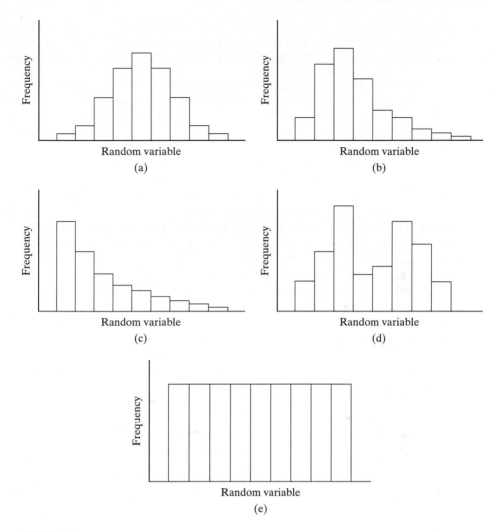

FIGURE 6.2

Some shapes of distributions: (a) symmetric; (b) skewed; (c) J-shaped; (d) bimodal; (e) uniform. (After Johnson, 1988.)

Sample. A *sample* is a representative subset of a population on which an experiment is performed and numerical data are obtained. For example, 10 light bulbs can be selected from a production batch of 10,000, or wind speed can be sampled once each hour over a 24-h period. Different samples of the same population may be chosen for experimentation.

Sample space. The set of all possible outcomes of an experiment is called the *sample space*. For example, there are six possible outcomes in casting a fair die. If the sample space is made of discrete values (such as the outcomes of casting a die or a coin, acceptable and unacceptable products), it is a *discrete sample space*. A random variable of a discrete sample space is discrete. If a sample space is a continuum, we have a *continuous sample space* and also continuous random variables. The sample space for the temperature measurements of gases coming out of a furnace is continuous.

    Random variable.   Engineering experiments and any associated measurement are influenced by many factors that cannot be totally controlled, and as a result, the outcome of the measurement and the experiment are not unique. Two examples of such experiments are the measurement of temperature of a hot gas flowing through a duct and the life expectancy of light bulbs. The value of temperature is a function of many factors, including the operation of the heating source, duct insulation and environment, and, more important, the nature of the flow itself and the measurement device. In the case of the bulb, variation in material properties, manufacturing process, and measurement process can influence the measured life of the bulbs. In each of the experiments mentioned, no matter how well we control the influencing parameters, given enough time, if the experiment is repeated, there will be variations in the values of the measured variables. The variables being measured (temperature and lifetime in these cases) are considered *random variables*. Mathematically, a random variable is a numerically valued function defined for the population. This means that, for every possible experimental outcome (sample point), there is a corresponding numerical value.

    A random variable can be continuous or discrete. The variables in the light bulb and duct temperature examples are continuous random variables. In principle, they can assume any real value. Other random variables, such as the outcome of tossing a die and fail/no-fail of products in a quality control process, are discrete random variables. Discrete random variables have a countable number of possible values.

    Distribution function.   A *distribution function* is a graphical or mathematical relationship that is used to represent the values of the random variable.

    Parameter.   A *parameter* is a numerical attribute of the entire population. As an example, the average value of a random variable in a population is a parameter for that population.

    Event.   An *event* is the outcome of a random experiment.

    Statistic.   A *statistic* is a numerical attribute of the sample. For example, the average value of a sample property is a statistic of the sample.

    Probability.   *Probability* is the chance of occurrence of an event in an experiment. The probability is obtained by dividing the number of successful occurrences by the total number of trials. For example, there is a 50% chance that the outcome of tossing a coin will be heads. This concept is elaborated further in Section 6.3.

6.2.2 Measures of Central Tendency

The most common parameter used to describe the central tendency is the *mean*. This is the everyday concept of average and for a sample is defined by

$$\bar{x} = \frac{x_1 + x_2 + \cdots + x_n}{n} = \sum_{i=1}^{n} \frac{x_i}{n} \tag{6.1}$$

where the x_i's are the values of the sample data and n is the number of measurements. For a population with a finite number of elements, N, with values x_i, the mean is denoted by the symbol μ and is given by

$$\mu = \frac{x_1 + x_2 + \cdots + x_n}{N} = \sum_{i=1}^{N} \frac{x_i}{N} \tag{6.2}$$

For a population with an infinite number of elements, the mean can be obtained from a more general definition presented later.

Two other common parameters describing central tendency are the median and the mode. If the measurands are arranged in ascending or descending order, the *median* is the value at the center of the set. If the set contains an even number of elements, the median is the average of the two central values. The *mode* is the value of the variable that corresponds to the peak value of the probability of occurrence of the event. In a discrete sample space, the mode can easily be identified as the most frequently occurring value. In a continuous sample space, the mode is taken as the midpoint of the data interval (bin) with the highest frequency. For some distributions (e.g., a uniform distribution), the mode may not exist at all; for other distributions (e.g., a bimodal distribution), there may be more than one peak frequency and more than one mode. When a distribution has more than one mode, the frequencies of occurrence of each mode need not be the same. While it is common for the mean, median, and mode to have close to the same value (although they will generally not have exactly the same value), in some data sets they may have significantly different values.

6.2.3   Measures of Dispersion

Dispersion is the spread or variability of the data. For example, a set of measurements ranging from 90 to 110 has a greater dispersion than a set of measurements ranging from 95 to 105. The following quantities are the most common ones used for representing the extent of dispersion of random variables around their mean value:

The *deviation* of each measurement is defined as

$$d_i = x_i - \overline{x} \tag{6.3}$$

The *mean deviation* is defined as

$$\overline{d} = \sum_{i=1}^{n} \frac{|d_i|}{n} \tag{6.4}$$

For a population with a finite number of elements, the *population standard deviation* is defined as

$$\sigma = \sqrt{\sum_{i=1}^{N} \frac{(x_i - \mu)^2}{N}} \tag{6.5}$$

As with the mean, for populations with an infinite number of elements, a more general definition will be presented at a later point.

The *sample standard deviation* is defined as

$$S = \sqrt{\sum_{i=1}^{n} \frac{(x_i - \overline{x})^2}{(n - 1)}} \tag{6.6}$$

The sample standard deviation is used when the data of a sample are used to estimate the population standard deviation.

The *variance* is defined as

$$\text{variance} = \begin{cases} \sigma^2 & \text{for the population} \\ S^2 & \text{for a sample} \end{cases} \tag{6.7}$$

For the temperature measurements in Table 6.1, the preceding statistics have the following values:

Mean	$\overline{x}$	$= 1103°C$
Median	x_m	$= 1104°C$
Standard deviation	S	$= 5.79°C$
Variance	S^2	$= 33.49°C^2$
Mode	m	$= 1104°C$

6.3   PROBABILITY

Probability is a numerical value expressing the likelihood of occurrence of an event relative to all possibilities in a sample space. As an example, if two dice are tossed, the probability of the event of two 1's is 1/36, since there are 36 possible outcomes for two dice and only one of these outcomes represents the event desired. If the desired event occurs, the outcome is considered successful. The prediction of the probability of an event is one purpose of statistical analysis. The probability of occurrence of an event A is defined as the number of successful occurrences (m) divided by the total number of possible outcomes (n) in a sample space, evaluated for $n \gg 1$:

$$\text{probability of event } A = \frac{m}{n} \tag{6.8}$$

The event may be represented by a continuous random variable x, in which case the probability will be represented by $P(x)$. For a discrete random variable x_i, the probability will be shown by $P(x_i)$.

The following are some standard properties associated with probability:

1. Probability is always a positive number with a maximum of 1: $0 \leq P(x \text{ or } x_i) \leq 1$.
2. If an event A is certain to occur, $P(A) = 1$.
3. If an event A is certain not to occur, $P(A) = 0$.
4. If event $\overline{A}$ is the complement of event A (this means that if event A occurs, event $\overline{A}$ cannot occur), then

$$P(\overline{A}) = 1 - P(A) \tag{6.9}$$

5. If the events A and B are mutually exclusive (i.e., the probability of simultaneous occurrence of A and B is zero), then the probability of occurrence of event A or event B is

$$P(A \text{ or } B) = P(A) + P(B) \tag{6.10}$$

For example, in the case of tossing a fair die, the probability of having a 3 or a 6 is

$$P(3 \text{ or } 6) = \frac{1}{6} + \frac{1}{6} = \frac{1}{3}$$

6. If the events A and B are independent of each other (this means that their occurrences do *not* depend on each other), the probability that both A and B will occur together is

$$P(AB) = P(A)P(B) \tag{6.11}$$

As an example, consider an assembly of two pieces, A and B, in a machine, such as a computer and a monitor. A and B are manufactured by two different companies. We are told that there is a 5% chance that A is defective and a 2% chance that B is defective. The probability that both A and B are defective is then

$$P(AB) = 0.05 \times 0.02 = 0.001 \text{ or } 0.1\%$$

7. The probability of occurrence of A or B or both, represented by $P(A \cup B)$ (read "the probability of A union B"), is

$$P(A \cup B) = P(A) + P(B) - P(AB) \tag{6.12}$$

In the preceding example, the probability of having A or B or both be defective will be

$$P(A \cup B) = 0.05 + 0.02 - 0.001 = 0.069 \text{ or } 6.9\%$$

6.3.1 Probability Distribution Functions

An important function of statistics is to use information from a sample to predict the behavior of a population. For example, after testing a sample from a batch of light bulbs for time to failure, we may want to know the probability that an additional bulb selected from the batch will have a time to failure of less than a certain value. In our duct temperature data in Table 6.1, we might ask the question, "What is the probability that our next measurement will be between 1105 and 1110°C?" One approach would be to use the sample data directly. This approach is called *use of an empirical distribution*. We can replot the data of Figure 6.1 in the form of relative frequency versus temperature, as shown in Figure 6.3. Relative frequency is the number of samples in each bin, divided by the total number of samples. Since the relative frequency of the bin for 1105 to 1110 is 0.23, we might estimate that the probability that an additional measurement will fall in the selected bin is approximately 0.23. However, such an approach has decided limitations. Due to the small sample size, the relative frequencies for the individual bins are quite uncertain; hence, the estimates of probabilities are only approximate. This is particularly true for the bins with only one or two samples. Although, intuitively, we expect some finite, small probability for temperatures less than 1085 or greater than 1119, the use of the sample data directly would not result in reasonable estimates of these probabilities.

For particular situations, experience has shown that the distribution of the random variable follows certain mathematical functions. Sample data are used to compute parameters in these mathematical functions, and then we use the mathematical functions to predict properties of the parent population. For discrete random variables,

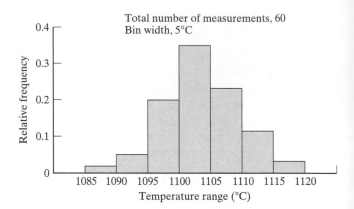

FIGURE 6.3

Relative frequencies of duct temperatures.

these functions are called *probability mass functions*. For continuous random variables, the functions are called *probability density functions*. For most experimental situations, the appropriate distribution function can be determined from experience. However, a technique known as *goodness of fit* can be used to determine the suitability of a distribution function for a given situation. This technique is presented in many statistical texts, such as Harnett and Murphy (1975).

Probability Mass Function   If a discrete random variable can have values $x_1, x_2, \ldots, x_n$, then the probability of occurrence of a particular value of x_i is $P(x_i)$, where P is the probability mass function for the variable x. As an example, consider a single die that has six possible states, each with equal probability 1/6. Then $P(x_i) = 1/6$ for each value of x_i. As another example, consider a biased coin that has a probability of heads of 2/3 and a probability of tails of 1/3. Considering heads to be x_1 and tails to be x_2, the complete function $P(x)$ is represented by $P(x_1) = 2/3$ and $P(x_2) = 1/3$.

The sum of the probabilities of all possible values of x must be 1:

$$\sum_{i=1}^{n} P(x_i) = 1 \tag{6.13}$$

The mean of the population for a discrete random variable is given by

$$\mu = \sum_{i=1}^{N} x_i P(x_i) \tag{6.14}$$

The quantity μ is also called the *expected value of x*, $E(x)$. The variance of the population is given by

$$\sigma^2 = \sum_{i=1}^{N} (x_i - \mu)^2 P(x_i) \tag{6.15}$$

Probability Density Function   For a continuous random variable, a function $f(x)$, called a *probability density function*, is defined such that the probability of occurrence of the random variable in an interval between x_i and $x_i + dx$ is given by

$$P(x_i \le x \le x_i + dx) = f(x_i)\,dx \tag{6.16}$$

To evaluate the probability that x will occur in a finite interval from $x = a$ to $x = b$, this equation can be integrated to obtain

$$P(a \le x \le b) = \int_a^b f(x)\,dx \tag{6.17}$$

For a continuous random variable, the probability of x having a single unique value is zero. If the limits of integration are extended to negative and positive infinity, we can be sure that the measurement is in that range and the probability will be $P(-\infty \le x \le \infty) = 1$.

The definition of $f(x)$ now allows us to define the mean of a population with probability density function $f(x)$:

$$\mu = \int_{-\infty}^{\infty} x f(x)\,dx \tag{6.18}$$

This is also the *expected value* of the random variable, $E(x)$. The *variance* of the population is given by

$$\sigma^2 = \int_{-\infty}^{\infty} (x - \mu)^2 f(x)\,dx \tag{6.19}$$

Example 6.1

The life of a given type of ball bearing can be characterized by a probability distribution function

$$f(x) = \begin{cases} 0 & x < 10\,\text{h} \\ \dfrac{200}{x^3} & x \ge 10\,\text{h} \end{cases}$$

$f(x)$ is shown in Figure E6.1.

(a) Calculate the expected life of the bearings.

(b) If we pick a bearing at random from this batch, what is the probability that its life (x) will be less than 20 h, greater than 20 h, and finally, exactly 20 h?

Solution:

(a) Using Eq. (6.18), we have

$$E(x) = \mu = \int_0^{\infty} x f(x)\,dx = \int_{10}^{\infty} x \frac{200}{x^3}\,dx = -\frac{200}{x}\Big|_{10}^{\infty} = 20\,\text{h}$$

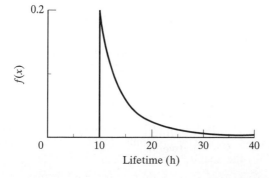

FIGURE E6.1

(b) The probability that the lifetime is less than 20 h is given by

$$P(x < 20) = \int_{-\infty}^{20} f(x)\, dx = \int_{-\infty}^{10} 0\, dx + \int_{10}^{20} \frac{200}{x^3}\, dx = 0.75$$

Also,

$$P(x \geq 20) = 1 - P(x \leq 20) = 0.25$$
$$P(x = 20) = 0$$

Cumulative Distribution Function   The cumulative distribution function is another method of presenting data for the distribution of a random variable. It is used to determine the probability that a random variable has a value less than or equal to a specified value. The cumulative distribution function for a continuous random variable (rv) is defined as

$$F(rv \leq x) = F(x) = \int_{-\infty}^{x} f(x)\, dx = P(rv \leq x) \tag{6.20}$$

For a discrete random variable, it is defined as

$$F(rv \leq x_i) = \sum_{j=1}^{i} P(x_j) \tag{6.21}$$

The next two relations follow from the definition of the cumulative distribution function:

$$P(a < x \leq b) = F(b) - F(a)$$
$$P(x > a) = 1 - F(a) \tag{6.22}$$

The use of the cumulative distribution function is demonstrated in Example 6.2.

Example 6.2

Find the probability that the lifetime of one of the ball bearings in Example 6.1 will have a lifetime of

(a) less than 15 h and

(b) less than 20 h, using the cumulative distribution function.

Solution:

(a) Using Eq. (6.20), we obtain, for the cumulative distribution function,

$$F(x) = \int_{-\infty}^{x} f(x)\, dx = \int_{-\infty}^{x} 0\, dx = 0 \qquad \text{for } x \leq 10$$

$$= 0 + \int_{10}^{x} \frac{200}{x^3}\, dx = 1 - \frac{100}{x^2} \qquad \text{for } x > 10$$

This function is plotted in Figure E6.2. Either by substituting 15 into the equation or by reading from the graph, we find that the probability that the lifetime is less than 15 h is 0.55.

(b) Similarly, the probability that the lifetime is less than 20 h is 0.75.

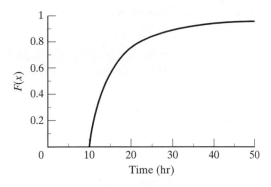

Time (hr)

FIGURE E6.2

6.3.2 Some Probability Distribution Functions with Engineering Applications

A number of distribution functions are used in engineering applications. In this section, some of the most common are described briefly and their application is discussed.

Binomial Distribution The *binomial distribution* is a distribution which describes discrete random variables that can have only two possible outcomes: "success" and "failure." This distribution has application in production quality control, when the quality of a product is either acceptable or unacceptable. The following conditions need to be satisfied for the binomial distribution to be applicable to a certain experiment:

1. Each trial in the experiment can have only the two possible outcomes of success or failure.
2. The probability of success remains constant throughout the experiment. This probability is denoted by p and is usually known or estimated for a given population.
3. The experiment consists of n independent trials.

The binomial distribution provides the probability (P) of finding exactly r successes in a total of n trials and is expressed as

$$P(r) = \binom{n}{r} p^r (1 - p)^{n-r} \tag{6.23}$$

In this relation, r is an integer and is always less than or equal to n, and

$$\binom{n}{r} = \frac{n!}{r!(n - r)!} \tag{6.24}$$

is called n *combination* r, which is the number of ways that we can choose r identical items from n items.

The expected number of successes in n trials for binomial distribution is

$$\mu = np \tag{6.25}$$

The standard deviation of the binomial distribution is

$$\sigma = \sqrt{np(1-p)} \qquad (6.26)$$

In some engineering applications, one is interested in finding the probability that the event will happen less than or equal to a certain number of times. If k is this number of occurrences $(k \leq n)$, then

$$P(r \leq k) = \sum_{i=0}^{k} P(r = i) = \sum_{i=0}^{k} \binom{n}{i} p^i (1-p)^{n-i} \qquad (6.27)$$

Example 6.3

A manufacturer of a brand of computer claims that his computers are reliable and that only 10% of the computers require repairs within the warranty period. Determine the probability that in a batch of 20 computers, 5 will require repair during the warranty period.

Solution: We can apply the binomial distribution because of the pass/fail outcome of the process. Success will be defined as not needing repair within the warranty period. In this case, based on the manufacturer's tests, $p = 0.9$. Other assumptions underlying the application of this distribution are that all trials are independent and that the probabilities of success and failure are the same for all computers. The problem amounts to determining the probability (P) of having 15 successes (r) out of 20 machines (n). Using Eqs. (6.23) and (6.24) yields

$$\binom{n}{r} = \binom{20}{15} = \frac{20!}{15!(20-15)!} = 15{,}504$$

$$P = \binom{20}{15} 0.9^{15}(1-0.9)^5 = 0.032$$

The conclusion here is that there is a fairly small chance (3.2%) that there will be exactly 5 computers out of a batch of 20 computers requiring repair.

Example 6.4

A light-bulb manufacturing company has discovered that, for a given batch, 10% of the light bulbs are defective. If we buy four of these bulbs, what are the probabilities of finding that four, three, two, one, and none of the bulbs are defective?

Solution: Again, we can use the binomial distribution. The number of trials is 4, and if we define success as bulb failure, $p = 0.1$. The probability of having four, three, two, one, and zero defective light bulbs can be calculated by using Eq. (6.23). The probability of finding four defective bulbs is determined from

$$P(r = 4) = \binom{4}{4} 0.1^4 (1-0.1)^{4-4}$$

$$= 0.0001$$

Note:

$$\binom{4}{4} = \left(\frac{4!}{4!(4-4)!}\right) = 1 \qquad \text{since } 0! = 1$$

Similarly,

$$P(r = 3) = 0.0036$$
$$P(r = 2) = 0.0486$$

$P(r = 1) = 0.2916$
$P(r = 0) = 0.6561$

The total probability of all five possible outcomes is

$$P = P(r = 4) + P(r = 3) + P(r = 2) + P(r = 1) + P(r = 0)$$
$$= 1.0000$$

Example 6.5

For the data of Example 6.4, calculate the probability of finding up to and including two defective light bulbs in the sample of four.

Solution: We use Eq. (6.27) for this purpose:

$$P(r \leq 2) = P(r = 0) + P(r = 1) + P(r = 2)$$
$$= 0.6561 + 0.2916 + 0.0486$$
$$= 0.9963$$

This means that the probability of finding two or fewer defective light bulbs in four is 99.63%.

Poisson Distribution The *Poisson distribution* is used to estimate the number of random occurrences of an event in a specified interval of time or space if the average number of occurrences is already known. For example, if it is known that, on average, 10 customers visit a bank per five-minute period during the lunch hour, the Poisson distribution can be used to predict the probability that 8 customers will visit during a particular five-minute period. The Poisson distribution can also be used for spatial variations. For instance, if it is known that there are, on average, two defects per square meter of printed circuit boards, the Poisson distribution can be used to predict the probability that there will be four defects in a square meter of boards.

The following two assumptions underline the Poisson distribution:

1. The probability of occurrence of an event is the same for any two intervals of the same length.
2. The probability of occurrence of an event is independent of the occurrence of other events.

The probability of occurrence of x events is given by

$$P(x) = \frac{e^{-\lambda}\lambda^x}{x!} \tag{6.28}$$

where λ is the expected or mean number of occurrences during the interval of interest.

The expected value of x for the Poisson distribution, the same as the mean, μ, is given by

$$E(x) = \mu = \lambda \tag{6.29}$$

The standard deviation is given by

$$\sigma = \sqrt{\lambda} \tag{6.30}$$

In some cases, the goal is to find the probability that a certain number of events or fewer will occur. To compute the probability that the number of occurrences is less than or equal to k, the sum of the probability of $k, k - 1 \ldots 0$ events must be computed. This is given by

$$P(x \le k) = \sum_{i=0}^{k} \frac{e^{-\lambda}\lambda^i}{i!} \tag{6.31}$$

Example 6.6

In a binary data stream, it is known that there is an average of three errors per minute. Find the probability that there are exactly zero errors during a one-minute period.

Solution: For this problem, λ is 3 and x is 0. Substituting into Eq. 6.28 yields

$$P(x) = \frac{e^{-3}3^0}{0!} = 0.050$$

The probability that there are no errors in a one-minute interval is thus 0.050.

Example 6.7

In Example 6.6, what is the probability that there will be
  (a) three or fewer errors.
  (b) more than three errors.

Solution: First we need to compute the probability of $0, 1, 2,$ and 3 errors using Eq. 6.28:

$$P(3) = \frac{e^{-3}3^3}{3!} = 0.224$$

Similarly,

$$P(2) = 0.224$$
$$P(1) = 0.149$$
$$P(0) = 0.050$$

Thus,

$$P(x \le 3) = 0.050 + 0.149 + 0.224 + 0.224 = 0.647$$

The probability that x is greater than 3 is then

$$P(x > 3) = 1 - P(x \le 3) = 1 - 0.647 = 0.353$$

Example 6.8

It has been found in welds joining pipes that there is an average of five defects per 10 linear meters of weld (0.5 defects per meter). What is the probability that there will be
  (a) a single defect in a weld that is 0.5 m long or
  (b) more than one defect in a weld that is 0.5 m long.

Solution: λ, the average number of defects in 0.5 m, is $0.5 \times 0.5 = 0.25$. Using Eq. 6.28, we find that the probability of one defect is then

$$P(1) = \frac{e^{-0.25}0.25^1}{1!} = 0.194$$

There is thus a probability of 0.194 that there would be a single defect. The probability that there will be more than one defect is

$$P(x > 1) = 1 - P(0) - P(1)$$

$P(0)$ can be computed from Eq. 6.28 as 0.778. The probability is

$$P(x > 1) = 1 - 0.778 - 0.194 = 0.028$$

Thus, the probability of more than one defect is only 0.028.

Normal (Gaussian) Distribution   The *normal* (or *Gaussian*) *distribution function* is a simple distribution function that is useful for a large number of common problems involving continuous random variables. The normal distribution has been shown to describe the dispersion of the data for measurements in which the variation in the measured value is due totally to random factors and occurrences of both positive and negative deviations are equally probable. The equation for the normal probability density function is

$$f(x) = \frac{1}{\sigma\sqrt{2\pi}}e^{-(x-\mu)^2/2\sigma^2} \tag{6.32}$$

In this equation, x is the random variable. The function has two parameters: the population standard deviation, σ, and the population mean, μ. A plot of $f(x)$ versus x for different values of σ and a fixed value of the mean is shown in Figure 6.4. As the figure shows, the distribution is symmetric about the mean value, and the smaller the standard deviation, the higher is the peak value of $f(x)$ at the mean.

According to the definition of the probability density function [Eq. (6.17)], for a given population, the probability of having a single value of x between a lower limit of

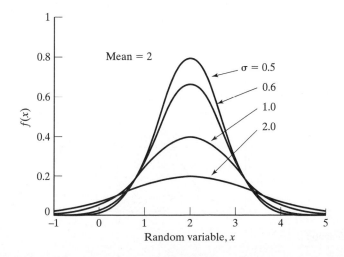

FIGURE 6.4

Normal distribution function.

x_1 and an upper limit of x_2 is

$$P(x_1 \leq x \leq x_2) = \int_{x_1}^{x_2} f(x)\, dx = \int_{x_1}^{x_2} \frac{1}{\sigma\sqrt{2\pi}} e^{-(x-\mu)^2/2\sigma^2}\, dx \qquad (6.33)$$

Since $f(x)$ is in the form of an error function, the preceding integral cannot be evaluated analytically, and as a result, the integration must be performed numerically. To simplify the numerical integration process, the integrand is usually modified with a change of variable so that the numerically evaluated integral is general and useful for all problems. A nondimensional variable z is defined as

$$z = \frac{x - \mu}{\sigma} \qquad (6.34)$$

It is now possible to define the function

$$f(z) = \frac{1}{\sqrt{2\pi}} e^{-z^2/2} \qquad (6.35)$$

which is called the *standard normal density function*. It represents the normal probability density function for a random variable z with mean μ equal to zero and $\sigma = 1$. This normalized function is plotted in Figure 6.5.

Taking the differential of Eq. (6.34), we have $dx = \sigma\, dz$. Equation (6.33) will then transform to

$$P(x_1 \leq x \leq x_2) = \int_{z_1}^{z_2} f(z)\, dz \qquad (6.36)$$

The probability that x is between x_1 and x_2 is the same as the probability that the transformed variable z is between z_1 and z_2:

$$P(x_1 \leq x \leq x_2) = P(z_1 \leq z \leq z_2) = P\left(\frac{x_1 - \mu}{\sigma} \leq \frac{x - \mu}{\sigma} \leq \frac{x_2 - \mu}{\sigma}\right) \qquad (6.37)$$

The probability $P(z_1 \leq z \leq z_2)$ has a value equal to the crosshatched area shown on Figure 6.5. The curve shown in the figure is symmetric with respect to the vertical axis at $z = 0$, which indicates that, with this distribution, the probabilities of positive and

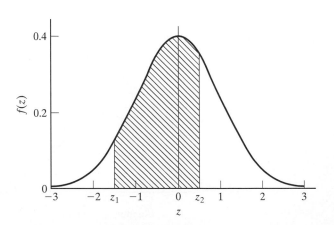

FIGURE 6.5

Standard normal distribution function.

negative deviations from $z = 0$ are equal. Mathematically, this can be stated as

$$P(-z_1 \le z \le 0) = P(0 \le z \le z_1) = \frac{P(-z_1 \le z \le z_1)}{2.0} \tag{6.38}$$

As mentioned, the integral in Eq. (6.33) has two parameters (μ and σ), and as a practical matter, it would have to be integrated numerically for each application. On the other hand, the integral in Eq. (6.36) has no parameters. If z_1 in Eq. (6.36) is chosen to be zero, it is practical to perform the integration numerically and tabulate the results as a function of z_2, or simply z. The results of this integration are shown in Table 6.3. Since the standard normal distribution function is symmetric about $z = 0$, the table can then be used to predict the probability of occurrence of a random variable anywhere in the range $-\infty$ to $+\infty$ if the population follows the basic characteristics of a normal distribution. The table presents the value of the probability that the random variable has a value between 0 and z for z values shown in the left column and the top row. The top row serves as the second decimal point of the first column. The process of using Table 6.3 to predict probabilities is demonstrated in Example 6.9.

Example 6.9

The results of a test that follows a normal distribution have a mean value of 10.0 and a standard deviation of 1. Find the probability that a single reading is

(a) between 9 and 12.
(b) between 8 and 9.55.
(c) less or equal to 9.
(d) greater than 12.

Solution:

(a) Using Eq. (6.34), we have $z_1 = (9 - 10)/1 = -1$ and $z_2 = (12 - 10)/1 = 2$. We are looking for the area under the standard normal distribution curve from $z = -1$ to $z = 2$ [Figure E6.9(a)]. This will be broken into two parts: from $z = -1$ to 0 and from $z = 0$ to $z = 2$. The area under the curve from $z = -1$ to 0 is the same as the area from $z = 0$ to $z = 1$. From Table 6.3, this value is 0.3413. The area from $z = 0$ to $z = 2$ is 0.4772. That is,

$$P(-1 \le z \le 0) = P(0, 1) = 0.3413 \quad \text{and} \quad P(0 \le z \le 2) = 0.4772$$

Hence,

$$P(-1 \le z \le 2) = 0.3413 + 0.4772 = 0.8185 \quad \text{or} \quad 81.85\%$$

(b) $z_1 = (8 - 10)/1 = -2$ and $z_2 = (9.55 - 10)/1 = -0.45$. We are looking for the area between $z = -2$ and $z = -0.45$ [Figure E6.9(b)]. From Table 6.3, the area from $z = -2$ to $z = 0$ is 0.4772 and the area from $z = -0.45$ to $z = 0$ is 0.1736. The result we seek is the difference between these two areas:

$$P(-2 \le z \le -0.45) = 0.4772 - 0.1736 = 0.3036 = 30.36\%$$

(c) $z_1 = -\infty$ and $z_2 = -1$ [Figure E6.9(c)]. From Table 6.3, we find that $P(-1 \le z \le 0)$ is 0.3413. We also know that $P(-\infty \le z \le 0)$ is 0.5. Hence,

$$P(-\infty \le z \le -1) = 0.5 - 0.3413 = 0.1587 \quad \text{or} \quad 15.87\%$$

TABLE 6.3   Area Under the Normal Distribution From $z = 0$ to z

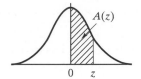

z	0.00	0.01	0.02	0.03	0.04	0.05	0.06	0.07	0.08	0.09
0.0	.0000	.0040	.0080	.0120	.0160	.0199	.0239	.0279	.0319	.0359
0.1	.0398	.0438	.0478	.0517	.0557	.0596	.0636	.0675	.0714	.0753
0.2	.0793	.0832	.0871	.0910	.0948	.0987	.1026	.1064	.1103	.1141
0.3	.1179	.1217	.1255	.1293	.1331	.1368	.1406	.1443	.1480	.1571
0.4	.1554	.1591	.1628	.1664	.1700	.1736	.1772	.1808	.1844	.1879
0.5	.1915	.1950	.1985	.2019	.2054	.2088	.2123	.2157	.2190	.2224
0.6	.2257	.2291	.2324	.2357	.2389	.2422	.2454	.2486	.2517	.2549
0.7	.2580	.2611	.2642	.2673	.2704	.2734	.2764	.2794	.2823	.2852
0.8	.2881	.2910	.2939	.2967	.2995	.3023	.3051	.3078	.3106	.3133
0.9	.3159	.3186	.3212	.3238	.3264	.3289	.3315	.3340	.3365	.3389
1.0	.3413	.3438	.3461	.3485	.3508	.3531	.3554	.3577	.3599	.3621
1.1	.3643	.3665	.3686	.3708	.3729	.3749	.3770	.3790	.3810	.3830
1.2	.3849	.3869	.3888	.3907	.3925	.3944	.3962	.3980	.3997	.4015
1.3	.4032	.4049	.4066	.4082	.4099	.4115	.4131	.4147	.4162	.4177
1.4	.4192	.4207	.4222	.4236	.4251	.4265	.4279	.4292	.4306	.4319
1.5	.4332	.4345	.4357	.4370	.4382	.4394	.4406	.4418	.4429	.4441
1.6	.4452	.4463	.4474	.4484	.4495	.4505	.4515	.4525	.4535	.4545
1.7	.4554	.4564	.4573	.4582	.4591	.4599	.4608	.4616	.4625	.4633
1.8	.4641	.4649	.4656	.4664	.4671	.4678	.4686	.4693	.4699	.4706
1.9	.4713	.4719	.4726	.4732	.4738	.4744	.4750	.4756	.4761	.4767
2.0	.4772	.4778	.4783	.4788	.4793	.4798	.4803	.4808	.4812	.4817
2.1	.4821	.4826	.4830	.4834	.4838	.4842	.4846	.4850	.4854	.4857
2.2	.4861	.4864	.4868	.4871	.4875	.4878	.4881	.4884	.4887	.4890
2.3	.4893	.4896	.4898	.4901	.4904	.4906	.4909	.4911	.4913	.4916
2.4	.4918	.4920	.4922	.4925	.4927	.4929	.4931	.4932	.4934	.4936
2.5	.4938	.4940	.4941	.4943	.4945	.4946	.4948	.4949	.4951	.4952
2.6	.4953	.4955	.4956	.4957	.4959	.4960	.4961	.4962	.4963	.4964
2.7	.4965	.4966	.4967	.4968	.4969	.4970	.4971	.4972	.4973	.4974
2.8	.4974	.4975	.4976	.4977	.4977	.4978	.4979	.4979	.4980	.4981
2.9	.4981	.4982	.4982	.4983	.4984	.4984	.4985	.4985	.4986	.4986
3.0	.4987	.4987	.4987	.4988	.4988	.4989	.4989	.4989	.4990	.4990
3.1	.4990	.4991	.4991	.4991	.4992	.4992	.4992	.4992	.4993	.4993
3.2	.4993	.4993	.4994	.4994	.4994	.4994	.4994	.4995	.4995	.4995
3.3	.4995	.4995	.4995	.4996	.4996	.4996	.4996	.4996	.4996	.4997
3.4	.4997	.4997	.4997	.4997	.4997	.4997	.4997	.4997	.4997	.4998
3.5	.4998	.4998	.4998	.4998	.4998	.4998	.4998	.4998	.4998	.4998
3.6	.4998	.4998	.4999	.4999	.4999	.4999	.4999	.4999	.4999	.4999
3.7	.4999	.4999	.4999	.4999	.4999	.4999	.4999	.4999	.4999	.4999
3.8	.4999	.4999	.4999	.4999	.4999	.4999	.4999	.4999	.4999	.4999
3.9	.5000	.5000	.5000	.5000	.5000	.5000	.5000	.5000	.5000	.5000
4.0	.5000	.5000	.5000	.5000	.5000	.5000	.5000	.5000	.5000	.5000

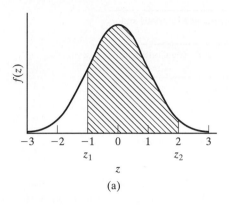

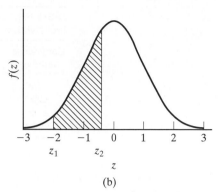

FIGURE E6.9(a, b)

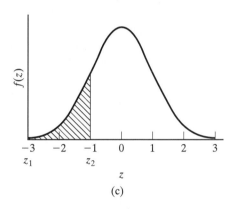

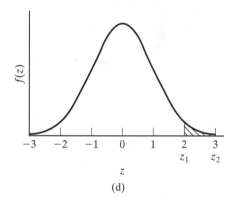

FIGURE E6.9(c,d)

(d) $z_1 = (12 - 10)/1 = 2$ and $z_2 = \infty$. We are seeking the area from $z = 2$ to $z = \infty$ [Figure E6.9(d)]. From Table 6.3, the area from $z = 0$ to $z = 2$ is 0.4772. The area from $z = 0$ to $z = \infty$ is 0.5. The required probability is then the difference between these two areas:

$$P(2 \le z \le \infty) = P(0, \infty) - P(0, 2) = 0.5 - 0.4772 = 0.0228 \quad \text{or} \quad 2.28\%$$

It is also of interest to determine the probability that a measurement will fall within one or more standard deviations (σ's) of the mean. The value for $1\sigma(z = 1)$ is evaluated from Table 6.3. For $z = 1$, the probability is 0.3413. Since we are looking for $\pm 1\sigma$, the probability we seek is twice this value, or 0.6826 (68.26%). This means that if a measurement is performed in an environment where the deviation from the mean value is totally influenced by random variables, we can be 68.26% confident that it will fall within one standard deviation from the mean. For 2σ, the probability is 95.44%. This value is close to 95%; hence, in practical usage, 2σ and 95% are often used interchangeably.

TABLE 6.4 Confidence Intervals and Probabilities

Confidence interval	Confidence level (%)
$\pm 1\sigma$	68.26
$\pm 2\sigma$	95.44
$\pm 3\sigma$	99.74
$\pm 3.5\sigma$	99.96

Table 6.4 shows the probability (confidence level) for different deviation limits (confidence intervals). The concepts of confidence intervals and confidence level are discussed in greater detail later in the chapter.

Example 6.10

In the mass production of engine blocks for a single-cylinder engine, the tolerance in the diameter (D) of the cylinders is normally distributed. The average diameter of the cylinders, μ, has been measured to be 4.000 in. and the standard deviation is 0.002 in. What are the probabilities of the following cases?

(a) A cylinder is measured with $D \leq 4.002$.
(b) A cylinder is measured with $D \geq 4.005$.
(c) A cylinder is measured with $3.993 \leq D \leq 4.003$.
(d) If cylinders with a deviation from the mean diameter of greater than 2σ are rejected, what percent of the blocks will be rejected?

Solution: The deviation in diameter in this case is $d = D - \mu$, where D is the diameter of the cylinder and μ is the mean value for diameter. It has been established that the diameter is normally distributed, so we can use the standard normal distribution function and Table 6.3 to solve this problem.

(a) $z = (D - \mu)/\sigma = (4.002 - 4.000)/0.002 = 1$. From Table 6.3, for $z = 1$, the area from $z = 0$ to $z = 1$ is 0.3413. The area from $z = -\infty$ to $z = 0$ is 0.5. Then

$$P(-\infty \leq z \leq 1) = 0.5 + 0.3413 = 0.8413 \text{ or } 84.3\%$$

(b) $z = (4.005 - 4.000)/0.002 = 2.5$. From Table 6.3, for $z = 2.5$, the area from $z = 0$ to $z = 2.5$ is 0.4938. The area from $z = 0$ to $z = \infty$ is 0.5. Then

$$P(z > 2.5) = 0.5 - 0.4938 = 0.0062 \quad \text{or} \quad 0.62\%$$

(c) $z_1 = (3.993 - 4.000)/0.002 = -3.5$ and $z_2 = (4.003 - 4.000)/0.002 = 1.5$. From Table 6.3, we obtain 0.4998 and 0.4332 for $z = 3.5$ and 1.5, respectively. Then

$$P(-3.5 \leq z \leq 1.5) = 0.4998 + 0.4332 = 0.933 \quad \text{or} \quad 93.3\%$$

(d) If the rejection criterion is 2σ, the fraction of rejected blocks will be

$$1 - 2 \times P(0 \leq z \leq 2) = 1 - 2 \times 0.4772 = 0.0456 \quad \text{or} \quad 4.56\%$$

TABLE 6.5   Summary of Important Distributions in Engineering

Distribution	Application
Binomial	Used for quality control (rejection of defective products), failure–success, good–bad type of observation. This is a discrete distribution.
Poisson	Used to predict the probability of occurrence of a specific number of events in a space or time interval if the mean number of occurrences is known.
Normal (Gaussian)	Continuous, symmetrical, and most widely used distribution in experimental analysis in physical science. Used for explanation of random variables in engineering experiments, such as gas molecule velocities, electrical power consumption of households, etc.
Student's t	Continuous, symmetrical, used for analysis of the variation of sample mean value for experimental data with sample size less than 30. For sample sizes greater than 30, Student's t approaches normal distribution.
χ^2	Continuous, nonsymmetrical, used for analysis of variance of samples in a population. For example, consistency of chemical reaction time is of prime importance in some industrial processes and χ^2 distribution is used for its analysis. This distribution is also used to determine goodness of fit of a distribution for a particular application.
Weibull	Continuous, nonsymmetrical, used for describing the life phenomena of parts and components of machines.
Exponential	Continuous, nonsymmetrical, used for analysis of failure and reliability of complete systems and assemblies.
Lognormal	Continuous, nonsymmetrical, used for life and durability studies of parts and components.
Uniform	Continuous, symmetrical, used for estimating the probabilities of random values generated by computer simulation.

Other Distribution Functions   Several other types of probability distribution functions are used in engineering experiments. Detailed discussion of all these functions are beyond the scope of this book. In Table 6.5, we summarize briefly important distributions that have significant applications in engineering. For details on the subject, the reader is referred to more comprehensive texts on the subject, such as Lipson and Sheth (1973) and Lapin (1990).

6.4   PARAMETER ESTIMATION

In most experiments, the sample size is small relative to the population, yet the principal intention of a statistical experiment is to estimate the parameters describing the entire population. While there are many population parameters that can be estimated, the most common estimated parameter is the population mean, μ. In some cases, it is also necessary to estimate the population standard deviation, σ. An estimate of the population mean, μ, is the sample mean, $\bar{x}$ [as defined by Eq. (6.1)], and an estimate of the population standard deviation, σ, is the sample standard deviation, S [as defined by Eq. (6.6)]. However, simple estimation of values for these parameters is not sufficient. Different samples from the same population will yield different values. Consequently, it is also necessary to determine uncertainty intervals for the estimated parameters. In the sections that follow, we discuss the determination of these uncertainty intervals. For further details, see Walpole and Myers (1998).

6.4.1   Interval Estimation of the Population Mean

We wish to make an estimate of the population mean, which takes the form

$$\mu = \bar{x} \pm \delta \quad \text{or} \quad \bar{x} - \delta \leq \mu \leq \bar{x} + \delta \tag{6.39}$$

where δ is an uncertainty and $\bar{x}$ is the sample mean. The interval from $\bar{x} - \delta$ to $\bar{x} + \delta$ is called the *confidence interval* on the mean. However, the confidence interval depends on a concept called the *confidence level*, sometimes called the *degree of confidence*. The confidence level is the probability that the population mean will fall within the specified interval:

$$\text{confidence level} = P(\bar{x} - \delta \leq \mu \leq \bar{x} + \delta) \tag{6.40}$$

We will have a higher confidence level that the mean will fall in a large interval than that it will fall within a small interval. For example, we might state that the mean is 11 ± 5 with a 95% confidence level or that the mean is 11 ± 3 with a 60% confidence level. The confidence level is normally expressed in terms of a variable α called the *level of significance*:

$$\text{confidence level} = 1 - \alpha \tag{6.41}$$

α is then the probability that the mean will fall outside the confidence interval.

The *central limit theorem* makes it possible to make an estimate of the confidence interval with a suitable confidence level. Consider a population of the random variable x with a mean value of μ and standard deviation σ. From this population, we could take several different samples, each of size n. Each of these samples would have a mean value $\bar{x}_i$, but we would not expect each of these means to have the same value. In fact, the $\bar{x}_i$'s are values of a random variable. The central limit theorem states that if n is sufficiently large, the $\bar{x}_i$'s follow a normal distribution and the standard deviation of these means is given by

$$\sigma_{\bar{x}} = \frac{\sigma}{\sqrt{n}} \tag{6.42}$$

The population need not be normally distributed for means to be normally distributed and for the standard deviation of the means to be given by Eq. (6.42). The standard deviation of the mean is also called the *standard error of the mean*. For the central limit theorem to apply, the sample size n must be large. In most cases, to be considered large, n should exceed 30 (Harnett and Murphy, 1975; Lapin, 1990).
The following are important conclusions from the central limit theorem:

1. If the original population is normal, the distribution for the $\bar{x}_i$'s is normal.
2. If the original population is not normal and n is large ($n > 30$), the distribution for the $\bar{x}_i$'s is normal.
3. If the original population is not normal and if $n < 30$, the $\bar{x}_i$'s follow a normal distribution only approximately.

If the sample size is large, we can use the central limit theorem directly to make an estimate of the confidence interval. Since $\bar{x}$ is normally distributed, we can use the statistic as defined by

$$z = \frac{\bar{x} - \mu}{\sigma_{\bar{x}}} \tag{6.43}$$

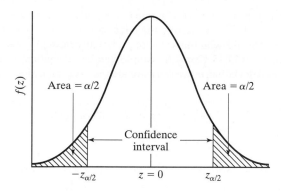

FIGURE 6.6

Concept of confidence interval of the mean.

Using the standard normal distribution function and the standard normal distribution tables, we can estimate the confidence interval on z. σ is the population standard deviation, which, in general, is not known. However, for large samples, the sample standard deviation, S, can be used as an approximation for σ.

The estimation of the confidence interval is shown graphically in Figure 6.6. If z has a value of zero, this means that the estimate $\bar{x}$ has a value that is exactly the population mean, μ. However, we only expect the true value of μ to lie somewhere in the confidence interval, which is $\pm z_{\alpha/2}$. The probability that z lies in the confidence interval, between $-z_{\alpha/2}$ and $z_{\alpha/2}$, is the area under the curve between these two z values and has the value $1 - \alpha$. The confidence level is this probability, $1 - \alpha$. The term α is equal to the sum of the areas of the left- and right-hand tails in Figure 6.6. These concepts can then be restated as

$$P[-z_{\alpha/2} \leq z \leq z_{\alpha/2}] = 1 - \alpha \tag{6.44}$$

Substituting for z, we obtain

$$P\left[-z_{\alpha/2} \leq \frac{\bar{x} - \mu}{\sigma/\sqrt{n}} \leq z_{\alpha/2}\right] = 1 - \alpha \tag{6.45a}$$

This can be rearranged to give

$$P\left[\bar{x} - z_{\alpha/2}\frac{\sigma}{\sqrt{n}} \leq \mu \leq \bar{x} + z_{\alpha/2}\frac{\sigma}{\sqrt{n}}\right] = 1 - \alpha \tag{6.45b}$$

It can also be stated as

$$\mu = \bar{x} \pm z_{\alpha/2}\frac{\sigma}{\sqrt{n}} \quad \text{with confidence level } 1 - \alpha \tag{6.46}$$

The next example demonstrates this method.

Example 6.11

We would like to determine the confidence interval of the mean of a batch of resistors made in a certain process. Based on 36 measurements, the average resistance is 25 Ω and the sample standard deviation is 0.5 Ω. Determine the 90% confidence interval of the mean resistance of the batch.

Solution: The desired confidence level is 90%, so $1 - \alpha = 0.90$ and $\alpha = 0.1$. Because the number of samples is greater than 30, we can use the normal distribution and Eq. (6.45b) to determine the confidence interval. Based on the nomenclature of Figure 6.6, we can determine the value of $z_{\alpha/2}$. Since the area between $z = 0$ and $z = \infty$ is 0.5, the area between $z = 0$ and $z_{\alpha/2}$ is $0.5 - \alpha/2 = 0.45$. We can now enter Table 6.3 with this value of probability (area) to find the corresponding value of $z_{\alpha/2}$. The value of $z_{\alpha/2}$ is 1.645. Using the sample standard deviation, S, as an approximation for the population standard deviation, σ, we can estimate the uncertainty interval on μ:

$$\bar{x} - \frac{z_{\alpha/2}S}{\sqrt{n}} \leq \mu \leq \bar{x} + \frac{z_{\alpha/2}S}{\sqrt{n}} = 25 - 1.645 \times \frac{0.5}{\sqrt{36}} \leq \mu \leq 25 + 1.645 \times \frac{0.5}{\sqrt{36}}$$

$$= (24.86 \leq \mu \leq 25.14) \ \Omega$$

This result can be stated as "The average resistance is expected to be $25 \pm 0.14 \ \Omega$ with a confidence level of 90%."

If the sample size is small ($n < 30$), the assumption that the population standard deviation can be represented by the sample standard deviation may not be accurate. Due to the uncertainty in the standard deviation, for the same confidence level, we would expect the confidence interval to be wider. In case of small samples, a statistic called *Student's t* is used:

$$t = \frac{\bar{x} - \mu}{S/\sqrt{n}} \tag{6.47}$$

In this formula, $\bar{x}$ is the sample mean, S is the sample standard deviation, and n is the sample size. In contrast to the normal distribution that is independent of sample size, there is a family of t-distribution functions that depend on the number of samples. The functional form of the t-distribution function is given by (see Lipson and Sheth, 1973)

$$f(t, v) = \frac{\Gamma\left(\dfrac{v + 1}{2}\right)}{\sqrt{v\pi}\,\Gamma\left(\dfrac{v}{2}\right)\left(1 + \dfrac{t^2}{v}\right)^{(v+1)/2}} \tag{6.48}$$

$\Gamma(x)$ is a standard mathematical function known as a gamma function and can be obtained from tables of mathematical functions, such as those of the U.S. Department of Commerce (1964). v is a parameter known as the *degrees of freedom*. It is given by the number of independent measurements minus the minimum number of measurements that are theoretically necessary to estimate a statistical parameter. For example, to make an estimate of the diameter of a shaft in a production batch, the minimum number of measurements is 1. If we make 10 measurements to make a more accurate estimate of the diameter, $v = n - 1 = 9$. It is unlikely that an engineer would actually use Eq. (6.48) to perform an analysis that uses the t distribution. The function is built into spreadsheet and statistical analysis programs. For purposes of the present course, all required values can be obtained from a single table.

Figure 6.7 [a plot of Eq. (6.48)] shows the Student's t-distribution for different values of degrees of freedom v. Like the normal distribution, these are symmetric curves. As the number of samples increases, the t-distribution approaches the normal distribution. For lesser values of v, the distribution is broader with a lower peak.

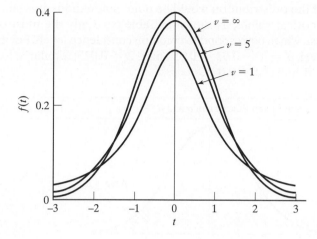

FIGURE 6.7

Probability density function using the Student's *t*-distribution.

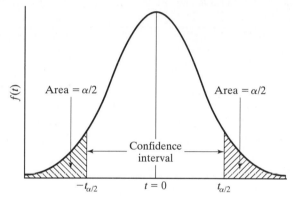

FIGURE 6.8

Confidence interval for the *t*-distribution.

The *t*-distribution can be used to estimate the confidence interval of a mean value of a sample with a certain confidence level for small sample sizes (less than 30). The *t*-distribution is used in much the same manner as the normal distribution, except that the curve for the appropriate value of v is selected as shown in Figure 6.8. The probability that t falls between $-t_{\alpha/2}$ and $t_{\alpha/2}$ is then $1 - \alpha$. This can be stated as

$$P[-t_{\alpha/2} \leq t \leq t_{\alpha/2}] = 1 - \alpha \tag{6.49}$$

Substituting for t, we obtain

$$P\left[-t_{\alpha/2} \leq \frac{\bar{x} - \mu}{S/\sqrt{n}} \leq t_{\alpha/2}\right] = 1 - \alpha \tag{6.50a}$$

This can be rearranged to give

$$P\left[\bar{x} - t_{\alpha/2}\frac{S}{\sqrt{n}} \leq \mu \leq \bar{x} + t_{\alpha/2}\frac{S}{\sqrt{n}}\right] = 1 - \alpha \tag{6.50b}$$

This equation can also be stated as

$$\mu = \bar{x} \pm t_{\alpha/2}\frac{S}{\sqrt{n}} \text{ with confidence level } 1 - \alpha \tag{6.51}$$

Since complete tables of the *t*-distribution would be quite voluminous, it is common to present only a table of critical values, as shown in Table 6.6. Only the most common values of *t* are used—those which correspond to common confidence levels. For example, for a 95% confidence level, $\alpha = 1 - 0.95 = 0.05$ and $\alpha/2 = 0.025$. Similarly, for a 99% confidence level, $\alpha = 0.01$ and $\alpha/2 = 0.005$.

TABLE 6.6   Student's *t* as a Function of α and v

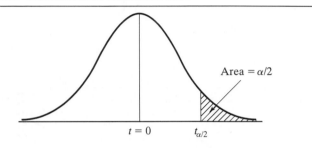

v	0.100	0.050	0.025	0.010	0.005
1	3.078	6.314	12.706	31.823	63.658
2	1.886	2.920	4.303	6.964	9.925
3	1.638	2.353	3.182	4.541	5.841
4	1.533	2.132	2.776	3.747	4.604
5	1.476	2.015	2.571	3.365	4.032
6	1.440	1.943	2.447	3.143	3.707
7	1.415	1.895	2.365	2.998	3.499
8	1.397	1.860	2.306	2.896	3.355
9	1.383	1.833	2.262	2.821	3.250
10	1.372	1.812	2.228	2.764	3.169
11	1.363	1.796	2.201	2.718	3.106
12	1.356	1.782	2.179	2.681	3.054
13	1.350	1.771	2.160	2.650	3.012
14	1.345	1.761	2.145	2.624	2.977
15	1.341	1.753	2.131	2.602	2.947
16	1.337	1.746	2.120	2.583	2.921
17	1.333	1.740	2.110	2.567	2.898
18	1.330	1.734	2.101	2.552	2.878
19	1.328	1.729	2.093	2.539	2.861
20	1.325	1.725	2.086	2.528	2.845
21	1.323	1.721	2.080	2.518	2.831
22	1.321	1.717	2.074	2.508	2.819
23	1.319	1.714	2.069	2.500	2.807
24	1.318	1.711	2.064	2.492	2.797
25	1.316	1.708	2.060	2.485	2.787
26	1.315	1.706	2.056	2.479	2.779
27	1.314	1.703	2.052	2.473	2.771
28	1.313	1.701	2.048	2.467	2.763
29	1.311	1.699	2.045	2.462	2.756
30	1.310	1.697	2.042	2.457	2.750
∞	1.283	1.645	1.960	2.326	2.576

Example 6.12

A manufacturer of VCR systems would like to estimate the mean failure time of a VCR brand with 95% confidence. Six systems are tested to failure, and the following data (in hours of playing time) are obtained: 1250, 1320, 1542, 1464, 1275, and 1383. Estimate the population mean and the 95% confidence interval on the mean.

Solution: Because the sample size is small ($n < 30$), we should use the t-distribution to estimate the confidence interval. But first we have to calculate the mean and standard deviation of the data.

$$\bar{x} = \frac{1250 + 1320 + 1542 + 1464 + 1275 + 1383}{6} = 1372 \text{ h}$$

$$S = ((1250 - 1372)^2 + (1320 - 1372)^2 + (1542 - 1372)^2$$

$$+ \frac{(1464 - 1372)^2 + (1275 - 1372)^2 + (1383 - 1372)^2}{5} \Bigg]^{1/2} = 114 \text{ h}$$

A 95% confidence level corresponds to $\alpha = 0.05$. From Table 6.6 of the t-distribution, for $v = n - 1 = 5$ and $\alpha/2 = 0.025$, $t_{\alpha/2} = 2.571$. From Eq. (6.51) the mean failure time will be

$$\mu = \bar{x} \pm t_{\alpha/2} \frac{S}{\sqrt{n}} = 1372 \pm 2.571 \times \frac{114}{\sqrt{6}} = 1372 \pm 120 \text{ h}$$

It should be noticed that if we were to increase the confidence level, the estimated interval will also expand, and vice versa.

Example 6.13

In Example 6.12, to reduce the 95% confidence interval to ±50 h from ±120 h, the VCR manufacturer decides to test more systems to failure. Determine how many more systems should be tested.

Solution: Since we do not know the number of samples in advance, we cannot select the appropriate t-distribution curve. Hence, the solution process is one of converging trial and error. To obtain a first estimate of the number of measurements n, we assume that $n > 30$, so that we can use the normal distribution. Then Eq. (6.46) can be applied, and the confidence interval becomes

$$\bar{x} \pm z_{\alpha/2} \frac{\sigma}{\sqrt{n}} = \bar{x} \pm 50$$

so

$$z_{\alpha/2} \frac{\sigma}{\sqrt{n}} = 50 \quad \text{and} \quad n = \left(z_{\alpha/2} \frac{\sigma}{50} \right)^2.$$

For a 95% confidence level, $\alpha/2 = 0.025$. Using the standard normal distribution, we find from Table 6.3 that $z_{0.025} = 1.96$. Using $S = 114$ (from Example 6.12) as an estimate for σ, we obtain a first estimate of n:

$$n = \left(1.96 \times \frac{114}{50} \right)^2 = 20$$

Because $n < 30$, we have to use the t-distribution instead of the standard normal distribution. We can use $n = 20$ for the next trial. For $v = n - 1 = 19$ and $\alpha/2 = 0.025$, from Table 6.6 we obtain $t = 2.093$. This value of t can be used with Eq. (6.51) to estimate a new value for n:

$$\mu = \bar{x} \pm t_{\alpha/2} \frac{S}{\sqrt{n}} = \bar{x} \pm 50$$

$$t_{\alpha/2} \frac{S}{\sqrt{n}} = 50$$

$$n = \left(t_{\alpha/2} \frac{S}{50}\right)^2 = \left(2.093 \times \frac{114}{50}\right)^2 = 23$$

We can use this number as a trial value to get a new value of t and recalculate n, but the result will be the same. Note that with the additional tests, the value of the sample mean, $\bar{x}$, may also change and may no longer have the value 1372 h.

6.4.2   Interval Estimation of the Population Variance

In many situations, the variability of the random variable is as important as its mean value. The best estimate of the population variance, σ^2, is the sample variance, S^2. As with the population mean, it is also necessary to establish a confidence interval for the estimated variance. For normally distributed populations, a statistic called χ^2 (pronounced "kye squared") is used for the purpose of establishing a confidence interval. Consider a random variable x with population mean value μ and standard deviation σ. If we assume that $\bar{x}$ is equal to μ, Eq. (6.6) can be written

$$S^2 = \frac{\sum_{i=1}^{n}(x_i - \mu)^2}{n - 1} \tag{6.52}$$

The random variable χ^2 is defined as

$$\chi^2 = \frac{\sum_{i=1}^{n}(x_i - \mu)^2}{\sigma^2} \tag{6.53}$$

Combining Eqs. (6.52) and (6.53), we obtain

$$\chi^2 = (n - 1)\frac{S^2}{\sigma^2} \tag{6.54}$$

That is, the variable χ^2 relates the sample variance to the population variance. χ^2 is a random variable, and it has been shown that in the case of a normally distributed population its probability density function is given by

$$f(\chi^2) = \frac{(\chi^2)^{(v-2)/2}e^{-\chi^2/2}}{2^{v/2}\Gamma(v/2)} \quad \text{for } \chi^2 > 0 \tag{6.55}$$

where v is the number of degrees of freedom and Γ is a gamma function, which can be obtained from standard tables such as those of the U.S. Department of Commerce (1964).

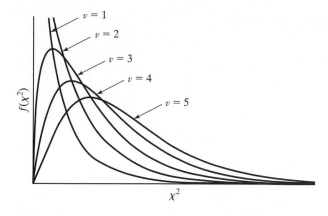

FIGURE 6.9

Chi-squared distribution function.

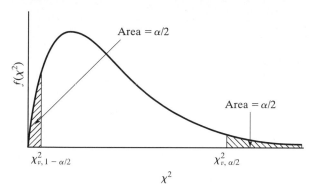

FIGURE 6.10

Confidence interval for the chi-squared distribution.

The number of degrees of freedom is the same as for the Student's t-distribution: the number of samples minus 1. $f(\chi^2)$ is plotted in Figure 6.9 for several values of the degrees of freedom.

As with other probability density functions, the probability that the variable χ^2 falls between any two values is equal to the area under the curve between those values (as shown in Figure 6.10). In equation form, this is

$$P(\chi^2_{v,1-\alpha/2} \leq \chi^2 \leq \chi^2_{v,\alpha/2}) = 1 - \alpha \qquad (6.56)$$

where α is the level of significance defined earlier and is equal to $(1 -$ confidence level). Substituting for χ^2 from Eq. (6.54), we obtain

$$P\left[\chi^2_{v,1-\alpha/2} \leq (n-1)\frac{S^2}{\sigma^2} \leq \chi^2_{v,\alpha/2}\right] = 1 - \alpha \qquad (6.57)$$

Since χ^2 is always positive, this equation can be rearranged to give a confidence interval on the population variance:

$$\frac{(n-1)S^2}{\chi^2_{v,\alpha/2}} \leq \sigma^2 \leq \frac{(n-1)S^2}{\chi^2_{v,1-\alpha/2}} \qquad (6.58)$$

The critical values of χ^2 used in this equation have been tabulated in Table 6.7. In Eq. (6.58), α is the total area of the two tails shown in Figure 6.10, so each tail has an

TABLE 6.7   Critical Values of the Chi-Squared Distribution

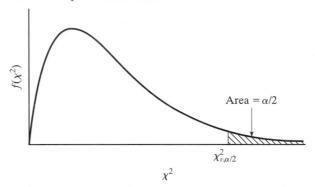

					Area of right hand tail		$1 - \alpha/2$ and $\alpha/2$			
$v\downarrow$	0.995	0.990	0.975	0.950	0.900	0.100	0.050	0.025	0.010	0.005
1	0.000039	0.000157	0.000982	0.003932	0.015791	2.706	3.841	5.024	6.635	7.879
2	0.0100	0.0201	0.0506	0.1026	0.2107	4.605	5.991	7.378	9.210	10.597
3	0.0717	0.1148	0.2158	0.3518	0.5844	6.251	7.815	9.348	11.345	12.838
4	0.2070	0.2971	0.4844	0.7107	1.0636	7.779	9.488	11.143	13.277	14.86
5	0.4118	0.5543	0.8312	1.1455	1.6103	9.236	11.070	12.832	15.086	16.750
6	0.6757	0.8721	1.2373	1.6354	2.2041	10.645	12.592	14.449	16.812	18.548
7	0.9893	1.2390	1.6899	2.1673	2.8331	12.017	14.067	16.013	18.475	20.278
8	1.3444	1.6465	2.1797	2.7326	3.4895	13.362	15.507	17.535	20.090	21.955
9	1.7349	2.0879	2.7004	3.3251	4.1682	14.684	16.919	19.023	21.666	23.589
10	2.1558	2.5582	3.2470	3.9403	4.8652	15.987	18.307	20.483	23.209	25.188
11	2.6032	3.0535	3.8157	4.5748	5.5778	17.275	19.675	21.920	24.725	26.757
12	3.0738	3.5706	4.4038	5.2260	6.3038	18.549	21.026	23.337	26.217	28.300
13	3.5650	4.1069	5.0087	5.8919	7.0415	19.812	22.362	24.736	27.688	29.819
14	4.0747	4.6604	5.6287	6.5706	7.7895	21.064	23.685	26.119	29.141	31.319
15	4.6009	5.2294	6.2621	7.2609	8.5468	22.307	24.996	27.488	30.578	32.801
16	5.1422	5.8122	6.9077	7.9616	9.3122	23.542	26.296	28.845	32.000	34.267
17	5.6973	6.4077	7.5642	8.6718	10.0852	24.769	27.587	30.191	33.409	35.718
18	6.2648	7.0149	8.2307	9.3904	10.8649	25.989	28.869	31.526	34.805	37.156
19	6.8439	7.6327	8.9065	10.1170	11.6509	27.204	30.144	32.852	36.191	38.582
20	7.4338	8.2604	9.5908	10.8508	12.4426	28.412	31.410	34.170	37.566	39.997
21	8.0336	8.8972	10.2829	11.5913	13.2396	29.615	32.671	35.479	38.932	41.401
22	8.6427	9.5425	10.9823	12.3380	14.0415	30.813	33.924	36.781	40.289	42.796
23	9.2604	10.1957	11.6885	13.0905	14.8480	32.007	35.172	38.076	41.638	44.181
24	9.8862	10.8563	12.4011	13.8484	15.6587	33.196	36.415	39.364	42.980	45.558
25	10.5196	11.5240	13.1197	14.6114	16.4734	34.382	37.652	40.646	44.314	46.928
26	11.1602	12.1982	13.8439	15.3792	17.2919	35.563	38.885	41.923	45.642	48.290
27	11.8077	12.8785	14.5734	16.1514	18.1139	36.741	40.113	43.195	46.963	49.645
28	12.4613	13.5647	15.3079	16.9279	18.9392	37.916	41.337	44.461	48.278	50.994
29	13.1211	14.2564	16.0471	17.7084	19.7677	39.087	42.557	45.722	49.588	52.335
30	13.7867	14.9535	16.7908	18.4927	20.5992	40.256	43.773	46.979	50.892	53.672
40	20.7066	22.1642	24.4331	26.5093	29.0505	51.805	55.758	59.342	63.691	66.766
50	27.9908	29.7067	32.3574	34.7642	37.6886	63.167	67.505	71.420	76.154	79.490
60	35.5344	37.4848	40.4817	43.1880	46.4589	74.397	79.082	83.298	88.379	91.952
70	43.2753	45.4417	48.7575	51.7393	55.3289	85.527	90.531	95.023	100.425	104.215
80	51.1719	53.5400	57.1532	60.3915	64.2778	96.578	101.879	106.629	112.329	116.321
90	59.1963	61.7540	65.6466	69.1260	73.2911	107.565	113.145	118.136	124.116	128.299
100	67.3275	70.0650	74.2219	77.9294	82.3581	118.498	124.342	129.561	135.807	140.170

area $\alpha/2$. In Table 6.7, $\alpha/2$ is the area of the right-hand tail only. For example, consider finding the interval boundary values of χ^2 for a 90% confidence level and $v = 10$. In this case, $\alpha = 1 - 0.9 = 0.1$, so the area of each tail is $\alpha/2 = 0.05$. We look up the upper value of χ^2 in Table 6.7 under the column labeled 0.05 and obtain $\chi^2_{10, 0.05} = 18.307$. For the lower value of χ^2, the area of the right-hand tail is $1 - \alpha/2 = 0.95$. Looking under the column labeled 0.95, we obtain $\chi^2_{10, 0.95} = 3.9403$. Although the χ^2 distribution and test have been developed for data with normal distributions, they are often used satisfactorily for populations with other distributions.

Example 6.14

To estimate the uniformity of the diameter of ball bearings in a production batch, a sample of 20 is chosen and carefully measured. The sample mean is 0.32500 in, and the sample standard deviation is 0.00010 in. Obtain a 95% confidence interval for the standard deviation of the production batch.

Solution: We will assume that the population is normal, and we have

$$v = n - 1 = 19 \quad \alpha = 1 - 0.95 = 0.05 \quad \alpha/2 = 0.025$$

Using Table 6.7 for $v = 19$ and $\alpha/2 = 0.025$ and $1 - \alpha/2 = 0.975$ yields

$$\chi^2_{19, 0.025} = 32.825 \quad \text{and} \quad \chi^2_{19, 0.975} = 8.907$$

Using Equation 6.58, we can determine the interval for the standard deviation:

$$(20 - 1)0.0001^2/32.825 \le \sigma^2 \le (20 - 1)0.0001^2/8.907 \text{ or } 0.000076 \le \sigma \le 0.00015 \text{ in.}$$

6.5 CRITERION FOR REJECTING QUESTIONABLE DATA POINTS

In some experiments, it happens that one or more measured values appear to be out of line with the rest of the data. If some clear faults can be detected in measuring those specific values, they should be discarded. But often the seemingly faulty data cannot be traced to any specific problem. There exist a number of statistical methods for rejecting these *wild* or *outlier* data points. The basis of these methods is to eliminate values that have a low probability of occurrence. For example, data values that deviate from the mean by more than two or more than three standard deviations might be rejected. It has been found that so-called two-sigma or three-sigma rejection criteria normally must be modified to account for the sample size. Furthermore, depending on how strong the rejection criterion is, good data might be eliminated or bad data included.

The method recommended in the document *Measurement Uncertainty* (ASME, 1998) is the *modified Thompson τ technique*. In this method, if we have n measurements that have a mean $\bar{x}$ and standard deviation S, the data can be arranged in ascending order $x_1, x_2, \ldots, x_n$. The extreme values (the highest and lowest) are suspected outliers. For these suspected points, the deviation is calculated as

$$\delta_i = |x_i - \bar{x}| \tag{6.59}$$

TABLE 6.8   Values of Thompson's τ

Sample size		Sample size	
n	τ	n	τ
3	1.150	22	1.893
4	1.393	23	1.896
5	1.572	24	1.899
6	1.656	25	1.902
7	1.711	26	1.904
8	1.749	27	1.906
9	1.777	28	1.908
10	1.798	29	1.910
11	1.815	30	1.911
12	1.829	31	1.913
13	1.840	32	1.914
14	1.849	33	1.916
15	1.858	34	1.917
16	1.865	35	1.919
17	1.871	36	1.920
18	1.876	37	1.921
19	1.881	38	1.922
20	1.885	39	1.923
21	1.889	40	1.924

Source: ASME (1998).

and the largest value is selected. The next step is to find a value of τ from a table (given here as Table 6.8). The largest value of δ_i should be compared with the product of τ and the standard deviation, S. If the value of δ exceeds τS, the data value can be rejected as an outlier. According to this method, only one data value should be eliminated. The mean and standard deviation of the remaining data values should then be recomputed and the process repeated. The process can be repeated until no more of the data can be eliminated.

It should be noted that eliminating an outlier is not an entirely positive event. The outlier probably resulted from a problem with the measurement system. One common cause of outliers, human error in recording the data, is less common now that most data are recorded automatically. Other methods for rejecting outlier data are presented by Lipson and Sheth (1973). A method to identify outliers in $y = f(x)$ data sets is presented later in this chapter.

Example 6.15

Nine voltage measurements in a circuit have produced the following values: 12.02, 12.05, 11.96, 11.99, 12.10, 12.03, 12.00, 11.95, 12.16 V. Determine whether any of the values can be rejected.

Solution: For the nine values, $\overline{V} = 12.03$ V and $S = 0.07$ V. We will use the Thompson τ test to determine possible outliers. The largest and the smallest values are suspected to be outliers. We have

$$\delta_1 = |V_{largest} - \overline{V}| = |12.16 - 12.03| = 0.13$$
$$\delta_2 = |V_{smallest} - \overline{V}| = |11.95 - 12.03| = 0.08$$

Using Table 6.8 for $n = 9$, we find that $\tau = 1.777$. Then $S\tau = 0.07 \times 1.77 = 0.12$. Since $\delta_1 = 0.13$, it exceeds $S\tau$ and should be rejected. We can now recalculate S and $\overline{V}$ to obtain 0.05 and 12.01, respectively. For $n = 8$, $\tau = 1.749$ and $S\tau = 0.09$, so none of the remaining data points should be rejected.

6.6   CORRELATION OF EXPERIMENTAL DATA

6.6.1   Correlation Coefficient

Scatter due to random errors is a common characteristic of virtually all measurements. However, in some cases the scatter may be so large that it is difficult to detect a trend. Consider an experiment in which an independent variable x is varied systematically and the dependent variable y is then measured. We would like to determine whether the value of y depends on the value of x. If the results appeared as in Figure 6.11(a), we would immediately conclude that there is a strong relationship between y and x. On the other hand, if the data appeared as shown in Figure 6.11(b), we would probably

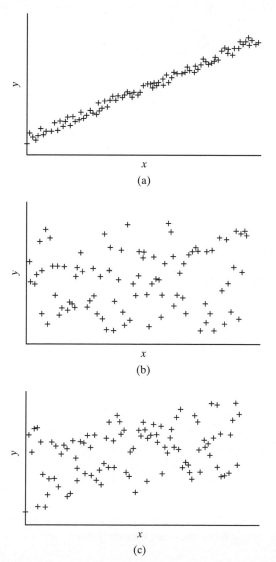

(a)

(b)

(c)

FIGURE 6.11

Data showing significant scatter.

conclude that there is no functional relationship between y and x. If the data appeared as shown in Figure 6.11(c), however, we might be uncertain. There appears to be a trend of increasing y with increasing x, but the scatter is so great that the apparent trend might be a consequence of pure chance. Fortunately, there exists a statistical parameter called the *correlation coefficient*, which can be used to determine whether an apparent trend is real or could simply be a consequence of pure chance.

The correlation coefficient, r_{xy}, is a number whose magnitude can be used to determine whether there in fact exists a functional relationship between two measured variables x and y. For example, one would not expect even a weak correlation between exam scores and the height of the student. On the other hand, we might expect a fairly strong correlation between the total electric power consumption in a region and the time of the day. If we have two variables, x and y, and our experiment yields a set of n data pairs $[(x_i, y_i), i = 1, n]$, we can define the linear correlation coefficient

$$r_{xy} = \frac{\sum_{i=1}^{n} (x_i - \bar{x})(y_i - \bar{y})}{\left[\sum_{i=1}^{n} (x_i - \bar{x})^2 \sum_{i=1}^{n} (y_i - \bar{y})^2 \right]^{1/2}} \tag{6.60}$$

where

$$\bar{x} = \frac{\sum_{i=1}^{n} x_i}{n} \qquad \bar{y} = \frac{\sum_{i=1}^{n} y_i}{n} \tag{6.61}$$

are the mean values of the x values and the y values obtained experimentally. The resulting value of r_{xy} will lie in the range from -1 to $+1$. A value of $+1$ would indicate a perfectly linear relationship between the variables with a positive slope (i.e., increasing x results in increasing y). A value of -1 indicates a perfectly linear relationship with negative slope (i.e., increasing x decreases y). A value of zero indicates that there is no linear correlation between the variables. Even if there is no correlation, it is unlikely that r_{xy} will be exactly zero. For any finite number of data pairs, pure chance means that a nonzero correlation coefficient is likely. For a given sample size, statistical theory can be used to determine whether a computed r_{xy} is significant or could be a consequence of pure chance. Harnett and Murphy (1975) discuss the process of determining whether a given correlation coefficient is significant, and Johnson (1988) also discusses the issue.

For practical problems, this process can be simplified to the form of a single table. Critical values for r have been established that can be compared with the computed r_{xy}. For two variables and n data pairs, the appropriate critical values of r, r_t, have been computed and are presented in Table 6.9.[†] r_t is a function of the number of samples and the level of significance, α. The values of r in this table are limiting values

---

[†]This approach assumes that r_{xy} could be either positive or negative. If we know that r_{xy} can be only positive or only negative, then the approach must be modified. See Johnson (1988).

TABLE 6.9   Minimum Values of the Correlation Coefficient for
Significance Level α

n	α				
	0.20	0.10	0.05	0.02	0.01
3	0.951	0.988	0.997	1.000	1.000
4	0.800	0.900	0.950	0.980	0.990
5	0.687	0.805	0.878	0.934	0.959
6	0.608	0.729	0.811	0.882	0.917
7	0.551	0.669	0.754	0.833	0.875
8	0.507	0.621	0.707	0.789	0.834
9	0.472	0.582	0.666	0.750	0.798
10	0.443	0.549	0.632	0.715	0.765
11	0.419	0.521	0.602	0.685	0.735
12	0.398	0.497	0.576	0.658	0.708
13	0.380	0.476	0.553	0.634	0.684
14	0.365	0.458	0.532	0.612	0.661
15	0.351	0.441	0.514	0.592	0.641
16	0.338	0.426	0.497	0.574	0.623
17	0.327	0.412	0.482	0.558	0.606
18	0.317	0.400	0.468	0.543	0.590
19	0.308	0.389	0.456	0.529	0.575
20	0.299	0.378	0.444	0.516	0.561
25	0.265	0.337	0.396	0.462	0.505
30	0.241	0.306	0.361	0.423	0.463
35	0.222	0.283	0.334	0.392	0.430
40	0.207	0.264	0.312	0.367	0.403
45	0.195	0.248	0.294	0.346	0.380
50	0.184	0.235	0.279	0.328	0.361
100	0.129	0.166	0.197	0.233	0.257
200	0.091	0.116	0.138	0.163	0.180

that might be expected by pure chance. For each r_t value in the table, there is only a probability α that an experimental value of r_{xy} will be larger by pure chance. Conversely, if the experimental value exceeds the table value, we can expect that the experimental value shows a real correlation with confidence level $1 - \alpha$. For common engineering purposes, the confidence level is often taken as 95%, which corresponds to a value of α of 0.05. For a given set of data, we obtain r_t from the table and compare it with the value of r_{xy} computed from the data. If $|r_{xy}| > r_t$, we can presume that y does depend on x in a nonrandom manner, and we can expect that a linear relationship will offer some approximation of the true functional relationship. A value of $|r_{xy}|$ less than r_t implies that we cannot be confident that a linear functional relationship exists.

It is not necessary for the functional relationship actually to be linear for a significant correlation coefficient to be calculated. For example, a parabolic functional relationship that shows little data scatter will often show a large correlation coefficient. On the other hand, some functional relationships (e.g., a multivalued circular function), while very strong will result in a very low value of r_{xy}.

Two additional precautions to be aware of in using the correlation coefficient should be mentioned. First, a single bad data point can have a strong effect on the value of r_{xy}. If possible, outliers should be identified before evaluating the coefficient. It is also a mistake to conclude that a significant value of the correlation coefficient implies that a change in one variable causes the other variable to change. Causality should be determined from other knowledge about the problem.

Example 6.16

It is thought that lap times for a race car depend on the ambient temperature. The following data for the same car with the same driver were measured at different races:

Ambient temperature (°F)	40	47	55	62	66	88
Lap time (s)	65.3	66.5	67.3	67.8	67	66.6

Does a linear relationship exist between these two variables?

Solution: First, we plot the data as shown in Figure E6.16. From the plot, it looks as though there might exist a weak positive correlation between lap time and ambient temperature, but we can compute the correlation coefficient to determine whether this correlation is real or might be due to pure chance. We can determine this coefficient using Eq. (6.60). The following computation table is prepared:

x	y	$x - \bar{x}$	$(x - \bar{x})^2$	$y - \bar{y}$	$(y - \bar{y})^2$	$(x - \bar{x})(y - \bar{y})$
40	65.3	−19.67	386.78	−1.45	2.10	28.52
47	66.5	−12.67	160.44	−0.25	0.06	3.17
55	67.3	−4.67	21.78	0.55	0.30	−2.57
62	67.8	2.33	5.44	1.05	1.10	2.45
66	67	6.33	40.11	0.25	0.06	1.58
88	66.6	28.33	802.78	−0.15	0.02	−4.25
$\sum = 358.00$	$\sum = 400.50$		$\sum = 1417.33$		$\sum = 3.66$	$\sum = 28.90$

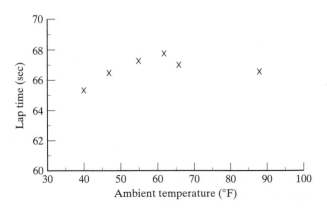

FIGURE E6.16

We can now compute the correlation coefficient, using Eq. (6.60):

$$r_{xy} = \frac{28.9}{(1417.33 \times 3.66)^{1/2}} = 0.4013$$

For a 95% confidence level, $\alpha = 1 - 0.95 = 0.05$. For six pairs of data, from Table 6.9, we obtain a value of r_t of 0.811. Since r_{xy} is less than r_t, we conclude that the apparent trend in the data is probably caused by pure chance. The calculation of r_{xy} is a feature of some spreadsheet programs. The user needs only to input two columns of numbers and then call the appropriate function.

6.6.2 Least-Squares Linear Fit

It is a common requirement in experimentation to correlate experimental data by fitting mathematical functions such as straight lines or exponentials through the data. One of the most common functions used for this purpose is the straight line. Linear fits are often appropriate for the data, and in other cases the data can be transformed to be approximately linear. As shown in Figure 6.12, if we have n pairs of data (x_i, y_i), we seek to fit a straight line of the form

$$Y = ax + b \tag{6.62}$$

through the data. We would like to obtain values of the constants a and b. If we have only two pairs of data, the solution is simple, since the points completely determine the straight line. However, if there are more points, we want to determine a "best fit" to the data. The experimenter can use a ruler and "eyeball" a straight line through the data, and in some cases this is the best approach.

A more systematic and appropriate approach is to use the method of *least squares* or *linear regression* to fit the data. Regression is a well-defined mathematical formulation that is readily automated. Let us assume that the test data consist of data pairs (x_i, y_i). For each value of x_i (which is assumed to be error free), we can predict a value of Y_i according to the linear relationship $Y = ax + b$. For each value of x_i, we then have an error

$$e_i = Y_i - y_i \tag{6.63}$$

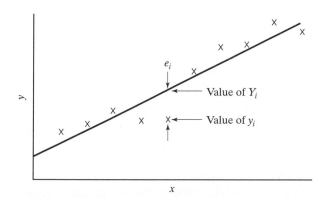

FIGURE 6.12

Fitting a straight line through data.

and the square of the error is

$$e_i^2 = (Y_i - y_i)^2 = (ax_i + b - y_i)^2 \tag{6.64}$$

The sum of the squared errors for all the data points is then

$$E = \sum (Y_i - y_i)^2 = \sum (ax_i + b - y_i)^2 \quad \text{where} \sum = \sum_{i=1}^{n} \tag{6.65}$$

We now choose a and b to minimize E by differentiating E with respect to a and b and setting the results to zero:

$$\frac{\partial E}{\partial a} = 0 = \sum 2(ax_i + b - y_i)x_i$$

$$\frac{\partial E}{\partial b} = 0 = \sum 2(ax_i + b - y_i) \tag{6.66}$$

These two equations can be solved simultaneously for a and b:

$$a = \frac{n \sum x_i y_i - \left(\sum x_i \right)\left(\sum y_i \right)}{n \sum x_i^2 - \left(\sum x_i \right)^2} \tag{6.67a}$$

$$b = \frac{\sum x_i^2 \sum y_i - \left(\sum x_i \right)\left(\sum x_i y_i \right)}{n \sum x_i^2 - \left(\sum x_i \right)^2} \tag{6.67b}$$

The resulting line, $Y = ax + b$, is called the *least-squares best fit* to the data represented by (x_i, y_i).

When a linear regression analysis has been performed, it is desirable to determine how good the fit actually is. Some idea of this measure can be obtained by examining a plot, such as Figure 6.12, of the data and the best-fit line. However, it is desirable to have a mathematical expression of how well the best-fit line represents the actual data. A good measure of the adequacy of the regression model is called the *coefficient of determination*, given by

$$r^2 = 1 - \frac{\sum (ax_i + b - y_i)^2}{\sum (y_i - \bar{y})^2} \tag{6.68}$$

In the second term on the right-hand side of Eq. (6.68), the numerator is the sum of the square deviations of the data from the best fit. The denominator of this term is the sum of the squares of the variation of the y data about the mean, $\bar{y}$, which is the average of the y_i's. For a good fit, r^2 should be close to unity. That is, the best-fit line accounts for most of the variation in the y data. For engineering data, r^2 will normally be quite high (0.8–0.9 or higher), and a low value might indicate that there exists some important variable that was not considered, but that is affecting the result.

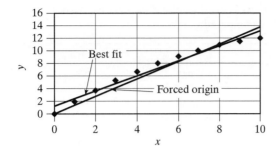

FIGURE 6.13

Least-squares line with forced origin.

Another measure of how well the best-fit line represents the data is called the *standard error of estimate*, given by

$$S_{y.x} = \sqrt{\frac{\sum y_i^2 - b\sum y_i - a\sum x_i y_i}{n - 2}} \tag{6.69}$$

This is, effectively, the standard deviation of the differences between the data points and the best-fit line, and it has the same units as y.

In some cases, a somewhat different form of linear regression is used in which the line is forced to go through the origin ($x = 0$, $y = 0$). This form is often employed to calibrate instruments in which the zero offset can be adjusted to zero prior to making measurements. This situation is shown in Figure 6.13. If a linear best fit of the form of Eq. (6.62) is used, the best-fit line will not go through the origin. If the line is forced to go through the origin, the fit is not quite as good, but the fitted curve will give the correct value when $x = 0$.

With the origin fixed at $(0, 0)$, the fitted line will have the form

$$Y = ax \tag{6.70}$$

The value of a is calculated from

$$a = \frac{\sum_{i=1}^{n} x_i y_i}{\sum_{i=1}^{n} x_i^2} \tag{6.71}$$

This form of the least-squares fit is a standard feature of common spreadsheet programs.

Example 6.17

The following table represents the output (volts) of a linear variable differential transformer (LVDT; an electric output device used for measuring displacement) for five length inputs:

L(cm)	0.00	0.50	1.00	1.50	2.00	2.50
V(V)	0.05	0.52	1.03	1.50	2.00	2.56

Determine the best linear fit to these data, draw the data on a (V, L) plot, and calculate the standard error of estimate as well as the coefficient of determination.

Solution: To solve this problem, we substitute the data into Eq. (6.67). The following table shows how the individual sums are computed:

x_i	x_i^2	y_i	$x_i y_i$	y_i^2
0	0	0.05	0	0.0025
0.5	0.25	0.52	0.26	0.2704
1	1	1.03	1.03	1.0609
1.5	2.25	1.5	2.25	2.25
2	4	2	4	4.0
2.5	6.25	2.56	6.4	6.5536
$\sum x_i = 7.5$	$\sum x_i^2 = 13.75$	$\sum y_i = 7.66$	$\sum x_i y_i = 13.94$	$\sum y_i^2 = 14.137$

Then,

$$a = \frac{6 \times 13.94 - 7.5 \times 7.66}{6 \times 13.75 - 7.5^2}$$
$$= 0.9977$$

and

$$b = \frac{13.75 \times 7.66 - 7.5 \times 13.94}{6 \times 13.75 - 7.5^2}$$
$$= 0.0295$$

The resulting best-fit line is then

$$Y = 0.9977x + 0.0295$$

where Y is the voltage and x is the displacement. The best-fit line, together with the data, is plotted in Figure E6.17.

The standard error for these data is obtained from Eq. (6.69):

$$S_{y.x} = \left(\frac{14.137 - 0.0295 \times 7.66 - 0.9977 \times 13.94}{6 - 2} \right)^{1/2} = 0.0278$$

This represents deviation of the y data about the values predicted by the best-fit line.

We can now compute the coefficient of determination with the use of Eq. (6.68). Noting that $\bar{y} = \frac{1}{n} \sum y_i = \frac{1}{6}(7.66) = 1.27666$, we compute the following table:

x_i	y_i	$(Y_i - y_i)^2$	$(y_i - \bar{y})^2$
0	0.05	0.000419	1.504711
0.5	0.52	7.02E-05	0.572544
1	1.03	7.63E-06	0.060844
1.5	1.5	0.000681	0.049878
2	2	0.000623	0.523211
2.5	2.56	0.00131	1.646944
		$\sum (Y_i - y_i)^2$	$\sum (y_i - \bar{y})^2$
		$= 0.00311$	$= 4.358133$

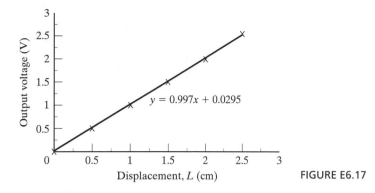

FIGURE E6.17

The figure shows a plot of Output voltage (V) versus Displacement, L (cm), with the line $y = 0.997x + 0.0295$.

r^2 is thus

$$r^2 = 1 - \frac{\sum(Y_i - y_i)^2}{\sum(y_i - \overline{y})^2} = 1 - \frac{.00311}{4.358133} = 0.999286$$

This is high, close to unity, and consistent with the plot of the results shown in Figure E6.17.

Comment: Linear regression is a standard feature of statistical programs and most spreadsheet programs. It is only necessary to input columns of the data, and the remaining calculations are performed immediately. In Excel®, regression can be found under Tools/Data Analysis. Values of b (Intercept Coefficient), a (x Variable 1 Coefficient), r^2 (R Square) and $S_{x,y}$ (Standard Error) will be output. A lot of other information is also supplied, which is used only in more advanced analyses.

There are some important considerations to bear in mind about the least-squares method:

1. Variation in the data is assumed to be normally distributed and due to random causes.
2. In deriving the relation $Y = ax + b$, we are assuming that random variation exists in y values, while x values are error free.
3. Since the error has been minimized in the y direction, an erroneous conclusion may be made if x is estimated based on a value for y. Linear regression of x in terms of $y(X = cy + d)$ cannot simply be derived from $Y = ax + b$.

6.6.3   Outliers in x–y Data Sets

In Section 6.5, the Thompson τ technique was presented as a method for eliminating bad data when several measurements are made of a single variable. When a variable y is measured as a function of an independent variable x, in most cases there is only one value of y measured for each value of x. As a result, at each x there are insufficient data to determine a meaningful standard deviation. One way to identify outliers is to plot the data

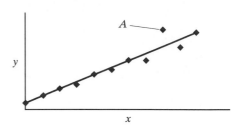

FIGURE 6.14

x–*y* data showing an outlier.

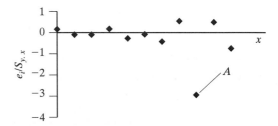

FIGURE 6.15

Plot of standardized residuals.

and the best-fit line as shown in Figure 6.14. A point such as *A*, which shows a much larger deviation from the line than the other data, might be identified as an outlier.

Montgomery, et al. (1998) suggest a more sophisticated method of identifying outliers in *x*–*y* data sets. The method involves computing the ratio of the residuals [e_i, Eq. (6.63)] to the standard error of estimate [$S_{y \cdot x}$, Eq. (6.69)] and making a graph. These ratios, called the standardized residuals, can be plotted as a function of the independent variable, *x*, the dependent variable, *y*, or the time or sequence in which the *x*–*y* pair was measured. Figure 6.15 shows a plot of the standardized residuals as a function of *x* for the same data shown in Figure 6.14.

If we assume that the residuals are normally distributed, then we would expect that 95% of the standardized residuals would be in the range ±2 (that is, within two standard deviations from the best-fit line). We might consider data values with standardized residuals exceeding 2 to be outliers, since they only have a 5% chance of occurring, due to simple randomness in the measurement. In the example, point *A* has a value of −3, and hence deviates from the best-fit line by three standard deviations. Since it is not consistent with its neighbors, it is almost certainly an outlier.

However, there are several complications, and the process of determining outliers will require considerable judgment on the part of the experimenter.

The first problem occurs when the data set is relatively small. In this case, the potential outlier itself has a major effect on the value of $S_{y \cdot x}$. As a result, it is unlikely that the standardized residual for this suspect data point will have a large value. The Thompson τ test in Section 6.4.3 accounts for this effect by reducing the value of τ for a small sample size. No similar guidance exists for the *x*–*y* outlier determination for small samples in common statistics practice.

A second problem occurs when the data themselves are not linear. In this case, correct data may result in high standardized residuals. Consider the non-linear data plotted in Figure 6.16. The standardized residuals are plotted in Figure 6.17.

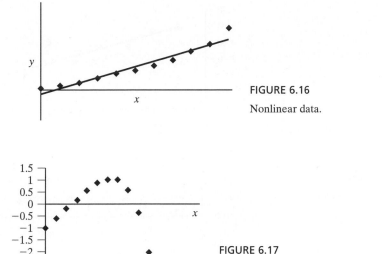

FIGURE 6.16
Nonlinear data.

FIGURE 6.17
Normalized residuals for nonlinear data.

Although the magnitude of the normalized residual at the highest value of x is greater than 2, this is not an outlier. The plot of the normalized residuals is smooth, and these data simply reflect nonlinear behavior. Potential outliers at the extreme ends of the data are likely real data. It is also possible that there is a change in the character of the physical phenomenon represented at the high or low values of the independent variable.

In summary, outliers in x–y data sets cannot be determined by simple mechanistic rules. The experimenter can make use of plots of the data and plots of the standardized residuals, but ultimately, it is a judgment call as to whether to reject any data point. It may in fact be necessary to consider the actual data-gathering process in order to make a decision.

Example 6.18

The following data were taken from a small water-turbine experiment:

rpm	torque (N-m)
100	4.89
201	4.77
298	3.79
402	3.76
500	2.84
601	4.00
699	2.05
799	1.61

Fit a least-squares straight line to these data, and determine whether any of the torque values appear to be outliers.

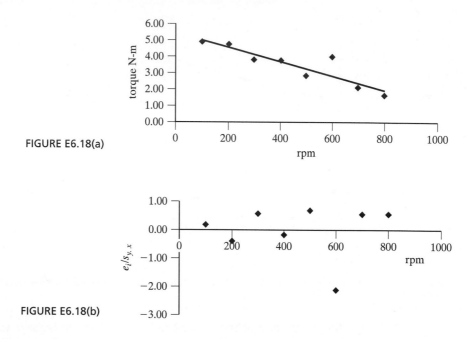

FIGURE E6.18(a)

FIGURE E6.18(b)

Solution: We will use a spreadsheet program to fit the straight line. This is shown as Figure E6.18(a).

Examining Figure E6.18(a), we find that the torque value at about 600 rpm appears to be inconsistent with the other data and is probably an outlier. To confirm this, we can prepare a plot of the standardized residuals. For these data, the best-fit equation is

$$\text{torque} = 5.433 - 0.0044 \text{ rpm}$$

and $S_{y.x} = 0.5758$. This information can be used to prepare the following table:

rpm	torque (measured)	torque (predicted)	$e/S_{x,y}$
100	4.89	4.99	0.18
201	4.77	4.55	−0.38
298	3.79	4.12	0.58
402	3.76	3.66	−0.16
500	2.84	3.23	0.68
601	4.00	2.79	−2.10
699	2.05	2.36	0.53
799	1.61	1.92	0.54

The standardized residuals are plotted against the rpm in Figure E6.18(b).

Clearly, the torque value at 600 rpm has a high probability of being an outlier. The standardized residual has a magnitude greater than 2. Furthermore, the standardized residual plot appears random and shows no trend, so the high standard residual is not likely caused by nonlinearity.

6.6.4 Linear Regression Using Data Transformation

Commonly, test data do not show an even approximately linear relationship between the dependent and independent variables, and linear regression is not directly useful. In some cases, however, the data can be transformed into a form that is linear, and a straight line can then be fitted by linear regression. Examples of data relationships that can easily be transformed to linear form are $y = ax^b$ and $y = ae^{bx}$. In these relations, a and b are constant values and x and y are variables. For example, we examine the equation $y = ae^{bx}$. Taking the natural logarithms of each side, we obtain $\ln(y) = bx + \ln(a)$. Since $\ln(a)$ is just a constant, we now have a form in which $\ln(y)$ is a linear function of x.

Example 6.19

In a compression process in a piston–cylinder device, air pressure and temperature are measured. Table E6.19 presents these data. Determine an explicit relation of the form $T = F(P)$ for the given data.

Solution: The data are plotted in Figure E6.19(a). As can be seen, the temperature–pressure relationship is curved. In a compression process of this type, it is known from thermodynamics that the temperature and the pressure are related by an equation of the form

$$\frac{T}{T_0} = \left(\frac{P}{P_0}\right)^{\frac{n-1}{n}}$$

where T is the *absolute* temperature, P is the absolute pressure, and n is a number, called the *polytropic exponent*, that depends on the actual compression process. For this system of units, $T_{abs} = T_F + 460.0$. If we take natural logarithms of both sides of the equation for T/T_0, we obtain

TABLE E6.19

Pressure (psia)	Temperature (°F)
20.0	44.9
40.4	102.4
60.8	142.3
80.2	164.8
100.4	192.2
120.3	221.4
141.1	228.4
161.4	249.5
181.9	269.4
201.4	270.8
220.8	291.5
241.8	287.3
261.1	313.3
280.4	322.3
300.1	325.8
320.6	337.0
341.1	332.6
360.8	342.9

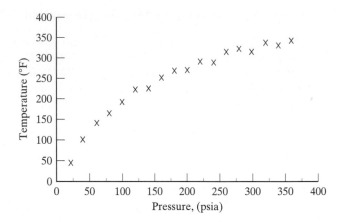

FIGURE E6.19(a)

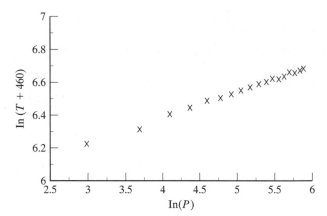

FIGURE E6.19(b)

an equation of the form

$$\ln(T) = a \ln(P) + b$$

where $a = (n - 1)/n$ and b is another constant. In this form, $\ln(T)$ is a linear function of $\ln(P)$. Natural logarithms of the pressure and absolute temperature have been plotted in Figure E6.19(b). The data now appear to follow an approximate straight line. Using the method of least squares, the equation of the best-fit straight line is

$$\ln(T + 460) = 0.1652 \ln(P) + 5.7222$$

This is the functional form required; T is in °F and P is in psia.

6.6.5 Multiple and Polynomial Regression

Regression analysis is much more general than the methods presented so far. Best-fit functions can be determined for situations with more than one independent variable (multiple regression) or for polynomials of the independent variable (polynomial regression). The calculations for these methods can be quite tedious, but they are standard features of statistical-analysis programs, and many features are available in common spreadsheet programs.

In multiple regression, we seek a function of the form

$$Y = a_0 + a_1\hat{x}_1 + a_2\hat{x}_2 + a_3\hat{x}_3 + \ldots + a_k\hat{x}_k \tag{6.72}$$

We can have several independent variables, $\hat{x}_1 \ldots \hat{x}_k$. At first, it would appear that the dependent variable will be a linear function of each of the independent variables, but in fact the method is more general, similar in concept to the transformation of variables demonstrated in Section 6.5.4. It is for this reason that $\hat{x}$ rather than x is used in Eq. (6.72). The $\hat{x}$'s can be independent variables, or they can be functions of the independent variables. As an example, consider a situation with two independent variables, x_1 and x_2. We could use three $\hat{x}$'s:

$$\hat{x}_1 = x_1$$
$$\hat{x}_2 = x_2$$
$$\hat{x}_3 = x_1 x_2$$

In this case, $\hat{x}_3$ is the product of the two independent variables.

The theoretical basis is the same as for simple linear regression (Section 6.5.2). For each data point, the error is

$$e_i = Y_i - y_i = a_0 + a_1\hat{x}_{1i} + a_2\hat{x}_{2i} + a_3\hat{x}_{3i} + \ldots + a_k\hat{x}_{ki} - y_i$$

The sum of the squares of the errors is then

$$E = \sum (a_0 + a_1\hat{x}_{1i} + a_2\hat{x}_{2i} + a_3\hat{x}_{3i} + \ldots + a_k\hat{x}_{ki} - y_i)^2 \quad \text{where} \quad \sum = \sum_{i=1}^{n}$$

E is then minimized by partially differentiating with respect to each a and then setting each resulting equation to zero. The set of equations can then be solved simultaneously for values of the a's.

One important aspect of multiple regression is that evaluating the adequacy of the best fit is considerably more difficult than it is for simple linear regression. This is because it may not be possible to generate suitable plots (since the function has three or more dimensions) and because it is more difficult to interpret the statistical parameters, such as the coefficient of determination (r^2). [Consult texts such as Devore (1991) and Montgomery (1998) for greater detail on interpreting multiple-regression results.]

Example 6.20

The following data represent the density of an oil mixture as a function of the temperature and mass fraction of three different component oils:

T(K)	m_1	m_2	m_3	ρ_{mixt}
300	0	1	0	879.6
320	0	0.5	0.5	870.6
340	0	0	1	863.6
360	0.5	0	0.5	846.4
380	0.5	0.25	0.25	830.8
400	0.5	0.5	0	819.1
420	1	0	0	796
440	1	0	0	778.2

Find the coefficients for a multiple regression of the form

$$\rho_{\text{mixt}} = a_0 + a_1 T + a_2 m_1 + a_3 m_2 + a_4 m_3$$

where ρ_{mixt} is the mixture density, T is the temperature, and m_1, m_2, and m_3 are the mass fractions of the component oils.

Solution: In an Excel® spreadsheet, the input table is as follows:

T(K)	m1	m2	m3	rhomixt
300	0	1	0	879.6
320	0	0.5	0.5	870.6
340	0	0	1	863.6
360	0.5	0	0.5	846.4
380	0.5	0.25	0.25	830.8
400	0.5	0.5	0	819.1
420	1	0	0	796
440	1	0	0	778.2

After calling the regression function in the spreadsheet program, the following output is obtained:

Summary Output	
	Coefficients
Intercept	4636.991971
X Variable 1	−0.592043796
X Variable 2	−3592.60073
X Variable 3	−3578.259367
X Variable 4	−3570.666667

"Intercept" is a_0, "X Variable 1" is a_1, "X Variable 2" is a_2, "X Variable 3" is a_3 and "X Variable 4" is a_4. Thus, the regression equation is

$$\rho_{\text{mixt}} = 4636.99 - 0.592044T - 3592.60 m_1 - 3578.26 m_2 - 3570.67 m_3$$

One should show great care in rounding the coefficients. The regression equation often involves differences between large numbers and is hence very sensitive to the significant figures of the coefficients.

Many physical relationships cannot be represented by a simple straight line, but can be easily fit with a polynomial. Polynomial regression involves only a single independent variable, but will involve terms with powers higher than unity. The form of a polynomial regression equation is

$$Y = a_0 + a_1 x + a_2 x^2 + \ldots + a_k x^k \tag{6.73}$$

where k, the exponent of the highest-order term, is the degree of the polynomial. Polynomial regression can be performed in statistical programs by simply inputting the data and the order of the polynomial desired. It can also be treated as a subset of multiple regression by making the $\hat{x}$'s in Eq. (6.72) be x, x^2, x^3, etc.

Example 6.21

Consider the following x-y data:

x	y
0	4.997
1	6.165
2	6.950
3	8.218
4	9.405
5	10.404
6	10.425
7	10.440
8	9.393
9	7.854
10	5.168

Perform a polynomial regression for a third-order polynomial on these data.

Solution: We will use the multiple-regression function in Excel®. The input field is

x	x^2	x^3	y
0	0	0	4.997
1	1	1	6.165
2	4	8	6.95
3	9	27	8.218
4	16	64	9.405
5	25	125	10.404
6	36	216	10.425
7	49	343	10.44
8	64	512	9.393
9	81	729	7.854
10	100	1000	5.168

Note that columns for x^2 and x^3 have been created, and these, together with x, will be used to generate the regression model.

The spreadsheet output is as follows:

	Coefficients
Intercept	5.023965035
X Variable 1	0.836644911
X Variable 2	0.158657343
X Variable 3	−0.024113442

Thus, the regression equation is

$$Y = 5.02396 + 0.836645x + 0.158657x^2 - 0.0241134x^3$$

The regression curve and the data are shown in Figure E6.21.

In the Excel® spreadsheet program, a simpler approach is available. A polynomial-regression feature is built into the plotting function. It is only necessary to plot the data, select the trendline function, and specify the order of the polynomial.

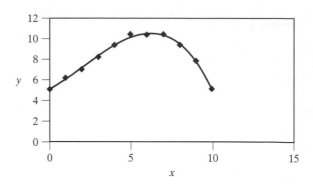

FIGURE E6.21

Comment: The use of high-order polynomials to match data can sometimes have unforeseen complications. Slight irregularities in the data can produce nonphysical wiggles in the best fit. There is normally no problem with second-order fits, but third- or higher order polynomials may produce unreasonable results. To ensure that this doesn't happen, it is necessary to examine the fit by comparing it with the data.

6.7 LINEAR FUNCTIONS OF RANDOM VARIABLES

A variable that is a function of other random variables is itself a random variable. Consider the simplest case, in which the variable y is a linear function of variables $x_1, x_2, \ldots x_n$:

$$y = a_0 + a_1 x_1 + a_2 x_2 \ldots + a_n x_n \tag{6.74}$$

Devore (1991) shows that if the x's are independent of each other, the mean value and standard deviation of y can be evaluated as

$$\mu_y = a_0 + a_1 \mu_{x_1} + a_2 \mu_{x_2} \ldots + a_n \mu_{x_n} \tag{6.75}$$

and

$$\sigma_y = [(a_1 \sigma_{x_1})^2 + (a_2 \sigma_{x_2})^2 \ldots (a_n \sigma_{x_n})^2]^{1/2} \tag{6.76}$$

These equations are useful in evaluating tolerances in fabricated parts and other applications.

Example 6.22

Consider a shaft in a bearing, as shown in Figure E6.22. The shaft diameter, D_s, is 25.400 mm, and the bearing inside diameter, D_b, is 25.451 mm. The standard deviation of the shaft diameter is 0.008 mm, and the standard deviation of the bearing diameter is 0.010 mm. For satisfactory operation, the difference in diameters (clearance) between the bearings must be between 0.0381 mm and 0.0635 mm. What fraction of the final assemblies will be unsatisfactory?

Solution: The difference in diameters, ΔD, is a linear function of D_b and D_s: $\Delta D = D_b - D_s$. Equations (6.75) and (6.76) can then be used to evaluate the mean and standard deviation of the clearance:

$$\mu_{\Delta D} = 25.451 - 25.400 = 0.051 \text{ mm}$$
$$\sigma_{\Delta D} = [.010^2 + .008^2]^{1/2} = 0.013 \text{ mm}$$

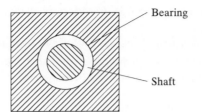

Bearing

Shaft

FIGURE E6.22

The fraction of acceptable parts would equal the area under the standard normal distribution function for

$$z_1 = \frac{0.0381 - 0.051}{0.013} = -0.99 \quad z_2 = \frac{0.0635 - 0.051}{0.013} = 0.096$$

Using the same method as in Example 6.9, we find that the area under the curve is 0.67, so 67% of the assemblies will be acceptable and 33% will be rejected.

Comment: Such a high rejection rate would not be acceptable. There are two things that could be done to reduce the rejection rate. The standard deviations of the dimensions of the parts could be reduced by improving the manufacturing process. Alternatively, the parts could be paired on a selective basis—large shafts mated with large bearings, small shafts with small bearings.

6.8   APPLYING COMPUTER SOFTWARE FOR STATISTICAL ANALYSIS OF EXPERIMENTAL DATA

Statistical analysis and data presentation have become a necessary feature of many engineering and business projects. Most modern spreadsheet programs contain some statistical functions, and some spreadsheet programs contain extensive statistical capabilities. The better programs contain not only calculations for the mean and the standard deviation, data sorting into bins, and histogram plotting, but also calculations of linear-regression coefficients and correlation coefficients. They also have built-in tables for the common distribution functions (normal, *t*-distribution, and chi-squared). There are also several specialized software packages to perform customized statistical-analysis tasks. The reader should check the field for available packages if major statistical analysis is planned.

REFERENCES

[1]   ANDERSON, DAVID R., SWEENEY, DENNIS J., AND WILLIAMS, THOMAS A. (1991). *Introduction to Statistics, Concepts and Applications*, West Publishing Co, St. Paul, Minnesota.

[2]   ASME (1998). *Measurement Uncertainty*, Part I, ASME PTC 19.1-1998.

[3]   BLAISDELL, ERNEST A. (1998). *Statistics in Practice*, Saunders College, Fort Worth, Texas.

[4]   CROW, E., DAVIS, F., AND MAXFIELD, M. (1960). *Statistics Manual*, Dover Publications, New York.

[5]   DEVORE, JAY L. (1991). *Probability and Statistics for Engineering and the Sciences*, Brookes/Cole, Pacific Grove, California.

[6]   DUNN, OLIVE JEAN, AND CLARK, VIRGINIA A. (1974). *Applied Statistics: Analysis of Variance and Regression*, John Wiley, New York.

[7] HARNETT D., AND MURPHY, J. (1975). *Introductory Statistical Analysis*, Addison-Wesley, Reading, Massachusetts.

[8] JOHNSON, R. (1988). *Elementary Statistics*, PWS-Kent, Boston.

[9] LAPIN, L. (1990). *Probability and Statistics for Modern Engineering*, PWS-Kent, Boston.

[10] LIPSON, C., AND SHETH, N. (1973). *Statistical Design and Analysis of Engineering Experiments*, McGraw-Hill, New York.

[11] Montgomery, Douglas C., Runger, George C., and Hubele, Norma F. (1998). *Engineering Statistics*, John Wiley, New York.

[12] REES, D. (1987). *Foundations of Statistics*, Chapman and Hall, London and New York.

[13] Scheaffer, Richard L., and McClave, James T. (1995). *Probability and Statistics for Engineers*, Duxbury Press, Belmont, California.

[14] U.S. DEPARTMENT OF COMMERCE (1964). *Handbook of Mathematical Functions*. U.S. Dept. of Commerce, Washington, D.C.

[15] WALPOLE, R., AND MYERS, R. (1998). *Probability and Statistics for Engineers and Scientists*, Prentice Hall, Upper Saddle River, New Jersey.

PROBLEMS

Note: An asterisk (*) denotes a spreadsheet problem.

6.1 A certain length measurement is made with the following results:

Reading	1	2	3	4	5	6	7	8	9	10
x (cm)	49.3	50.1	48.9	49.2	49.3	50.5	49.9	49.2	49.8	50.2

(a) Arrange the data into bins with width 2 mm.
(b) Draw a histogram of the data.

6.2 The air pressure (in psi) at a point near the end of an air supply line is monitored every hour in a 12-h period, and the following readings are obtained:

Reading	1	2	3	4	5	6	7	8	9	10	11	12
P (psi)	110	104	106	94	92	89	100	114	120	108	110	115

Draw a histogram of the data with a bin width of 5 psi.

6.3 Calculate the standard deviation, mean, median, and mode of the values of the data in Problem 6.1.

6.4 Calculate the standard deviation, mean, median, and mode of the values of the data in Problem 6.2.

6.5 Calculate the probability of having a 6 and a 3 in tossing two fair dice.

6.6 At a certain university, 15% of the electrical engineering students are women and 80% of the electrical engineering students are undergraduates. What is the probability that an electrical engineering student is an undergraduate woman?

6.7 A distributor claims that the chance that any of the three major components of a computer (CPU, monitor, and keyboard) is defective is 3%. Calculate the chance that all three will be defective in a single computer.

6.8 In filling nominally 12-oz beer cans, the probability that a can has 12 or more ounces is 99%. If five cans are filled, what is the probability that all five will have 12 or more ounces. What is the probability that all of the cans will have less than 12 ounces?

6.9 It has been found that the probability that a light bulb will last longer than 3000 hours is 90%. If a room contains six of these bulbs, what is the probability that all six will last longer than 3000 hours?

6.10 The probability that a certain electronic component will fail in less than 1000 hours is 0.2. There are two of these components in an instrument. What is the probability that either one or both will fail before 1000 hours?

6.11 Consider the following probability distribution function for a continuous random variable:

$$f(x) = \frac{3x^2}{35} \quad -2 < x < 3$$
$$= 0 \quad elsewhere$$

   (a) Show that this function satisfies the requirements of a probability distribution function.
   (b) Calculate the expected (mean) value of x.
   (c) Calculate the variance and the standard deviation of x.

6.12 In Problem 6.11, calculate the probability of the following cases:

   (a) $x \le 0$.
   (b) $0 < x \le 1$.
   (c) $x = 2$.

6.13 In Problem 6.11, determine the cumulative distribution of the random variable x. Calculate the numerical values of the cumulative distribution for $x = -2$, 0, and 3. Interpret your results.

6.14 An electronic production line yields 95% satisfactory parts. Assume that the quality of each part is independent of the others. Four parts are tested. Determine the probability that

   (a) all parts tested are satisfactory.
   (b) at least two parts are satisfactory.

6.15 In Example 6.3, what is the chance that at most two computers will fail in the warranty period?

6.16 Repeat Example 6.3 for the case in which between two and five computers will fail in the warranty period.

6.17 Repeat Problem 6.14 if six parts are tested and the other information is the same.

6.18 A utility company has determined that the probability of having one or more power failures is 5% in a given month. Assuming that each power failure is an independent event, find the probability that

   (a) there is no power failure during a three-month period.
   (b) there is exactly one month involving power failure during the next four months.
   (c) there is at least one power failure during the next five months.

6.19 A tire company has determined that its tires have a 1% premature rate of failure on the road. The failure mechanisms seem not to be related to each other. What is the chance that none of the 16 tires shipped to a retailer will have premature failure?

6.20 In the manufacturing of an electronic part, the probability of a part being defective is 5%.

(a) What are the probabilities that, in a batch of 100 parts,
- 2 will be defective?
- 5 will be defective?
- 10 will be defective?

(b) What should be the probability (p) of a part being defective if we do not want more than 1 out of 10 parts to fail with a chance (probability) of 99% (i.e., the probability of 0 or 1 part failing is 99%)? This part of the problem may need a trial-and-error solution.

6.21 An airplane has 175 seats. Because some ticketed passengers do not show up, the airline sells 180 seats. The probability that a passenger will not show up is 5%. Find the probability that more than 175 passengers will show up, causing a shortage of seats.

6.22 A manufacturer needs to make 50 products. The probability that the main component is defective is 5%. If the manufacturer orders 55 components, what is the probability that there are 50 or more good components so that all the products can be completed?

6.23 Automobiles visit an inspection station an average of 40 times in an eight-hour workday. What is the probability that more than five automobiles will visit the inspection station in a one-hour period?

6.24 Defects in a particular diode occur at a rate of 1 in 1000. What is the probability that a circuit board using four of the diodes contains a single defective diode?

6.25 It has been found that a particular computer system crashes, on average, once every three days. What is the probability that the system will not crash at all during a five-day period?

6.26 A plastic material is used to cover the cabinets in a kitchen. The occurrence of a noticeable surface flaw is 1 for 50 m^2 of the material. A typical kitchen uses 10 m^2 of the material. What is the probability of one or more flaws in a typical kitchen?

6.27 In an office building, light bulbs fail at an average rate of two per day. The company that replaces the bulbs will not do so unless at least three bulbs have been reported failed by the end of the previous day. The company came yesterday. A bulb in your office fails today. What is the probability that it will be replaced the tomorrow?

6.28 In connecting overseas phone calls, there is an average of 5 failures in 100 calls.

(a) What is the probability that there will be 5 failures in 50 calls?

(b) What is the probability of having 5 or fewer failures in 50 calls?

(c) What is the probability of having more than 5 failures in 50 calls?

6.29 During the one-hour lunch period in the first five days of the month, a bank expects, on average, 20 customers. What is the probability that, during the lunch period in the first five days of the month, the bank

(a) will have 25 customers?

(b) will have between 20 and 25 (inclusive) customers?

(c) will have 10 or fewer customers?

(d) will have more than 10 customers?

6.30 In the manufacturing of circuit boards, multiple boards are manufactured on a single base (e.g., metal sheet) and tested. Six units of one type of board are manufactured on a single sheet. On average, there are 3 defects per sheet. What is the chance that one will encounter

(a) ten defects in a single sheet?

(b) no defects in a single sheet?

(c) one defect in a single board?

(d) more than one defect in a single board?

6.31 In Problem 6.30, the manufacturer wants to reduce the probability of having one or more defects per board to 1%. Calculate the required average value of defects per sheet (λ).

6.32 A voltmeter is used to measure a known voltage of 100 V. Forty percent of the readings are within 0.5 V of the mean value. Assuming a normal distribution for the error, estimate the standard deviation for the voltmeter. What is the probability that a single reading will have an error greater than 0.75 V?

6.33 A certain length measurement is performed 100 times. The mean value is 6.832 m, and the standard deviation is 1 cm. Assuming a normal distribution for the error in the measurement, how many readings fall within

(a) 0.5 cm,

(b) 2 cm,

(c) 5 cm, and

(d) 0.1 cm of the mean value?

6.34 Consider the following scores of a class of 15 students:

$$95, 86, 83, 79, 79, 78, 75, 70, 70, 68, 63, 63, 55, 55, 50$$

The scores (X) are out of 100 in descending order.

(a) Calculate the mean (X_{av}), median (X_m), and standard deviation (S) of the scores.

(b) Assign letter grades to the students on the basis of the following criteria:

A	$X \geq X_{av} + 2S$	A–	$X \geq X_{av} + 1.5S$		
B+	$X \geq X_{av} + 1S$	B	$X \geq X_{av} + 0.5S$	B–	$X \geq X_{av}$
C+	$X \geq X_{av} - 0.5S$	C	$X \geq X_{av} - 1S$	C–	$X \geq X_{av} - 1.5S$
D	$X \geq X_{av} - 2S$	F	$X < X_{av} - 2S$		

(c) If the grade distribution were normal, with the same X_{av} and S, how many students would you expect to get A's, B's, C's, D's, and F's? Compare your results with Part (b) in a table.

6.35 The lighting department of a city has installed 2000 electric lamps with an average life of 1000 h and a standard deviation of 200 h.

(a) After what period of lighting hours would we expect 10% of the lamps to fail?

(b) How many lamps may be expected to fail between 900 and 1300 h?

6.36 On an assembly line that fills 8-ounce cans, a can will be rejected if its weight is less than 7.90 ounces. In a large sample, the mean and the standard deviation of the weight of a can is measured to be 8.05 and 0.05 oz, respectively.

(a) Calculate the percentage of the cans that is expected to be rejected on the basis of the given criterion.

(b) If the filling equipment is adjusted so that the average weight becomes 8.10 oz, but the standard deviation remains 0.05 oz, calculate the rejection rate (% of cans being rejected).

(c) If the filling equipment is adjusted so that the average weight remains 8.05 oz, but the standard deviation is reduced to 0.03 oz, calculate the rejection rate.

6.37 In the production of a certain type of plastic part, the diameter of the part is expected to be 5.00 mm with a standard deviation of 0.05 mm. If the part diameter deviates by more than two standard deviations, it will be rejected.

(a) Calculate the probability of rejection of a part.

(b) In a sample of 20 parts, what are the probabilities that 2, 4, and 10 will be rejected?

6.38 In manufacturing steel shelves, the column pieces are cut automatically with a standard deviation of 0.2 in. The average length is 7.25 ft. A variation in length of 0.3 in. can be tolerated. Calculate the percentage of the rejected columns. To reduce the rejection rate by one-half, calculate the required standard deviation. The tolerance will remain the same.

6.39 In the mass production of a certain piece of a machine, the average length is 10.500 in. and the standard deviation is 0.005 in. Calculate the probabilities for the following cases:

(a) The length of the piece is less than 10.520.

(b) The length of the piece is between 10.485 and 10.515.

(c) If the manufactured pieces with length deviations of more than 2.5 standard deviations are rejected, what percentage of the products will be rejected?

6.40 In a test measuring the life span of a certain brand of tire, 100 tires are tested. The results showed an average lifetime of 50,000 miles, with a standard deviation of 5000 miles. One hundred thousand of these tires have been sold and are on the road.

(a) After what mileage would you expect 10% of the tires to have worn out?

(b) How many tires are expected to wear out between 60,000 and 70,000 miles?

(c) How many tires are expected to have a life of less than 20,000 miles?

(d) What are your major assumptions in these calculations?

6.41 A particular brand of light bulbs is used in a large manufacturing plant. This brand is known to have a normally distributed life with a mean of 3600 h and a standard deviation of 160 h. After how many hours should all of the bulbs be simultaneously replaced to ensure that no more than 10% of the bulbs fail while in use?

6.42 A manufacturer of PVC plastic pipes measures the weight of ten 10-ft-long pipes to determine some statistical data for each production batch. The following data have been produced for the weight (in pounds) of a sample of 10 pipes:

6.75; 6.68, 6.70, 6.64, 6.69, 6.66, 6.77, 6.71, 6.62, 6.64

Calculate the mean and standard deviation for the sample, and estimate the standard deviation of the mean value.

6.43 To determine the wind speed in a certain location, 40 samples are taken in a limited period of time. The average value of the measurements is 30 miles per hour, and the standard deviation of the sample is 2 miles per hour. Determine a 95% confidence interval for the mean value of the wind speed.

6.44 To determine the confidence interval on the mean weight of filled buttercups, 40 samples are measured and found to have a mean value of 8 oz and a standard deviation of 0.2 oz. Determine the confidence interval on the mean weight of buttercups with confidence levels of 90%, 95%, and 99%.

6.45 In problem 6.44, if the sample size was 20 with the same statistics, determine the confidence intervals on the mean of the weight of buttercups for the given confidence levels.

6.46 To estimate the average weight of a yogurt-cup batch, the weights, in ounces, of a sample of 12 cups are determined to be as follows:

16.05, 16.14, 16.10, 16.04, 15.95, 16.08, 15.92, 16.02, 16.10, 15.90, 16.08, 16.12

(a) Calculate the mean and standard deviation of the sample, and estimate the standard deviation of the mean.

(b) Determine the 95% confidence interval on the mean of the weight of the yogurt cup.

6.47 For the data in Problem 6.40, estimate the 95% confidence interval on the mean.

6.48 To estimate the failure rate of a brand of VCR, 10 systems are tested to failure. The average life and standard deviation are calculated to be 1500 h and 150 h, respectively.

(a) Estimate the 95% confidence interval on the mean of the life of these systems.

(b) How many of the systems should be tested to obtain a 95% confidence interval on the mean of 50 h?

6.49 A batch of machine parts is considered acceptable if the average tensile strength exceeds $41.0 \times 10^6 \text{ N/m}^2$. In a manufactured batch, 10 samples were tested and found to have an average strength of $41.25 \times 10^6 \text{ N/m}^2$, with a standard deviation of $0.30 \times 10^6 \text{ N/m}^2$.

(a) Determine whether, with a 99% confidence level, the manufacturer can state that the product has an average strength exceeding 41.0 million N/m^2.

(b) Supposing that 20 samples are tested and yield the same statistics, answer the question in part (a).

6.50 To determine the efficiency of a batch of production of electric motors, 10 motors are tested. The average efficiency of the sample is calculated to be 91.0%, and the standard deviation of the sample is 0.8 percentage point.

(a) Determine a 95% confidence interval for the mean efficiency of the given batch of motors.

(b) If we want to cut the confidence interval by one-half, how many more motors should be tested?

6.51 A clothing manufacturer is setting up an experiment to obtain reliable (99% confidence level) estimates or the mean height and mean sleeve length of a target population. On the basis of previous studies, we can assume that the standard deviations are 2.44 in. for height and 0.34 in. for sleeve length. The mean value of the height and length are 68 in. and 33 in., respectively. How many members of the target population should the manufacturer include in his sample to arrive at the desired estimates for the confidence interval on the mean height and hand length of 1 in. and 0.2 in., respectively'?

6.52 To have a reasonable estimate of NO_x emissions from a boiler, the exhaust is tested 15 times. The mean value of the emissions is 25 ppm (parts per million) and the standard deviation is 3 ppm. Determine a 95% confidence interval for the standard deviation of the NO_x emissions from the boiler.

6.53 Rivet holes are punched in steel beams. To ensure that the rivets will fit and that the joint will have adequate strength, it is necessary to control the standard deviation of the diameter, and measurements are made periodically. Ten measurements are made of nominally 1-inch-diameter holes, and the standard deviation is found to be 0.002 inches. What is the 95% confidence interval on the standard deviation?

6.54 With eight tests, the standard deviation of the tensile strength of a steel alloy is found to be 5500 psi. Estimate the 99% uncertainty interval on the standard deviation.

6.55 In a certain circuit design, it is important that the standard deviation of the current be less than 15 mA. In a test on 12 parts, the sample standard deviation is found to be 10 mA. Can we be 99% confident that the standard deviation will not exceed 15 mA?

6.56 To avoid any high-speed imbalance in a rotating shaft, it needs to have a variance of its diameter below 0.0004 mm². The shaft diameter has been measured at 10 equally distributed locations, with a standard deviation of 0.01 mm.

(a) Determine the confidence interval on the variance of the diameter along the shaft with a 95% confidence level.

(b) With a confidence level of 95%, is this shaft acceptable to the customer? (That is, is the variance in diameter along the shaft below 0.0004 mm²?)

(c) Answer the questions in Parts (a) and (b) if five measurements are done and the same standard deviation is calculated.

6.57 In problem 6.56, if we want to increase the confidence level to 99%,

(a) determine the confidence interval on the variance of the diameter along the shaft, based on 10 measurements with the same standard deviation.

(b) is the shaft acceptable to the customer? (That is, is the variance in diameter along the shaft below 0.0004 mm²?)

6.58 In problem 6.57(b), it turns out that the customer cannot be assured of the shaft balance. (In other words, the variance in diameter along the shaft is below 0.0004 mm².) As a result, it is proposed that the number of measurements be increased to reduce the interval on the variance to an acceptable level. Assuming the same standard deviation, determine the number of measurements needed to accomplish this aim. (A trial-and-error solution may be required.)

6.59 Apply the Thompson τ test to the data of Problem 6.1 to see if any of the data points can be rejected.

6.60 Apply the Thompson τ test to the data of Problem 6.2 to see if any of the data points can be rejected.

6.61 Thermocouples (temperature-sensing devices) are usually approximately linear devices in a limited range of temperature. A manufacturer of a brand of thermocouple has obtained the following data for a pair of thermocouple wires:

$T(°C)$	20	30	40	50	60	75	100
$V(mV)$	1.02	1.53	2.05	2.55	3.07	3.56	4.05

Determine the linear correlation coefficient between T and V.

6.62 For the data of Problem 6.61,

(a) determine the least-squares-fit line between T and V.

(b) plot the data on a linear graph paper, and show the experimental data points and the least-squares-fit line.

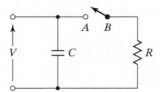

FIGURE P6.63

6.63 Consider the circuit in Figure P6.63. The capacitor C is initially charged. At time 0, the switch is connected from A to B and the voltage across the capacitor is measured. The following data are recorded by a data acquisition system:

t(s)	0.0	0.1	0.2	0.3	0.4	0.5
V(V)	4.98	1.84	0.68	0.25	0.09	0.03

(a) Determine the linear correlation coefficient between t and V.

(b) Determine the linear correlation coefficient between t and $\ln(V)$.

*6.64 The following data were obtained in order to calibrate a linear load cell with a digital output:

Pounds	0	5	10	15	20	25	30
Reading	−1.50	4.34	9.52	14.64	19.20	26.60	29.55

Determine the following:

(a) Coefficients of linear regression.

(b) Coefficients of linear regression found by forcing the line through zero.

(c) Standard error of estimate.

(d) Coefficient of determination.

(e) Any data points that are outliers.

Plot the data points and the lines, using a computer plotting routine.

*6.65 The following measurements were obtained in the calibration of a pressure transducer:

Voltage	$\Delta P \; ''H_2O$
0.31	1.96
0.65	4.20
0.75	4.90
0.85	5.48
0.91	5.91
1.12	7.30
1.19	7.73
1.38	9.00
1.52	9.90

(a) Determine the best-fit straight line.

(b) Find the coefficient of determination for the best fit.

*6.66 For the data of Problem 6.63,

(a) determine the best-fit lines for V and $\ln(V)$ with respect to t.

(b) determine the standard error of estimate for both cases, and discuss your results.

*6.67 The following data points show the flow rate versus measured pressure drop of a liquid in a venturi flowmeter:

ΔP (psi)	0.05	0.07	0.09	0.12	0.15	0.17	0.19	0.21	0.23	0.25
Q (CFM)	2.00	2.35	2.70	3.12	3.50	3.72	3.85	4.10	4.35	4.45

Let $Q = $ Constant $\times (\Delta P)^{1/2}$, plot Q as a function of ΔP, and plot $\log(Q)$ as a function of $\log(\Delta P)$. Observe the form of both curves, and determine the following quantities:

(a) Coefficients of linear regression.

(b) Standard error of estimate.

(c) Coefficient of determination.

(d) Any of the data points that are outliers.

Compare the results for both cases and comment on them.

*6.68 The following data are from the test for the flow rate of an engine oil pump as a function of engine rpm:

rpm	$T(°C)$	$Q(l/s)$
2000	90	0.50
3000	90	0.85
4000	90	1.13
5000	90	1.31
6000	90	1.40
2000	120	0.50
3000	120	0.86
4000	120	1.15
5000	120	1.35
6000	120	1.46

(a) Plot these data as two curves, one for each of the two temperatures.

(b) Find a suitable multiple regression model $Q = f(\text{rpm}, T)$. Consider terms of rpm, T, rpm^2, and $\text{rpm} \times T$.

(c) Using the regression formula, compute values of Q for each rpm and temperature. Calculate the deviations from the measured values.

*6.69 The following data are from the test of a water-pumping windmill:

V_{wind} (m/s)	Q (l/s)
4	0.94
5	1.69
6	2.44
7	3.08
8	3.72

(a) Plot these data.

(b) Fit a linear regression line to the data.

(c) Fit a second-degree polynomial to the data.

Which do you think is the better regression function?

*6.70 The following data are taken from the calibration of a fluid-velocity measuring device called a hot-film anemometer:

Voltage	Velocity (f/s)
0.01	0.07
0.115	0.08
0.29	0.11
0.48	0.135
0.59	0.145
0.81	0.185
0.88	0.19
1.02	0.22
1.12	0.24
1.325	0.285
1.4	0.295

Perform a second-degree polynomial regression on the data [Velocity $= f(\text{Voltage})$].

*6.71 The following table shows high-quality data taken from an experiment to measure the tensile strength of a steel as a function of temperature:

T (°F)	$\sigma(\text{ksi})$[a]
200	41.5
300	40.3
400	39.2
500	37.8
600	35
700	32.5
800	28.5
900	23
1000	16
1100	9.5
1200	6.5
1300	5

[a]Thousand pounds per square inch

(a) Plot these data at $\sigma = f(T)$.

(b) What do you expect would be the minimum degree of polynomial necessary to fit these data? Explain.

(c) Find the coefficients for a fourth-degree fit to the data.

6.72 In a kitchen-cabinet shop, the cabinets and the shelves are manufactured separately. In a particular cabinet model, the distance between the sides is 24 inches, with a standard deviation of 0.06 inch. The length of the shelves can be cut with a standard deviation of 0.03 inch. What is the maximum length of each shelf such that 99% of the shelves will fit? (The clearance is greater than zero.)

C H A P T E R   7

Experimental Uncertainty Analysis

Uncertainty analysis is a vital part of any experimental program or measurement system design. Common sources of experimental uncertainty were defined in Chapters 2 through 4. In this chapter, methods are provided to combine uncertainties from all sources to estimate the uncertainty of the final results of an experiment.

7.1   INTRODUCTION

Any experimental measurement will involve some level of uncertainty that may originate from causes such as inaccuracy in measurement equipment, random variation in the quantities measured, and approximations in data—reduction relations. All these uncertainties in individual measurements eventually translate into uncertainty in the final results. This "propagation of uncertainty" is an important aspect of any engineering experiment. Uncertainty analysis is performed during the design stage of an experiment to assist in the selection of measurement techniques and devices. It is also performed after data have been gathered, in order to demonstrate the validity of the results. Uncertainty analysis is a useful tool for identifying corrective actions while validating and performing experiments. Basic aspects of uncertainty analysis will be presented in this chapter. For additional details, see ASME (1998), Taylor and Kuyatt (1994), ISO (1993), Coleman and Steele (1989), and Taylor (1990).

7.2   PROPAGATION OF UNCERTAINTIES—GENERAL CONSIDERATIONS

Consider a test in which readings from a voltmeter and an ammeter are used to determine the power, $P = VI$. We would like to state the results in the form $P = A \pm w_P$, where A is the numerical value of the power and w_P is an estimated uncertainty. If the voltmeter and ammeter readings have uncertainty intervals of w_V and w_A, we need to estimate the uncertainty in w_P. In the equation $P = VI$, the uncertainties in the measurands V and I propagate into the uncertainty of the result, P.

In most experiments, it is possible for the experimenter to estimate the uncertainty of each of the measurands. Uncertainty estimates should take into account the imprecision in the measurements (*random uncertainty*, sometimes called *precision uncertainty*) and an estimated maximum fixed error (*systematic uncertainty*, sometimes called *bias uncertainty*). All uncertainty estimates should be made to the same confidence level—95%, for example. This confidence level implies that 95% of the times an uncertainty is estimated, the actual error will be less than the estimated uncertainty. Generally, if a higher confidence level is required, say, 99%, the estimated uncertainty will be larger.

In the general case, consider the result, R, to be a function of n measured variables $x_1, x_2, \ldots, x_n$; that is,

$$R = f(x_1, x_2, \ldots x_n) \tag{7.1}$$

We can relate a small change δR in R to small changes δx_i's in the x_i's through the differential equation

$$\delta R = \delta x_1 \frac{\partial R}{\partial x_1} + \delta x_2 \frac{\partial R}{\partial x_2} + \cdots + \delta x_n \frac{\partial R}{\partial x_n} = \sum_{i=1}^{n} \delta x_i \frac{\partial R}{\partial x_i} \tag{7.2}$$

This equation is exact if the δ's are infinitesimal; otherwise, it is an approximation. If R is a calculated result based on measured x_i's, then the values of the δx_i's can be replaced by the uncertainties in the variables, denoted by w_{xi}'s, and δR can be replaced by the uncertainty in the result, denoted by w_R. The partial derivative, $\partial R / \partial x_i$, is called the sensitivity coefficient of result R with respect to variable x_i.

Each of the terms on the right-hand side of Eq. (7.2) may be positive or negative, and because we designate the w's as a plus-or-minus range for the most probable errors, Eq. (7.2) will not produce a realistic value for w_R. In principle, it would be possible for the terms to cancel out, leading to a zero value for the uncertainty in R. However, we can estimate the maximum uncertainty in R by forcing all terms on the right-hand side of Eq. (7.2) to be positive. In mathematical form, this becomes

$$w_R = \sum_{i=1}^{n} \left| w_{x_i} \frac{\partial R}{\partial x_i} \right| \tag{7.3}$$

It is not very probable that all terms in Eq. (7.2) will become positive simultaneously or that errors in the individual x's will all be at the extreme value of the uncertainty interval. Consequently, Eq. (7.3) will produce an unreasonably high estimate for w_R. A better estimate for the uncertainty is given by

$$w_R = \left(\sum_{i=1}^{n} \left[w_{x_i} \frac{\partial R}{\partial x_i} \right]^2 \right)^{1/2} \tag{7.4}$$

The conceptual basis for Eq. (7.4) is discussed in Coleman and Steel (1989) and Taylor (1982). It is known as the *root of the sum of the squares* (RSS). When Eq. (7.4) is used, the confidence level in the uncertainty in the result, R, will be the same as the confidence levels of the uncertainties in the x_i's. Hence, it is important that all uncertainties used in Eq. (7.4) be evaluated at the same confidence level.

There is a significant restriction on the use of Eq. (7.4): Each of the measured variables should be independent of each other. That is, an error in one variable must not correlate with an error in another. If the variables are not independent, then the formulation is slightly different. [See ASME (1998) and Coleman and Steele (1989).] The use of Eqs. (7.3) and (7.4) is demonstrated in Example 7.1.

If the result R is dependent only on the product of the measured variables—that is,

$$R = Cx_1{}^a x_2{}^b \ldots x_n{}^N \tag{7.5}$$

it can readily be shown that Eq. (7.4) takes the simpler form

$$\frac{w_R}{R} = \left\{ \left(a\frac{w_1}{x_1} \right)^2 + \left(b\frac{w_2}{x_2} \right)^2 + \cdots + \left(N\frac{w_n}{x_n} \right)^2 \right\}^{1/2} \tag{7.6}$$

This formula is easier to use, since the fractional uncertainty in the result, R, is directly related to the fractional uncertainties in the individual measurements. Each of the exponents, [in Eq. (7.5)] or coefficients [in Eq. (7.6)] $a, b, \ldots, N$ can be positive or negative.

One important feature of Eqs. (7.4) and (7.6) is that, since the individual terms are squared before they are added, the larger uncertainties tend to dominate the result. In general, far more can be gained by reducing a single large error source than by reducing several small error sources. Eq. (7.4) can also be used in the design phase of an experiment to determine the required accuracy of instruments and other components and the suitability of the experimental approach. Example 7.2 is a simple demonstration of the use of Eq. (7.4) in the design phase.

Example 7.1

To calculate the power consumption in a resistive electric circuit ($P = VI$), the voltage and current have been measured and found to be

$$V = 100 \pm 2 \text{ V}$$

$$I = 10 \pm 0.2 \text{ A}$$

Calculate the maximum possible error [Eq. (7.3)] and also the best-estimate uncertainty [Eq. (7.4)] in the computation of the power. Assume that the confidence levels for the uncertainties in V and I are the same.

Solution: We are going to use Eqs. (7.3) and (7.4) to calculate the maximum uncertainty and the best-estimate uncertainty in the power ($P = VI$). To do this, we have to calculate the partial derivatives of P with respect to V and I:

$$\frac{\partial P}{\partial V} = I = 10.0 \text{ A and } \frac{\partial P}{\partial i} = V = 100.0 \text{ V}$$

Then,

$$(w_P)_{\text{max}} = \left|\frac{\partial P}{\partial V}w_V\right| + \left|\frac{\partial P}{\partial I}w_I\right| = 10 \times 2 + 100 \times 0.2 = 40 \text{ W}$$

$$w_P = \left(\left(\frac{\partial P}{\partial V}w_V\right)^2 + \left(\frac{\partial P}{\partial I}w_I\right)^2\right)^{1/2} = ((10 \times 2)^2 + (100 \times 0.2)^2)^{1/2} = 28.3 \text{ W}$$

Comment: The maximum uncertainty of 40 W is 4% of the power ($P = VI = 100 \times 10 = 1000$ W), whereas the uncertainty estimate of 28.3 W is 2.8% of the power. The maximum error estimate is too high in most circumstances.

Example 7.2

Manometers are pressure-measuring devices that determine a pressure by measuring the height of a column of fluid. We would like to achieve an accuracy of 0.1% of the maximum reading, 10 kPa. This is to be done by using a type of manometer called a well manometer, which has an uncertainty of 1/10 mm in reading the scale. Estimate the uncertainty that can be tolerated in the density of a gage fluid, which has a nominal value of 2500 kg/m^3.

Solution: The relationship between the pressure and the measured column height for a manometer is

$$P = \rho g h$$

where ρ is the manometer fluid density, g is the acceleration due to gravity, and h is the height of the column of fluid. It is assumed that the value of g is known to a much higher degree of accuracy than the rest of the parameters. The value of h for the maximum pressure reading of this device is

$$h = \frac{P}{(\rho g)} = \frac{10,000}{(2500 \times 9.81)} = 0.408 \text{ m}$$

The maximum uncertainty in pressure is 0.1% of 10 kPa, which is 10 Pa. Eq. (7.4) can be used to estimate the uncertainty of the gage fluid density:

$$w_P^2 = \left(\frac{\partial P}{\partial \rho}\right)^2 w_\rho^2 + \left(\frac{\partial P}{\partial h}\right)^2 w_h^2 = (10.0)^2 = 100 \text{ Pa}^2$$

$$\frac{\partial P}{\partial \rho} = gh = 9.81 \times 0.408 = \frac{4.00 \text{ m}^2}{\text{s}^2}$$

$$\frac{\partial P}{\partial h} = \rho g = 2500.0 \times 9.81 = \frac{24525.0 \text{ kg}}{\text{m}^2 - \text{s}^2}$$

$$w_h = 0.0001 \text{ m}$$

Substituting these values into the uncertainty equation yields

$$100 = (4.00 \times w_\rho)^2 + (24525.0 \times 0.00010)^2$$

$$w_\rho = 2.4 \text{ kg/m}^3$$

The maximum allowable uncertainty in the gage fluid density is $\dfrac{2.4}{2500} = 0.1\%$.

7.3   CONSIDERATION OF SYSTEMATIC AND RANDOM COMPONENTS OF UNCERTAINTY

In the early phases of the design of an experiment, it is often not practical to separate the effects of systematic and random uncertainties. For example, manufacturer-specified instrument accuracies inherently combine random and systematic uncertainties. Eq. (7.4) will then be used with these combined uncertainties in measured variables to estimate the uncertainty of the result. In more detailed uncertainty analyses, it is usually desirable to keep separate track of the systematic uncertainty, denoted B, and the random uncertainty, denoted P, associated with the measurement. Estimating random uncertainty depends on the sample size. A systematic error does not vary during repeated readings and is independent of the sample size. It is this distinction that makes it desirable to handle the terms separately in uncertainty analysis.

The random uncertainty is estimated using the t-distribution, introduced in Chapter 6. If the variable x is measured n times, then the standard deviation of the sample can be determined from

$$S_x = \left(\sum_{i=1}^{n} \frac{(x_i - \bar{x})^2}{(n-1)} \right)^{1/2} \tag{7.7}$$

and the mean can be determined from

$$\bar{x} = \frac{\sum_{i=1}^{n} x_i}{n} \tag{7.8}$$

For a given confidence level (usually 95% for uncertainty analysis), a value of t can be obtained from Table 6.6 for the degrees of freedom ($v = n - 1$), and the random uncertainty in the mean of x is determined from

$$P_{\bar{x}} = \pm t \frac{S_x}{\sqrt{n}} \tag{7.9}$$

where $\dfrac{S_x}{\sqrt{n}}$ is the estimate of the standard deviation of the mean.

It is not actually necessary for S_x to be determined in the same tests that are used to determine the final mean value of the variable x. In fact, it is common to obtain S_x from auxiliary tests run prior to taking the final data. In other cases, S_x is obtained by combining standard deviations for various components of the measurement system (elemental uncertainties), as discussed in Section 7.4. In case the mean value is obtained from a different set of tests, the final value of the mean of x, denoted by $\bar{x}_{\text{final}}$ is determined from M measurements of x and is given by

$$\bar{x}_{\text{final}} = \frac{\sum_{i=1}^{M} x_i}{M} \tag{7.10}$$

The random uncertainty in $\bar{x}_{\text{final}}$ is given by

$$P_{\bar{x}_{\text{final}}} = \pm t \frac{S_x}{\sqrt{M}} \qquad (7.11)$$

In Eq. (7.11), t is determined on the basis of the degrees of freedom, $v = n - 1$ for the tests used to determine S_x and the desired confidence level. M in Eqs. (7.10) and (7.11) is the number of samples used to obtain $\bar{x}_{\text{final}}$.

It is common in very complicated tests to measure the final value of x only once—that is, to take a single reading. In this case, $M = 1$ in Eqs. 7.10 and 7.11, and the random uncertainty in this single measurement is given by

$$P_{x_{\text{final}}} = t S_x \qquad (7.12)$$

Eq. (7.12) not only estimates the random uncertainty of x in a test in which a single final measurement is made, but also estimates the random uncertainty in each individual measurement of x in the final tests if more than one value of x is measured.

Usually, S_x is determined in tests with a large sample size ($n > 30$). In addition, a confidence level of 95% is commonly used to determine the random uncertainty interval. For these two conditions, the appropriate value of t is 2.0. A large sample size simplifies an uncertainty analysis considerably, and for this reason, the ASME Standard [ASME, (1998)] suggests the use of large sample sizes in most uncertainty analyses. While ASME presents the theory for smaller samples ($n < 30$), it relegates most of that material to appendices. This text will emphasize the simplified approach, but it provides enough information and examples that the engineer can deal with small sample sizes.

B_x, the systematic uncertainty in the measured variable x, remains constant if the test is repeated under the same conditions. Systematic uncertainties include those errors which are known, but have not been eliminated through calibration, as well as other fixed errors that can be estimated, but not eliminated from the measurement process.

While estimating systematic uncertainty involves considerable judgment, manufacturers' specifications, calibration tests, comparisons with independent measurements, and mathematical modeling can lead to rational values of this kind of uncertainty.

Unlike random uncertainty, which has a well-established statistical basis, if the measurement conditions remain the same, the systematic uncertainty does not change and does not lend itself to rigorous statistical analysis. According to ASME (1998), systematic uncertainty is the 95%-confidence-level estimate of the limits of a true systematic error. In the context of systematic uncertainty, a confidence level of 95% implies that the actual error will be less than the estimate 95% of the time an estimate is made. According to Taylor and Kuyatt (1994), other confidence levels may also be used for estimating systematic uncertainty, but 95% is the recommended one. The term "coverage" has sometimes been used to mean "confidence level" for systematic uncertainty.

Systematic and random uncertainties need to be evaluated with the same confidence level to be combined—usually the aforementioned 95% [ASME (1998) and ISO (1993)].

Figure 7.1 shows graphically how systematic and random uncertainties affect a measured result. In Figure 7.1(a), the bell curve is the frequency distribution, which would be expected if a very large number of measurements were made. The curve peaks at the

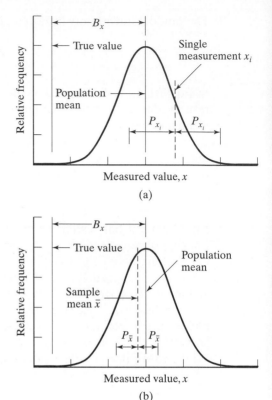

FIGURE 7.1

Graphical display of random and systematic uncertainties. (Based on W. Coleman and W. Steele, *Experimental Uncertainty Analysis for Engineers*, Wiley, 1989. Reprinted by permission of John Wiley and Sons, Inc.)

population mean, which differs from the true value of x by the systematic uncertainty, B_x. An individual reading, x_i, also deviates from the population mean and has an uncertainty interval (i.e., the interval about the reading) of $\pm P_{x_i}$. If the confidence level is 95%, the value of the population mean should be within this $\pm P_{x_i}$ interval 95% of the times the measurement is made. Figure 7.1(b) shows the situation if the measurement of x is made several times. The sample mean of these measurements, $\bar{x}$, also differs from the population mean and has an uncertainty interval of $\pm P_{\bar{x}}$ that is narrower than the interval $\pm P_{x_i}$, but that should include the population mean 95% of the time that a sample is taken.

Finally, systematic and random uncertainties are combined to obtain the *total uncertainty*, using RSS. For the mean of x,

$$W_{\bar{x}} = (B_x^2 + P_{\bar{x}}^2)^{1/2} \tag{7.13}$$

while for a single measurement of x,

$$W_x = (B_x^2 + P_x^2)^{1/2} \tag{7.14}$$

P_x is computed from Eq. (7.12) while $P_{\bar{x}}$ is computed from Eq. (7.9) or Eq. (7.11). The confidence level in the uncertainty, w, is the same as the confidence level for B and P. Examples 7.3 and 7.4 show how the random uncertainty is evaluated, and Examples 7.5 and 7.6 show how the random and systematic uncertainties are combined to produce estimates of the total errors in measurements.

Example 7.3

In a chemical-manufacturing plant, load cells are used to measure the mass of a chemical mixture during a batch process. From 10 measurements, the average of the mass is measured to be 750 kg. From a large number of previous measurements, it is known that the standard deviation of the measurements is 15 kg (which implies that $t = 2.0$ for 95% confidence level). Assuming that the load cells do not introduce any random uncertainty into the measurement, calculate for 95% confidence level

(a) the standard deviation and random uncertainty of each measurement.

(b) the standard deviation and random uncertainty of the mean value of the ten measurements.

Solution: In this problem,

$$S_x = 15 \text{ kg}$$

obtained from a large number of previous measurements, and

$$M = 10,$$

the number of measurements used for determining the average.

(a)   For each (single) measurement, the standard deviation using Eq. (7.12) is

$$S_x = 15 \text{ kg}$$

and the uncertainty of a single measurement is

$$P_x = tS_x = 2S_x = 30 \text{ kg}$$

(b)   For the average value of the measurement, $\bar{x} = 750$ kg, the standard deviation is

$$S_{\bar{x}} = \frac{S_x}{(M)^{1/2}}$$
$$= \frac{15}{(10)^{1/2}}$$
$$= 4.7 \text{ kg}$$

The random uncertainty of the mean value is

$$P_{\bar{x}} = 2S_{\bar{x}}$$
$$= 2 \times 4.7 = 9.4 \text{ kg}$$

Example 7.4

In estimating the heating value of natural gas from a gas field, 10 samples are taken and the heating value of each sample is measured by a calorimeter. The measured values of the heating value, in kJ/kg, are as follows:

48530, 48980, 50210, 49860, 48560, 49540, 49270, 48850, 49320, 48680

Assuming that the calorimeter itself does not introduce any random uncertainty into the measurement, calculate (for 95% confidence level)

(a) the random uncertainty of each measurement.

(b) the random uncertainty of the mean of the measurements.

(c) the random uncertainty of the mean of the measurements, assuming that S was calculated on the basis of a large sample ($n > 30$); but has the same value as computed in parts (a) and (b).

Solution: With x_i being the heating value, the mean value is

$$\bar{x} = \frac{\Sigma x_i}{n} = 49180 \text{ kJ/kg}$$

The standard deviation of the sample is

$$S_x = \left[\frac{\Sigma(x_i - \bar{x})^2}{(n-1)}\right]^{1/2} = 566.3 \text{ kJ/kg}$$

(a) Using the Student's t-distribution (Table 6.6) for a confidence level of 95% and degrees of freedom of $10 - 1 = 9$, we find that $t = 2.26$.

From Eq. (7.12) the random uncertainty of each sample will be

$$P_i = tS_x = 2.26 \times 566.3 = 1280 \text{ kJ/kg}$$

(b) Since $S_{\bar{x}} = \dfrac{S_x}{(n)^{1/2}}$, the random uncertainty of the mean value will be

$$P_{\bar{x}} = \frac{tS_x}{n^{\frac{1}{2}}} = \frac{2.26 \times 566.3}{10^{1/2}} = 404.7 \text{ kJ/kg}$$

(c) If we assume large n, but the same value of S_x, t will be 2.0, and the random uncertainty of the mean will be

$$P_{\bar{x}} = \frac{tS_x}{n^{\frac{1}{2}}} = \frac{2.0 \times 566.3}{10^{1/2}} = 358.2 \text{ kJ/kg}$$

Example 7.5

The manufacturer's specification for the calorimeter in Example 7.4 states that the device has an accuracy of 1.5% of the full range from 0 to 100,000 kJ/kg. Estimate the total uncertainty of (a) the mean value of the measurement in Example 7.4 and (b) a single measurement of heating value of 49,500 kJ/kg, obtained later than the measurements in Example 7.4.

Solution: The available data are as follows:

Mean value	$\bar{x} = 49180$ kJ/kg
Random uncertainty of the mean	$P_{\bar{x}} = 404.7$ kJ/kg (95% confidence)
Random uncertainty of a single value	$P_x = 1280$ kJ/kg
Systematic uncertainty	$B_x = 0.015 \times 100000 = 1500$ kJ/kg (95% confidence assumed)

We have assumed that the "accuracy" is entirely a systematic uncertainty. This point is discussed in Section 7.8.

(a) The random uncertainty of the mean is 404.7 kJ/kg. According to Eq. (7.13), the total uncertainty of measurement with a confidence level of 95% will be

$$w_{\bar{x}} = (B_x^2 + P_{\bar{x}}^2)^{1/2} = (1500^2 + 404.7^2)^{1/2} = 1554 \text{ kJ/kg}$$

which is 3.1% of the mean value.

(b) The random uncertainty of a single measurement is 1280 kJ/kg. According to Eq. (7.14), the uncertainty of a single measurement with 95% confidence will then be

$$w_x = (B_x^2 + P_x^2)^{1/2} = (1500^2 + 1280^2)^{1/2} = 1972 \text{ kJ/kg}$$

which is 4.0% of the measured value.

It is not necessary that the systematic uncertainty interval be symmetric: The uncertainty in the upward direction might be different from the uncertainty in the downward direction. In such asymmetric cases, Eq. (7.13) must be applied twice—once to obtain the total uncertainty in the positive direction and again to obtain the total uncertainty in the negative direction. These asymmetric uncertainties must then be carried through the remainder of the uncertainty analysis.

There are many situations in which a large systematic error results after the measurement device is installed into a particular application. An example is the measurement of the temperature of a hot gas when the gas is held in a cooler container. Radiation heat transfer between the cooler container walls and the measurement sensor will result in a measured value that is lower than the true gas temperature. Another source of systematic error is the location of an instrument when it is supposed to represent a variable in a space. Nonuniformity of the variable in the intended space is considered spatial error and results in systematic uncertainty. An example of this type of uncertainty occurs when a single thermometer measures the temperature in a box oven. Dynamic errors can also introduce a large systematic error. In many cases, it is possible to reduce this systematic error by analytically correcting the data. This correction process can significantly reduce the magnitude of the systematic uncertainty but, since the correction process is itself uncertain, the process cannot reduce the systematic error to zero. The analytical reduction of a systematic error is demonstrated in Example 7.6.

Example 7.6

As shown in Figure E7.6, a temperature-measuring sensor is used to measure the temperature, T_g, of a hot gas in a duct. The sensor reading, T_s, is 773 K, and the wall temperature, T_w, is 723 K. The sensor is expected to have a lower reading than the gas temperature, because the sensor is cooled by radiation heat transfer to the cooler duct wall. The following formula (derived in Chapter 9) can be used to correct for the measurement error due to radiation:

$$\Delta T_{\text{corr}} = T_g - T_s = \frac{\varepsilon}{h}\sigma(T_s^4 - T_w^4)$$

In this equation, σ is the Stefan–Boltzmann constant, which has the value 5.669×10^{-8} W/m²-K; h is the heat-transfer coefficient between the gas and the temperature sensor; and ε is the emissivity of the surface of the sensor. The temperatures must be in kelvin. The value of ε is

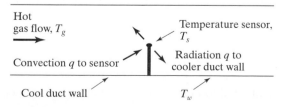

Hot gas flow, T_g

Temperature sensor, T_s

Convection q to sensor

Radiation q to cooler duct wall

Cool duct wall T_w

FIGURE E7.6

0.9 +(0.1/−0.2), and the value of h is 50 ± 10 W/m^2-K. The measured temperatures can be assumed to have negligible uncertainties. The uncertainties are systematic. Determine

(a) the temperature correction and

(b) the uncertainty in the correction.

Solution:

(a) Substituting into the given formula, we obtain

$$\Delta T_{corr} = \frac{\sigma \varepsilon}{h}(T_s^4 - T_w^4) = \frac{5.669 \times 10^{-8} \times 0.9}{50}(773^4 - 723^4) = 86 \text{ K}$$

(b) Eq. (7.6) can be used to estimate the uncertainties. Note that the positive uncertainty interval is different from the negative uncertainty interval due to the asymmetrical uncertainty in the emissivity. Using Eq. (7.6), we find that

$$\frac{w_{\Delta T}^+}{\Delta T} = \left[\left(\frac{w_\varepsilon^+}{\varepsilon}\right)^2 + \left(\frac{w_h}{h}\right)^2\right]^{1/2} = \left[\left(\frac{0.1}{0.9}\right)^2 + \left(\frac{10}{50}\right)^2\right]^{1/2} = 0.23$$

$$w_{\Delta T}^+ = 0.23 \times 86 = 20 \text{ K}$$

$$\frac{w_{\Delta T}^-}{\Delta T} = \left[\left(\frac{w_\varepsilon^-}{\varepsilon}\right)^2 + \left(\frac{w_h}{h}\right)^2\right]^{1/2} = \left[\left(\frac{0.2}{0.9}\right)^2 + \left(\frac{10}{50}\right)^2\right]^{1/2} = 0.30$$

$$w_{\Delta T}^- = 0.30 \times 86 = 26 \text{ K}$$

Thus, our best estimate of the gas temperature is

$$T_g = 773 + 86 = 859 + 20/-26 \text{ K}$$

7.4   SOURCES OF ELEMENTAL ERROR

In a typical measurement system, there are a large number of error sources. For example, an A/D converter can contribute quantization errors, sensitivity errors, and linearity errors. Each component in the chain can also contribute a number of different types of error. These error sources are known as *elemental error sources*, and each can generate either a systematic or a random error. Normally, there will be several elemental error sources in the measurement of each variable x, and the uncertainty in x will be a combination of the uncertainties due to these sources.

In order to estimate the systematic uncertainty of the measurement system with regard to a variable x, combining elemental uncertainties of the system components is usually the most practical method. By contrast, with regard to the random uncertainty of the measured variable x, there are three possible approaches to determining the standard deviation, S_x:

(a) Run the entire test a sufficient number of times so that the standard deviation can be obtained from the final test data. This was what was done in Example 7.4.

(b) Perform auxiliary tests on the system for each measured variable x, and use the data thereby obtained to determine S_x. A test matrix for a particular experiment might involve a large number of runs, with certain parameters being varied between the runs. For example, the test of an electric motor may involve varying the shaft speed, input voltage, and ambient temperature. An auxiliary test for the shaft speed might be performed for a single combination of speed, input voltage, and ambient temperature. However, the measurement itself will be repeated many times, so that S_x can be determined.

(c) Combine elemental random uncertainties of the system components, as discussed in this section.

The choice of approach will be determined by the specifics of the planned experiment.

For purposes of identifying and comparing measurement uncertainties, ASME (1998) suggests grouping the elemental errors into five categories: calibration uncertainties, data-acquisition uncertainties, data-reduction uncertainties, uncertainties due to methods, and other uncertainties. *Calibration* uncertainties are uncertainties that originate in the calibration process. While the process, of course, is performed to minimize resulting systematic errors, some residual error will remain, resulting in calibration uncertainties. The causes include factors such as uncertainty in standards, uncertainty in the calibration process, and randomness in the process. It is often convenient to include known, but uncorrected, calibration uncertainties, such as hysteresis and nonlinearity. For details on calibration uncertainties, see Taylor and Kuyatt (1994). *Data-acquisition* uncertainties are uncertainties that arise when a specific measurement is made. Among the errors produced are random variation of the measurand, installation effects such as measurand loading, A/D conversion uncertainties, and uncertainties in recording or indicating devices. *Data-reduction* uncertainties are caused by a variety of errors and approximations that are used in the data-reduction process. Interpolating, curve fitting, and differentiating data curves are examples of analytical processes that introduce data-reduction uncertainties. *Uncertainties due to methods* are those that originate from techniques or methods inherent in the measurement process. Examples of these uncertainties are assumptions or constants in the calculation routines, disturbance effects caused by instrumentation, spatial effects, and uncertainties due to instability, nonrepeatability, and hysteresis of test processes. Since elemental uncertainties are combined by using the square root of the sum of the squares (RSS), it does not matter into which category an elemental error is placed. Rather, the categories are primarily a bookkeeping and diagnostic tool. If one category dominates the errors, then it is the category to which corrective action should be applied first.

To obtain the systematic and random uncertainties of a measurement, elemental systematic and random uncertainties must be combined. The process for doing so is described schematically in Figure 7.2. The systematic uncertainties and the standard deviations are each combined using RSS:

$$B_x = \left[\sum_{i=1}^{k} B_i^2 \right]^{1/2} \tag{7.15}$$

$$S_x = \left[\sum_{i=1}^{m} S_i^2 \right]^{1/2} \tag{7.16}$$

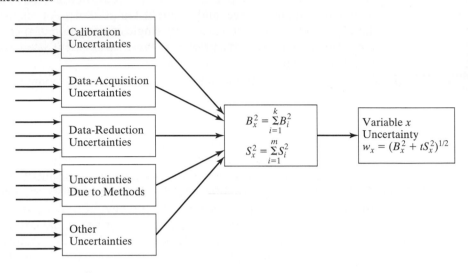

FIGURE 7.2

Sequence for calculating the uncertainty of a measured variable, x.

In these equations, there are k elemental systematic uncertainties and m elemental random uncertainties pertaining to the measured variable x.

The random uncertainty for a single measurement of variable x can then be obtained from

$$P_x = tS_x \tag{7.17}$$

and the random uncertainty in the mean can be obtained from

$$P_{\bar{x}} = \frac{tS_x}{\sqrt{M}} \tag{7.18}$$

where t is the value of the Student's t for the degrees of freedom v and the appropriate level of confidence and M is the number of times the final value of x was measured.

In order to apply the t-distribution to the resulting value of S, it is necessary to have a value for v, the degrees of freedom. If the sample sizes are greater than 30 for all of the elemental random uncertainties that were combined by Eq. (7.16), then the t is only a function of the confidence level. In some cases, the input elemental random uncertainties are based on different samples with different sample sizes. If any of the samples has a small size (less than 30), it is necessary to find a combined value of v. ASME (1998) suggests the use of the Welch–Satterthwaite formula,

$$v_x = \frac{\left[\sum\limits_{i=1}^{m} S_i^2\right]^2}{\sum\limits_{i=1}^{m} (S_i^4/v_i)} \tag{7.19}$$

where v_i is the degrees of freedom of the individual elemental error (number of data values in the sample, minus 1), m is the number of elemental random uncertainties combined in Eq. (7.16), and v_x is the value of the degrees of freedom for the variable x. The process of combining elemental errors is demonstrated in Examples 7.7 and 7.8.

If, in an experiment, the intention is to repeat the final measurement of the variable x a statistically significant number of times or to run an auxiliary test, the random uncertainty can be obtained from the test data and need not be propagated as shown in Figure 7.2. The systematic uncertainty, however, must still be derived from the elemental systematic uncertainties as shown here.

Example 7.7

In measuring the pressure in a chemical reaction tank, a high-quality pressure transducer is used. The following data are available from the manufacturer of the transducer:

Range	±3000 kPa
Sensitivity	±0.25% FS (full scale)
Linearity	±0.15% FS
Hysteresis	±0.10% FS

To determine the repeatability of the measured pressure, a large number of auxiliary tests at the desired average pressure of 1500 kPa were run. The standard deviation of these measurements is 10.0 kPa. In a separate test of the data-transmission system, a large number of tests for the same pressure input resulted in a standard deviation of 5.0 kPa. The A/D converter may produce a random uncertainty of 3.0 kPa for a 95% confidence level.

(a) Calculate the random uncertainty of the pressure measurement.
(b) Calculate the systematic uncertainty of the pressure measurement.
(c) Calculate the total uncertainty of the pressure measurement.

Solution: First, we need to convert the uncertainties that are in percentage of full scale to scaled values. The pressure-transducer uncertainties will be

Sensitivity	±0.25% FS	$= (0.25 \times 3000)/100 = 7.5$ kPa
Linearity	±0.15% FS	$= (0.15 \times 3000)/100 = 4.5$ kPa
Hysteresis	±0.10% FS	$= (0.10 \times 3000)/100 = 3.0$ kPa

The systematic uncertainties and the standard deviations of the measurements are shown in the following table:

Elemental Error	Systematic Uncertainty (kPa)	Standard Deviation (kPa)
Calibration	7.5	—
Hysteresis	3.0	—
Data Acquisition		
Repeatability	—	10
Transmission	—	5.0
A/D Conversion	—	3.0/2*
Data Reduction		
Linearity	4.5	—

*For a large number of measurements, the uncertainty is twice the standard deviation.

The standard deviation of the pressure measurement can be calculated from Eq. (7.16):

$$S_x = (10^2 + 5.0^2 + 1.5^2)^{1/2} = 11.3 \text{ kPa}$$

Because the number of measurements for determining uncertainty is large, the random uncertainty of the pressure measurement will be

$$P_x = tS_x = 2 \times 11.3 = 22.6 \text{ kPa}$$

The systematic uncertainty of the pressure measurement can be calculated from Eq. (7.15):

$$B_x = (7.5^2 + 4.5^2 + 3.0^2)^{1/2} = 9.2 \text{ kPa}$$

The total uncertainty of the pressure measurement with 95% confidence can be calculated from Eq. (7.14):

$$w_x = (B_x{}^2 + P_x{}^2)^{1/2} = (22.6^2 + 9.2^2)^{1/2} = 22.8 \text{ kPa}$$

Example 7.8

In the control of a chemical process, temperature is measured by a sensor that, according to the manufacturer, has a calibration uncertainty of $\pm0.5°C$. In 20 measurements in a test of the temperature sensor's repeatability, an average value of $150°C$ with a standard deviation of $1.5°C$ was determined. It is estimated that there can be a spatial variation of ±2 C and an installation effect of ±1 C. In a separate test of the data-transmission system, it was determined that, for 10 measurements, the standard deviation was $0.5°C$. The control program uses a linear relation between temperature and voltage, which can introduce an uncertainty of up to $\pm1°C$ in the applicable range.

(a) Calculate the random uncertainty, the number of degrees of freedom, and the total uncertainty of a single temperature measurement in this control process. Assume a 95% confidence level.

(b) Calculate the total uncertainty again, but assume that the combined standard deviation computed in Part (a) was determined with large samples ($n > 30$).

Solution:

(a) The systematic and random elemental uncertainties are identified in the following table, with an assumed confidence level of 95% for the systematic uncertainties:

Elemental Error	$B(°C)$	$S(°C)$	v
Calibration	0.5	—	—
Data Acquisition			
Repeatability	—	1.5	19
Spatial variation	2	—	—
Installation	1	—	—
Transmission	—	0.5	9
Data Reduction			
Linearity	1	—	—

The only random uncertainties in this process originate from the data-acquisition category. According to Eq. (7.16), the standard deviation of the temperature measurement will be

$$S_x = (1.5^2 + 0.5^2)^{1/2} = 1.58°C$$

From Eq. (7.19), the number of degrees of freedom will be

$$v = [0.0 + 1.5^2 + 0.5^2]^2/[0.0 + 1.5^4/19 + 0.5^4/9]$$
$$v = 22.9 \text{ or approximately } 23$$

For 23 degrees of freedom and a 95% confidence level, the Student's t-distribution (Table 6.6) gives $t = 2.07$. So the random uncertainty of each temperature measurement will be

$$P_x = tS = 2.07 \times 1.58 = 3.3 \text{ C}$$

From Eq. (7.15), the systematic uncertainty in the temperature measurement will be

$$B_x = (0.5^2 + 2^2 + 1^2 + 1^2)^{1/2} = 2.5 \text{ C}$$

The total uncertainty of measurement with 95% confidence can be calculated from Eq. (7.14):

$$w_x = (B_x{}^2 + P_x{}^2)^{1/2} = (2.5^2 + 3.3^2)^{1/2} = 4.1 \text{ C}$$

(b) If a large number of measurements is used to determine all the elemental standard deviations, t will have a value of 2.0, and the random uncertainty will be

$$P_x = tS_x = 2 \times 1.58 = 3.2°C$$

The systematic uncertainty will be the same as in Part (a), namely, $B_x = 2.5°C$, and the total uncertainty of measurement will be

$$w_x = (B_x^2 + S_x^2)^{1/2} = 4.1°C$$

Comment: In this case, accounting for the small sample size of the elemental errors did not change the uncertainty estimate significantly. This phenomenon is common: When there are several elemental errors with small sample size and small v, the combined v will be much larger. Here, one of the elemental S values had $v = 9$, but the combined S had $v = 23$.

7.5   UNCERTAINTY OF THE FINAL RESULTS FOR MULTIPLE-MEASUREMENT EXPERIMENTS

As discussed in Section 7.2, values of individual measured variables $x_1, \ldots x_n$ are combined to calculate a final result:

$$R = R(x_1, x_2, \ldots x_n) \qquad (7.20)$$

In some experimental situations, it is possible to repeat the test multiple times to determine M values of the result, $R_1 \ldots R_M$. The average value of R can then be computed, and the scatter in the measurements can be used to estimate the standard deviation of, and random uncertainty in, R. The mean of the results is given by

$$\bar{R} = \frac{1}{M}\sum_{j=1}^{M} R_j \qquad (7.21)$$

where there are M measurements of R.

The standard deviation of the result can be obtained from the formula

$$
S_R = \left[\frac{\displaystyle\sum_{j=1}^{M}(R_j - \overline{R})^2}{M - 1} \right]^{1/2}
\tag{7.22}
$$

As discussed in Section 6.4.1, the random uncertainty of the result is

$$
P_{\overline{R}} = t S_{\overline{R}} = t\frac{S_R}{\sqrt{M}}
\tag{7.23}
$$

The value of t is obtained from Table 6.6 for the selected level of confidence (usually 95%) and the degrees of freedom, $(v = M - 1)$.

In order to determine the systematic uncertainty of the result, the systematic uncertainties of each of the measured variables must be determined. This is usually done by combining elemental systematic uncertainties, as discussed in Section 7.4. The systematic uncertainty of the result is then determined from a variant of Eq. (7.4):

$$
B_R = \left\{ \sum_{i=1}^{n}\left(B_i \frac{\partial R}{\partial x_i} \right)^2 \right\}^{1/2}
\tag{7.24}
$$

Each partial derivative $(\partial R/\partial x_i)$ should be evaluated by using the average values of the measured variables.

The combined systematic and random uncertainty (total uncertainty) estimate of the mean value of R is then found from

$$
W_{\overline{R}} = (B_{\overline{R}}^2 + P_{\overline{R}}^2)^{1/2}
\tag{7.25}
$$

A test is considered to be a multiple measurement when enough independent data points are taken at each test condition to support a sound statistical determination of the random uncertainty. This means that the sample should represent the population in the statistical sense. In order to make the sample truly representative, the test should be replicated rather than simply repeated. Repeating a measurement in a short period of time does not allow for such effects as instrument thermal stability and drift. Replication, over a longer period of time, will better allow random effects to manifest themselves.

In multiple-measurement experiments, the result is obtained as a mean of several measurements and is hence less dependent on the random uncertainty inherent in individual measurements. In addition, the random-uncertainty estimate is obtained from the test results themselves and is often more reliable than an estimate obtained from auxiliary tests or combination of elemental uncertainties.

Performance and product quality tests are practically always multiple-measurement tests. Examples are measuring the life span of a certain brand of a tire, the life of a light bulb, the performance of a certain brand of engine, and the performance of a power plant upon its being commissioned. Jones (1991) and AGARD (1989) provide illustrations of such applications. Examples 7.9 and 7.10 demonstrate uncertainty estimates for a multiple-measurement experiment.

Example 7.9

In testing the power output of a large number of a certain brand of alternating-current motors, the average output power is measured to be 3.75 kW, with a standard deviation of the mean 0.05 kW. These results are obtained by measuring the rotational speed of the motor and its output torque through $P = \tau\omega$, where P is power in kW, τ is torque in N-m, and ω is angular speed in radians per second, related to the rpm (N, revolutions per minute) through $\omega = 2\pi N/60$.

The following information is available regarding the measurement of torque and speed based on the instrument's characteristics:

Parameter	Mean Value	Systematic Uncertainty
N (rpm)	1760	3.0
τ (Newton-m)	20.3	0.3

Calculate the total uncertainty in the measurement of the output power of the engine. Assume a 95% confidence level for the given systematic uncertainties.

Solution: We should propagate the systematic uncertainty of the given parameters (N and τ) to the result (in this case, power) and then calculate the total uncertainty.

Using Eq.(7.24), we can estimate the systematic uncertainty of the result, P:

$$B_P = \left[\left(\frac{\partial P}{\partial \tau} \times B_\tau \right)^2 + \left(\frac{\partial P}{\partial N} \times B_N \right)^2 \right]^{1/2}$$

$$\frac{\partial P}{\partial \tau} = \frac{2\pi N}{60} = 2 \times 3.14 \times \frac{1760}{60} = 184.31 \; 1/\text{sec}$$

$$\frac{\partial P}{\partial N} = \frac{2\pi\tau}{60} = 2 \times 3.14 \times \frac{20.3}{60} = 2.13 \; \text{N-m/(sec-rpm)}$$

$$B_P = [(184.31 \times 0.3)^2 + (2.13 \times 3)^2]^{1/2}$$

$$B_P = 55.7 \; \text{watts} = 0.06 \; \text{kW} \quad \text{(with 95\% confidence)}$$

For a 95% confidence level, and considering a large number of measurements, the random uncertainty in the measurement of the average power is

$$P_P = 2S_P$$
$$= 2 \times 0.05 = 0.10 \; \text{kW}$$

To obtain the total uncertainty in measurement of the mean value of power, we can use Eq. (7.25):

$$w_P = [B_P^2 + P_P^2]^{1/2} = [0.06^2 + 0.10^2]^{1/2}$$
$$= 0.12 \; \text{kW} \quad \text{(with 95\% confidence)}$$

As a result, the average motor power is $P = 3.75 \pm 0.12$ kW (with 95% confidence).

Example 7.10

The following relation is used to calculate the thermal efficiency (η) of a natural-gas internal combustion engine:

$$\eta = P/(m_f \cdot \text{HV})$$

Here, P is the power output in kW, m_f is the natural-gas mass flow rate in kg/s and HV is the heating value of the natural gas in kJ/kg.

To establish the mean (nominal) value for the efficiency of the engines from a production batch, five engines were tested under similar conditions, and the following results were obtained:

$$\text{Efficiency} \qquad 31.0, 30.5, 30.8, 30.6, 30.2$$

The average values of P, m_f, and HV are 50 kW, 0.2 kg/minute, and 49,180 kJ/kg, respectively. The systematic uncertainties have been evaluated as 0.2 kW, 0.003 kg/minute, and 1500 kJ/kg, respectively.

Calculate the mean value and the uncertainty of the efficiency of this production batch for a 95% confidence level.

Solution: First, we compute the average value of the efficiency:

$$\eta = \frac{(0.310 + 0.305 + 0.308 + 0.306 + 0.302)}{5} = 0.306$$

Next, we estimate the systematic uncertainty, using Eq. (7.20):

$$B_\eta = \left[\left(\frac{\partial \eta}{\partial P} \times B_P \right)^2 + \left(\frac{\partial \eta}{\partial m_f} \times B_{mf} \right)^2 + \left(\frac{\partial \eta}{\partial \text{HV}} \times B_{HV} \right)^2 \right]^{1/2}$$

$$\frac{\partial \eta}{\partial P} = \frac{1}{m_f \times \text{HV}} = \frac{1}{\left(\dfrac{0.200}{60} \times 49180 \right)} = 0.0061 \frac{1}{\text{kW}}$$

$$\frac{\partial \eta}{\partial m_f} = \frac{-P}{m_f^2 \times \text{HV}} = \frac{-50}{\left(\dfrac{0.200^2}{60^2} \times 49180 \right)} = -91.5 \frac{1}{\left(\dfrac{\text{kg}}{\text{sec}} \right)}$$

$$\frac{\partial \eta}{\partial \text{HV}} = \frac{-P}{m_f \times \text{HV}^2} = \frac{-50}{\left(\dfrac{0.200}{60} \times 49180^2 \right)} = 6.2 \times 10^{-6} \frac{1}{\left(\dfrac{\text{kJ}}{\text{kg}} \right)}$$

$$B_\eta = \left[(0.0061 \times 0.2)^2 + \left(91.5 \times \frac{0.003}{60} \right)^2 + (6.2 \times 10^{-6} \times 1500)^2 \right]^{1/2}$$

$$B_\eta = 10.43 \times 10^{-3} \quad \text{(dimensionless)}$$

Note that m_f was divided by 60 in each occurrence to convert from kg/minute to kg/s. The standard deviation of the efficiency is

$$S_\eta = \left\{ \frac{\begin{array}{c} [(0.310 - 0.306)^2 + (0.305 - 0.306)^2 + (0.308 - 0.306)^2 + \\ (0.306 - 0.306)^2 + (0.302 - 0.306)^2] \end{array}}{(5 - 1)} \right\}^{1/2}$$

$$= 0.0030 \quad \text{(dimensionless)}$$

The degrees of freedom are

$$v = M - 1 = 5 - 1 = 4$$

For a 95% confidence level and $v = 4$, the value of $t = 2.776$, from Table 6.6. From Eq. (7.19) the random uncertainty in the efficiency is then

$$P_\eta = \frac{tS}{M^{1/2}} = 2.776 \times \frac{0.0030}{5^{1/2}} = 0.0037$$

With the foregoing values of B_η and P_η, we can calculate the uncertainty of the mean value of the efficiency:

$$w_\eta = [B_\eta^2 + P_\eta^2]^{1/2} = [0.0104^2 + 0.0037^2]^{1/2}$$
$$= 0.0110$$

The mean efficiency of the engine will be

$$\eta = (30.6 \pm 1.1)\%$$

Comments: As the result shows, the systematic uncertainty is dominant in this measurement, and the multiple testing was of limited value.

7.6   UNCERTAINTY OF THE FINAL RESULT FOR SINGLE-MEASUREMENT EXPERIMENTS

An experiment is considered to be a single-measurement experiment if, for each test condition, it is run once or at most a few times. In single-measurement experiments, we do not have a large enough number of test results to calculate the standard deviation of the result reliably. In this type of test, the random uncertainty of each measured variable is usually determined through a secondary source, such as manufacturers' specifications or auxiliary tests. In a single-measurement experiment, we are dealing with the uncertainty of a single test result, rather than the uncertainty mean value of several test results.

A large number of engineering measurements and tests are categorized as single-measurement. These include practically all measurements used for process monitoring and control, such as measurements of the speed of your car, the exit temperature from a boiler, your blood pressure and heartbeat, and complicated experiments in engineering research.

The systematic uncertainty, B_R, is obtained with the use of Eq.(7.24), the same as for multiple-measurement experiments. In order to determine the random uncertainty of the result, the standard deviation of each of the measured variables, S_x, must be determined. This is usually done by running auxiliary tests, but it might be done by combining elemental random uncertainties as discussed in Section 7.4. The standard deviation of the result is given by

$$S_R = \left\{ \sum_{i=1}^{n} \left(S_i \frac{\partial R}{\partial x_i} \right)^2 \right\}^{1/2} \tag{7.26}$$

where the S_i's are the standard deviations of the individual measured variables. Each partial derivative, $\partial R/\partial x_i$, is evaluated at the measured or nominal values of the x_i's.

B_R and S_R can now be combined into the total uncertainty in the final result, given by

$$w_R = [B_R^2 + (tS_R)^2]^{1/2} \tag{7.27}$$

where t is the Student's t-value. In order to determine t, it is necessary to know the degrees of freedom of the standard deviation, S_R. If all the measured variables have v values greater than 30, t is independent of v (and has a value of 2.0 for a 95% confidence level). If S_x for one or more measured variable is based on a small sample, the appropriate value of v can be determined from another form of the Welch–Satterthwaite formula (ASME, 1998):

$$v_R = \frac{(S_R^2)^2}{\sum_{i=1}^{n}\left(\frac{1}{v_i}\left[\left(\frac{\partial R}{\partial x_i}S_i\right)^2\right]\right)} \tag{7.28}$$

Examples 7.11 and 7.12 demonstrate the calculation of the uncertainty in a single measurement with several measured variables.

Example 7.11

The power output of one of the motors of Example 7.9 is measured in a single test by measuring the motor's torque and the rotational speed. The following information is available regarding measurements of torque and speed based on instrument characteristics and a large number of auxiliary measurements:

Parameter	Value	Systematic Uncertainty	Standard Deviation
N (rpm)	1760	3.0	2.5
τ (Newton-m)	20.5	0.3	0.4

Calculate the power of the engine and the total uncertainty in the measurement of output power. Assume a 95% confidence level for the given systematic uncertainties.

Solution: The output power of the engine is

$$P = \tau \times \left(\frac{2\pi N}{60}\right) = 20.5 \times 2\pi \times \frac{1760}{60}$$
$$= 3778 \text{ watts} = 3.78 \text{ kW}$$

We should propagate the systematic and random uncertainties of the given parameters (N and τ) to the result and then calculate the total uncertainty.

The systematic uncertainty of the result (power) will be the same as the value calculated in Example 7.9, namely,

$$B_P = 55.7 \text{ watts} = 0.06 \text{ kW} \qquad \text{(with 95\% confidence)}$$

From Eq. (7.26), the standard deviation of the result (power) can be calculated on the basis of the standard deviations of torque and rotational speed:

$$S_P = \left[\left(\frac{\partial P}{\partial \tau} \times S_\tau\right)^2 + \left(\frac{\partial P}{\partial N} \times S_N\right)^2\right]^{1/2}$$
$$\frac{\partial P}{\partial \tau} = \frac{2\pi N}{60} = 2 \times 3.14 \times \frac{1760}{60} = 184.31 \text{ 1/sec}$$
$$\frac{\partial P}{\partial N} = \frac{2\pi \tau}{60} = 2 \times 3.14 \times \frac{20.5}{60} = 2.15 \text{ newton-m}$$
$$S_P = [(184.31 \times 0.4)^2 + (2.15 \times 2.5)^2]^{1/2} = 73.8 \text{ watts} = 0.07 \text{ kW}$$

For a 95% confidence level and considering a large number of measurements, the random uncertainty in a single measurement of power is

$$P_P = 2S_P$$
$$= 2 \times 0.07 = 0.14 \text{ kW}$$

To obtain the total uncertainty in measuring the power of the motor, we can use Eq. (7.27):

$$w_P = [B_P^2 + (2S_P)^2]^{1/2} = [0.06^2 + 0.14^2]^{1/2}$$
$$= 0.15 \text{ kW} \quad \text{(with 95\% confidence)}$$

As a result, the single measurement of motor power is $P = 3.78 \pm 0.15$ kW (with 95% confidence).

Example 7.12

The engine in Example 7.10 uses natural gas, for which the mean heating value and uncertainty have been previously determined. The power (P) and the fuel mass flow rate (m_f) were measured in a single test. The following information is available about the value, systematic uncertainties, standard deviations, and associated number of degrees of freedom of the these variables:

Parameter	x	B_x	S_x	v
P (kW)	50	0.2	0.3	15
m_f (kg/min)	0.200	0.003	—	—
HV (kJ/kg)	49,180	1500	167.6	9

Calculate the total uncertainty in a single measurement of the efficiency of the engine for a 95% confidence level.

Solution: The value of efficiency

$$\eta = P/(m_f \text{HV}) = \frac{50}{\left(\dfrac{0.200 \times 49180}{60} \right)}$$
$$= 0.305 \ (30.5\%)$$

The systematic uncertainty, B_η, was determined to be 10.43×10^{-3} in Example 7.10.
    Using the values of the partial derivatives computed in Example 7.10, we can determine the standard deviation of the result from Eq. (7.26):

$$S_\eta = \left[\left(\frac{\partial \eta}{\partial P} \times S_P \right)^2 + \left(\frac{\partial \eta}{\partial m_f} \times S_{m_f} \right)^2 + \left(\frac{\partial \eta}{\partial \text{HV}} \times S_{\text{HV}} \right)^2 \right]^{1/2}$$
$$S_\eta = [(0.0061 \times 0.3)^2 + (0)^2 + (6.2 \times 10^{-6} \times 167.6)^2]^{1/2}$$
$$S_\eta = 2.10 \times 10^{-3}$$

The number of degrees of freedom can be evaluated from Eq. (7.28):

$$v = \frac{(S)^4}{\left[\dfrac{\left(\dfrac{\partial \eta}{\partial P} \times S_P \right)^4}{v_P} + \dfrac{\left(\dfrac{\partial \eta}{\partial m_f} \times S_{m_f} \right)^4}{v_{mf}} + \dfrac{\left(\dfrac{\partial \eta}{\partial \text{HV}} \times S_{\text{HV}} \right)^4}{v_{\text{HV}}} \right]}$$

$$v = \frac{(2.10 \times 10^{-3})^4}{\left[\dfrac{(0.0061 \times 0.3)^4}{15} + 0 + \dfrac{(6.2 \times 10^{-6} \times 167.6)^4}{9} \right]}$$

$$v = 23$$

This is the number of degrees of freedom of the standard deviation of η. Table 6.6 tells us that the t value for the 95% confidence level is 2.074.

The total uncertainty of the result is obtained from Equation (7.27):

$$w_\eta = (B_\eta^2 + P_\eta^2)^{1/2} = [B_\eta^2 + (tS_\eta)^2]^{1/2}$$
$$= [(10.43 \times 10^{-3})^2 + (2.074 \times 2.10 \times 10^{-3})^2]^{1/2} = 0.0113$$

This corresponds to a 1.1-percentage-point uncertainty in the measurement of the efficiency of the engine:

$$\eta = (30.5 \pm 1.1)\% \text{ with a 95\% level of confidence.}$$

The uncertainty is essentially the same as for Example 7.10 because, for this situation, the systematic uncertainty dominates.

7.7   STEP-BY-STEP PROCEDURE FOR UNCERTAINTY ANALYSIS

Performing uncertainty analyses on tests can be complicated, requiring both documentation and a highly systematic bookkeeping approach. The following is an abridged step-by-step procedure for estimating the overall uncertainty of measurements as outlined by ASME (1998):

1. *Define the measurement process.*   This step involves reviewing test objectives, identifying all independent parameters and their nominal values, and defining the functional relationship between the independent parameters and the test results. For instance, in Example 7.12, the objective is to determine the nominal value, the random and systematic uncertainties, and the overall uncertainty of the efficiency measurement of a single engine. The functional relationship between the efficiency and the independently measured parameters (power, fuel mass flow rate, and heating value) are clearly identified.

2. *List all elemental error sources.*   This step includes making a complete and exhaustive list of all possible error sources for each measured parameter. To identify all uncertainties and maintain clear bookkeeping, group the uncertainties into categories based on their source: calibration, data acquisition, data reduction, methods, and other. Example 7.8 illustrates this step.

3. *Estimate the elemental errors.*   In this step, the systematic uncertainties and standard deviations must be estimated. If data are available to estimate the standard deviation of a parameter, or if the error is known to be random in nature, then it should be classified as random uncertainty; otherwise, classify it as systematic uncertainty. It is important to estimate all random uncertainties to the same confidence level and, for small samples, to identify the number of degrees of freedom associated with each standard deviation. Table 7.1 can be used as a guideline. Examples 7.7 and 7.8 illustrate this step.

4. *Calculate the systematic and random uncertainty for each measured variable.*   Systematic uncertainties and standard deviations for the variables identified in Step 1 are calculated at this step. The RSS formulation [Eqs. (7.15) and (7.16)] and the procedure

TABLE 7.1   Guidelines for assigning elemental error

ERROR	ERROR TYPE
accuracy	systematic
common-mode volt	systematic
hysteresis	systematic
installation	systematic
linearity	systematic
loading	systematic
noise	random*
repeatability	random*
resolution/scale/ quantization	random*
spatial variation	systematic
thermal stability (gain, zero, etc.)	random*

*Assume that the number of samples is greater than 30 unless specified otherwise.

outlined in Section 7.4 should be used. Data obtained in Step 3 are used in the calculations. Examples 7.7 and 7.8 illustrate this step.

5. *Propagate the systematic uncertainties and standard deviations all the way to the result(s).*   In this step, using the RSS relation, we propagate the systematic and random uncertainties of the measured variables to the final test result(s). Care should be taken so that the same confidence level is observed in all calculations. Procedures explained in Section 7.5 should be followed in this step. Examples 7.8 and 7.9 illustrate this process.

6. *Calculate the total uncertainty of the results.*   In this step, the RSS formula [Eq. (7.25) or (7.27)] is used to combine the systematic and random uncertainties to obtain the total uncertainty of the result(s). Examples 7.9 through 7.12 illustrate this step.

7.8   INTERPRETING MANUFACTURERS' UNCERTAINTY DATA

The required inputs to an uncertainty analysis from each error source are values for systematic uncertainty, standard deviation, and degrees of freedom of the standard deviation. Unfortunately, although manufacturers do supply uncertainty information, it is rarely in the desired form.

It is common for manufacturers to supply the uncertainty descriptor called *accuracy*. As defined in ANSI/ISA (1979), accuracy is a number that usually includes errors due to hysteresis, linearity, and repeatability, but assumes that the calibration is still valid and that zero errors (offsets) have been adjusted to close to zero. When accuracy is reported, it is treated as systematic uncertainty. This approach is conservative for two reasons: First, the accuracy probably has a level of confidence higher than 95%, since it bounds errors determined during the calibration process (as described in Chapter 2); second, in multiple-measurement experiments in which the random uncertainty is based on the actual measurement, the repeatability error embedded in accuracy will show its effect twice—once in the systematic uncertainty estimate and again in variations in measurement data used to determine the random uncertainty. When manufacturers report accuracy, they sometimes also report other elemental errors, such as thermal stability, that are not included in the accuracy and hence should be treated as separate elemental errors.

In some cases, manufacturers do list systematic and random uncertainties separately. For example, they may give a combined number for hysteresis and linearity and a separate number for repeatability. Hysteresis and linearity, whether separate or in combination, are considered to be systematic uncertainties. Repeatability, as defined in ANSI/ISA (1979), is the maximum observed difference in readings and, as such, is a bounding number rather than a 95% confidence estimate. As a result, combining the manufacturer's repeatability data with other statistically well-defined random elemental uncertainties is problematic. A possible approach is to assume that the manufacturer's repeatability number is a 95% confidence random uncertainty based on a sample size greater than 30. With these assumptions, a value for the standard deviation, S, can be obtained by dividing the manufacturer's repeatability numbers by 2, which is the value of Student's t for a 95% confidence level and a sample size greater than 30. This same approach can be used for other random uncertainty data given by manufacturers without a defined statistical basis.

7.9   APPLYING UNCERTAINTY ANALYSIS IN DIGITAL DATA-ACQUISITION SYSTEMS

Computer-based data-acquisition systems have been the norm in process-control applications for many years, and since the late 1970's they have become widespread in engineering experimentation. The basic components of computer systems are described in Chapter 4, and common aspects of electrical connections and signal conditioners are described in Chapter 3. Taylor (1990) is a useful reference on all aspects of these systems.

In this section, we apply uncertainty analysis techniques described earlier in the chapter to measurement systems that use electrical signal transducers and computerized data-acquisition systems. As discussed before, we can identify the following components in a generic measurement system using a digital data-acquisition system,

Sensor
Sensor Signal Conditioner
Amplifier
Filter
Multiplexer
Analog-to-digital (A/D) converter
Data reduction and analysis

Most of the errors associated with these components are discussed in Chapter 2, 3, and 4, although errors associated with some transducers and signal conditioners are

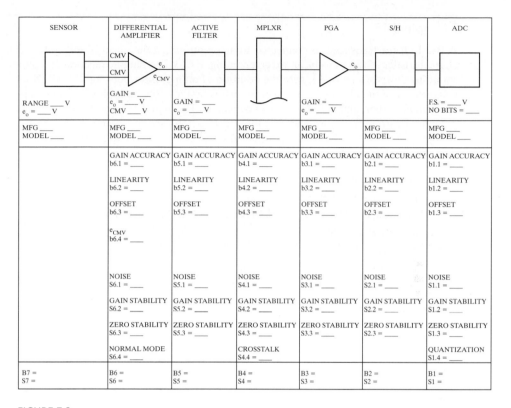

SENSOR	DIFFERENTIAL AMPLIFIER	ACTIVE FILTER	MPLXR	PGA	S/H	ADC
RANGE ___ V e_o = ___ V	GAIN = ___ e_o = ___ V CMV ___ V	GAIN = ___ e_o = ___ V		GAIN = ___ e_o = ___ V		F.S. = ___ V NO BITS = ___
MFG ___ MODEL ___	MFG ___ MODEL ___	MFG ___ MODEL ___	MFG ___ MODEL ___	MFG ___ MODEL ___	MFG ___ MODEL ___	MFG ___ MODEL ___
	GAIN ACCURACY b6.1 = ___	GAIN ACCURACY b5.1 = ___	GAIN ACCURACY b4.1 = ___	GAIN ACCURACY b3.1 = ___	GAIN ACCURACY b2.1 = ___	GAIN ACCURACY b1.1 = ___
	LINEARITY b6.2 = ___	LINEARITY b5.2 = ___	LINEARITY b4.2 = ___	LINEARITY b3.2 = ___	LINEARITY b2.2 = ___	LINEARITY b1.2 = ___
	OFFSET b6.3 = ___	OFFSET b5.3 = ___	OFFSET b4.3 = ___	OFFSET b3.3 = ___	OFFSET b2.3 = ___	OFFSET b1.3 = ___
	e_{CMV} b6.4 = ___					
	NOISE S6.1 = ___	NOISE S5.1 = ___	NOISE S4.1 = ___	NOISE S3.1 = ___	NOISE S2.1 = ___	NOISE S1.1 = ___
	GAIN STABILITY S6.2 = ___	GAIN STABILITY S5.2 = ___	GAIN STABILITY S4.2 = ___	GAIN STABILITY S3.2 = ___	GAIN STABILITY S2.2 = ___	GAIN STABILITY S1.2 = ___
	ZERO STABILITY S6.3 = ___	ZERO STABILITY S5.3 = ___	ZERO STABILITY S4.3 = ___	ZERO STABILITY S3.3 = ___	ZERO STABILITY S2.3 = ___	ZERO STABILITY S1.3 = ___
	NORMAL MODE S6.4 = ___		CROSSTALK S4.4 = ___			QUANTIZATION S1.4 = ___
B7 = S7 =	B6 = S6 =	B5 = S5 =	B4 = S4 =	B3 = S3 =	B2 = S2 =	B1 = S1 =

FIGURE 7.3

Worksheet for static error analysis of measurement using a computer data-acquisition system [adapted from Taylor (1990)].

not discussed until Chapters 8, 9, and 10. Taylor (1990) provides a useful worksheet, an adaptation of which is shown in Figure 7.3. This worksheet lists the type of uncertainty data that should be collected in order to perform a comprehensive error analysis of a measurement by means of a computerized data-acquisition system.

One problem that appears when a set of sequential components is to be analyzed is that not all components have the same range and the associated errors are usually quoted as a percent of full scale. One component will have the smallest range and will limit the range of the entire system. For example, a transducer might have an output range of ±5 volts, whereas the next stage, a lowpass filter, has an input range of ±10 volts. The actual range of the filter, when used with the transducer, will be only one-half of its maximum range. This arrangement requires some adjustment of uncertainty data that is expressed as a percent of full scale. In this example, if the stated linearity error of the filter is ±0.1% of full scale, this value will translate to ±0.01 V. In that case, the error will be ±0.01/5 × 100 = ±0.2% of the actual range of the filter.

Example 7.13 illustrates uncertainty analysis applied to a force-measuring system.

Example 7.13

A 0–5 lb force transducer is connected to a computerized data-acquisition system (DAS). The characteristics of the transducer are given in Fig. E7.13. The data-acquisition components correspond to those of Figure 4.1, with the addition of an amplifier between the MUX and the A/D converter. The signal from the transducer passes first into the multiplexer, then to a programmable gain amplifier, and next to the A/D converter. The signal conditioner shown in Figure 4.1 is contained within the transducer, and the specifications are for the combined device. Since the DAS combines the multiplexer, amplifier, and A/D converter, the following specifications are for the combined device:

> Number of bits: 12
>
> Programmable gain: −1, 10, 100, 1000
>
> Input range: ±10, ±1, ±0.1, ±0.01, 0–10, 0–1, 0–0.1, 0–0.01 V
>
> Gain error: ±2 LSB
>
> Linearity: ±2 LSB
>
> Thermal-stability errors ≪ 1 LSB, zero and gain
>
> CMRR error: 0.7 LSB/volt of common-mode voltage for $G = 100$; 7 LSB/volt for $G = 1000$

Estimate the uncertainty in a single measurement made by this force-measuring system for 95% confidence level.

Solution: In order to use the DAS, it is necessary to select a value for the gain and the input range. Since the transducer has an output of 2 mV/volt of the power supply and the power supply is 5 V, the maximum transducer output is 10 mV (and the minimum 0 V). We could select the 0–0.01-V range as the input range for the DAS; however, this choice would not allow any readings for slight overloads of the transducer. As a result, we select the 0–0.1-V range, which corresponds to a gain of 100. This means that when the transducer is at full scale, the DAS will be only

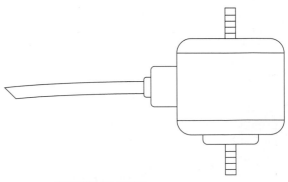

SPECIFICATION:
Linearity and Hysteresis: ±0.15% FS
Zero balance: ±2% max.
Repeatability: ±0.05% FS
Temp. effect on zero and span:
  ±0.005% FS/°F
Excitation voltage: 5 V dc
Signal output: 2 mV/V
Safe overload: 150% FS

FIGURE E7.13

at 0.01/0.1 = 10% of full scale. This low ratio adversely affects the accuracy of the DAS, but is necessary for the selected system.

In the analysis that follows, the errors associated with the transducer will be considered to be calibration errors and those associated with the DAS will be treated as data-acquisition errors.

CALIBRATION ERRORS

Systematic uncertainties.   There is only one systematic uncertainty: the linearity and hysteresis error of ±0.15% of full scale. This error translates to $B_1 = ±0.0075$ lb, since the full-scale output corresponds to 5 lb. A potential systematic uncertainty is due to the power supply, since output is proportional to that voltage, but we may assume that this error is negligible because the voltage is accurately adjusted. Another potential systematic uncertainty is due to the zero offset, but again, we assume that it has been accurately adjusted so that the error is negligible.

Random uncertainties.   There are two random uncertainties: the repeatability error and the thermal-stability error (the effect of temperature on zero and span). The repeatability error is ±0.05% of full scale (or ±0.0025 lb). The effect of the temperature error on span and zero is ±0.005% FS/°F. We will assume that the zero is adjusted and that only the temperature effect on span needs to be considered. If we assume that the environmental temperature varies by ±5 F, then this span error becomes ±0.025% FS (±0.00125 lb).

In estimating random uncertainty, the formulations use the standard deviation S, whereas the numbers in the previous paragraphs correspond more closely to the random uncertainty. The random uncertainty and the standard deviation are related by Eq. (7.17):

$$P = tS$$

To obtain a value for the standard deviation, we can assume that the sample on which S is based is larger than 30; hence, for a confidence level of 95%, $t = 2$. Thus, we can estimate the standard deviations as follows:

Repeatability $S_1 = 0.00125$ lb
Temperature $S_2 = 0.00063$ lb

DATA-ACQUISITION ERRORS

The number of possible output states for the DAS is $2^{12} = 4096$, since the device has a 12-bit A/D converter. The A/D converter is bipolar; thus, half the available states apply to the −0.1–0-V range. This leaves only 2048 bits available for the 0–0.1-V range. Furthermore, when the transducer is at full scale, its output will be only 0.010 V and the A/D will produce only 204 output states. As a result, all of the full-scale errors should be determined as percentages of 204, not 4096.

Systematic uncertainties.   The obvious systematic uncertainties are the linearity error and the gain error:

Linearity error $B_2 = 2/204 \times 100 = 0.98\%$ FS $= 0.049$ lb
Gain error $B_3 = 2/204 \times 100 = 0.49\%$ FS $= 0.049$ lb

Random uncertainties

The random uncertainties we have to consider for the DAS are the resolution (quantization error, ± 0.5 LSB) and the thermal-stability error. We should also consider noise in the signal, which appears between the transducer and the DAS. We assume that this noise has a 1% effect on the DAS output. Again, we have numbers for random uncertainty, not standard deviation, so we divide our values by a t value of 2. The treatment of the quantization error in this manner is decidedly an approximation, since that error is not a normally distributed random variable. However, it is not likely to introduce a major error in the final uncertainty estimate. The thermal-stability error is small enough to be neglected. Thus, we have the following uncertainties:

Quantization	$S_3 = (0.5/2)/204 \times 100 = 0.122\%$ FS $= 0.0061$ lb
Noise	$S_4 = 1/2\% = 0.025$ lb

The combined systematic and random uncertainties are

$$B_F = (0.0075^2 + 0.049^2 + 0.049^2)^{1/2} = 0.070 \text{ lb}$$
$$S_F = (0.00125^2 + 0.00063^2 + 0.0061^2 + 0.025^2)^{1/2} = 0.026 \text{ lb}$$

The uncertainty in the force can now be estimated. We have already assumed that the samples on which the random errors are based are larger than 30, so we assume that $t = 2$. The resulting estimated uncertainty in a single reading obtained from this measurement system is then

$$w_F = (B_F^2 + tS_F^2)^{1/2} = [0.070^2 + (2 \times 0.026)^2]^{1/2} = 0.087 \text{ lb}$$

Comments: The final uncertainty is 0.087 lb, which is 1.7% of the full-scale reading of 5 lb. Note that the data-acquisition errors are the most significant. One major problem is that the A/D converter in the DAS is operating over only 10% of its range. To allow for a slight increase in range with this transducer, it might be possible to reduce the power supply voltage to 4.5 V and then use an amplifier gain of 1000. This approach will substantially reduce the systematic uncertainty for the DAS. Noise is the other big contributor to the final error. It might be possible to reduce the noise level with better shielding or by using a filter between the transducer and the DAS.

7.10   ADDITIONAL CONSIDERATIONS FOR SINGLE-MEASUREMENT EXPERIMENTS

Moffat (1988) has a perspective on complicated multiparameter research experiments that is worth noting. Such experiments often require taking very large quantities of data over extended periods of time. It is then impractical to replicate all measurements a sufficient number of times to estimate random uncertainty on the basis of the measurements themselves. These experiments are single-measurement experiments by practical necessity. With this type of experiment in mind, Moffat (1989) suggests three levels of uncertainty estimation, which he calls zeroth order, the first order, and the Nth order. His suggestions are not incompatible with the ASME (1998) approach, but outline an alternative approach to uncertainty analysis that is suitable for single-measurement research experiments.

The zeroth-order uncertainty interval estimates the overall uncertainty arising from the instrumentation itself. This estimate is usually based on instrument manufacturers' specifications or previous experience and involves combining the elemental uncertainties discussed in Section 7.4. Zeroth-order uncertainty analysis is used in the design stage to evaluate the suitability of proposed instrumentation systems. For each system, this uncertainty is the smallest that could be achieved with the proposed instrumentation. It does not account for many sources of variability in the experiment, such as unsteadiness of the process. For a measured variable x_i, the zeroth-order uncertainty estimate will be

$$w_{i,0} = ((B_{i,0})^2 + (P_{i,0})^2)^{1/2} \qquad (7.29)$$

where $B_{i,0}$ and $P_{i,0}$ are the estimates of the systematic and random uncertainties in the instrumentation system.

The first-order uncertainty analysis estimates the scatter in the results of repeated trials using the same equipment, procedures, and instrumentation, but with the process running. The first-order uncertainty interval accounts for random uncertainty in the data, but does not include any systematic uncertainty. It differs from the random-uncertainty component of the zeroth-order uncertainty because the first-order estimate includes the process variability and other variability resulting from the measurement procedure. As an example, if a pump is used in an experiment in which the velocity of a fluid is measured, any random effects in the pump that influence the fluid flow will be a first-order uncertainty. This type of uncertainty cannot be evaluated at the zeroth-order level. In elaborate tests, the first-order uncertainty interval must be determined from an auxiliary experiment, before the final data are taken. These tests are usually run during the shakedown and debugging stages of the experiments. These variability measurements will be made by taking multiple measurements at selected operating conditions and at selected locations but will not cover the entire test matrix. These auxiliary tests result in estimates for $P_{i,1}$, the first-order random uncertainty in x_i.

The Nth-order uncertainty interval estimates the total uncertainty in the experiment as it is run. Nth-order uncertainty estimates include the zeroth-order systematic-uncertainty estimate, other sources of systematic uncertainties associated with the experiment (e.g., loading uncertainty), and the random-uncertainty estimate determined from the first-order analysis, namely,

$$w_{i,N} = ((B_{i,N})^2 + P_{i,1}^2)^{1/2} \qquad (7.30)$$

where $B_{i,N}$ is the RSS combination of the zeroth-order systematic uncertainty and other systematic uncertainties associated with the experiment.

In the preceding paragraphs, we computed the uncertainty of a measured value. To get the uncertainty in a result computed from a formula, it is better to compute the systematic uncertainty B_R and the standard deviation of the result, S_R by means of Eqs. (7.24) and (7.26) and then combine the two via Eq. (7.27). However, if the degrees-of-freedom value is large (greater than 30) for all computed standard deviations, we can use Eq. (7.4) to combine the uncertainties of the measured variables (the w_i's) to obtain the uncertainty of the result.

The foregoing analysis is demonstrated in Example 7.14. For further details, see Moffat (1988). Coleman and Steele (1988) provide additional examples.

Example 7.14

In the control of a process, it is necessary to measure the temperature of steam in a steam line. A temperature measuring device called a thermocouple is used for this purpose. The temperature-measurement system has been calibrated and found to have a systematic uncertainty of 0.5 C and a random uncertainty of 0.4 C. With the steam-line conditions held as constant as possible, several trial tests ($n > 30$) of the installed measuring system were run, and the data showed a standard deviation of 1.0 C. A loading uncertainty in the steamline measurement is expected to cause a systematic uncertainty of 1.0 C. No other measurement uncertainties are expected in this system. Estimate the zeroth-, first-, and Nth-order uncertainties for 95% confidence level.

Solution:

ZEROTH-ORDER ANALYSIS: MEASUREMENT SYSTEM UNCERTAINTY

$$B_0 = 0.5°C$$
$$P_0 = 0.4°C$$
$$w_0 = (0.5^2 + 0.4^2)^{1/2} = 0.64°C$$

FIRST-ORDER ANALYSIS: SCATTER IN THE MEASURED DATA

This uncertainty is caused by scatter in the steam-line system tests ($S = 1°C$). For $n > 30$ and a 95% confidence level, $t = 2.0$, and

$$P_1 = tS = 2 \times 1 = 2.0°C$$

NTH-ORDER ANALYSIS: OVERALL MEASUREMENT UNCERTAINTY

Here we include all systematic uncertainties and the first-order random uncertainty. We then have

$$B_0 = 0.5°C$$
$$B_L = 1.0°C \text{ (loading uncertainty)}$$
$$P_1 = 2.0°C$$
$$B_N = (B_0^2 + B_L^2)^{1/2} = (0.5^2 + 1.0^2)^{1/2} = 1.12°C$$
$$w_N = (B_N^2 + P_1^2)^{1/2} = (1.12^2 + 2.0^2)^{1/2} = 2.3°C$$

The estimated measurement uncertainty is thus ±2.3°C.

Comments:

1. Our analysis shows that, with the measuring system used, we cannot achieve an uncertainty less than ±0.64°C, no matter how the thermocouple is installed or how steady the process is.

2. The uncertainty in this measurement is caused primarily by unsteadiness and the installation loading effects of the steam line itself.

7.11   CLOSURE

This chapter presents a systematic method for evaluating uncertainty in measurements. Prior to the analysis of uncertainty, the experimenter must closely examine the system to ensure that components have a sufficient range and are compatible with each other.

The analysis described here cannot account for blunders in the measurement process. Nothing can replace human carefulness and patience in an experimental project!

REFERENCES

[1] Advisory Group for Aerospace Research & Development (AGARD) (1989). Measurement of Uncertainty within the Uniform Engine Test Program, AGARDograph No. 307, 1989.

[2] ASME (1998). *Test Uncertainty*, Part 1, ASME PTC 19.1-1998.

[3] ANSI/ISA (1979). *Process Instrumentation Terminology*, ANSI/ISA Standard S51.1—1979.

[4] COLEMAN H. AND STEELE, W. (1989). *Experimental Uncertainty Analysis for Engineers*, John Wiley, New York.

[5] ISO (1993). *Guide to the Expression of Uncertainty in Measurement*, International Standards Organization.

[6] JONES, P.A., SITKIEWITZ, S.D. AND FRIEDMAN, M.A. (1991). Application of Uncertainty Analysis to the Test Program at the Nucla Circulating Fluidized Bed Boiler Demonstration Project, *ASME Proceedings of the 1991 International Conference on Fluidized Bed Combustion*, Vol. 3, pp. 1339–1344.

[7] KLINE, S.J. AND McCLINTOCK, F.A. (1953). Uncertainty in Single Sample Experiments, *Mechanical Engineering*, Vol. 75, Jan 1953, pp. 3–8.

[8] MOFFAT, ROBERT J. (1988). Describing the Uncertainties in Experimental Results, *Experimental Thermal and Fluid Science*, Vol. 1, pp. 3–17.

[9] TAYLOR, B.W. AND KUYATT, C.E. (1994). *Guidelines for Evaluating and Expressing the Uncertainty of NIST Measurement Results*, U.S. National Institute of Standards and Technology (NIST), Technical Note 1297.

[10] TAYLOR, JAMES L. (1990). *Computer-Based Data Acquisition Systems—Design Techniques*, 2d ed., Instrument Society of America (ISA), Research Triangle Park, NC 27709.

[11] TAYLOR, JOHN R. (1982). *An Introduction to Error Analysis*, University Science Books, Mill Valley, CA 94941.

PROBLEMS

Note: An E after a problem number indicates that it is similar to the preceding problem except for the use of English units.

7.1 Two resistors, $R_1 = 100.0 \pm 0.2\ \Omega$ and $R_2 = 50.0 \pm 0.1\ \Omega$, are connected (a) in series and (b) in parallel. Calculate the uncertainty in the resistance of the resultant circuits. What is the maximum possible error in each case?

7.2 Orifice meters are used to measure the flow rate of a fluid. In an experiment, the flow coefficient K of an orifice is found by collecting and weighing water flowing through the orifice during a certain interval while the orifice is under a constant head. K is calculated from the following formula:

$$ K = \frac{M}{tA\rho(2g\,\Delta h)^{1/2}} $$

The values of the parameters have been determined to be as follows, with 95% confidence:

Mass $M = 393.00 \pm 0.03$ kg
Time $t = 600.0 \pm 1$ s
Density $\rho = 1000.0 \pm 0.1\%$ kg/m^3

Diameter $d = 1.270 \pm 0.0025$ cm (A is area)
Head $\Delta h = 366.0 \pm 0.3$ cm

Find the value of K, its uncertainty (with 95% confidence), and the maximum possible error.

7.2E Orifice meters are used to measure the flow rate of a fluid. In an experiment, the flow coefficient K of an orifice is found by collecting and weighing water flowing through the orifice during a certain interval while the orifice is under a constant head. K is calculated from the following formula:

$$K = \frac{M}{tA_\rho(2g\Delta h)^{1/2}} .$$

The values of the parameters have been determined to be as follows, with 95% confidence:

Mass $M = 865.00 \pm 0.05$ lbm
Time $t = 600.0 \pm 1$ s
Density $\rho = 62.36 \pm 0.1\%$ lbm/ft^3
Diameter $d = 0.500 \pm 0.001$ in. (A is area)
Head $\Delta h = 12.02 \pm 0.01$ ft

Find the value of K, its uncertainty (with 95% confidence), and the maximum possible error.

7.3 Variables R_1, R_2, R_3, and R_4 are related to three independent variables x_1, x_2, and x_3 by the formulas

$$R_1 = ax_1 + bx_2 + cx_3$$
$$R_2 = d(x_1)(x_2)(x_3)$$
$$R_3 = \frac{e(x_1)(x_2)}{x_3}$$
$$R_4 = f(x_1)^g(x_2)^h(x_3)^i$$

where $a, \ldots, i$ are constants. For each case, derive the uncertainty of the result, w_R, in terms of the uncertainties of individual variables (the w_x's). Generalize your results to take account of the effect of summation and multiplication on the propagation of errors.

7.4 A simple spring is used to measure force. The spring is considered to be linear, so that $F = kx$, where F is the force in newtons, k is the spring constant in newtons per centimeter, and x is the displacement in cm. If $x = 12.5 \pm 1.25$ cm and $k = 700 \pm 18$ N/cm, calculate the maximum possible error and the uncertainty of the measured force in absolute (dimensional) and relative (%) terms.

7.4E A simple spring is used to measure force. The spring is considered to be linear, so that $F = kx$, where F is force in pounds, k is the spring constant in pounds per inch, and x is the displacement in inches. If $x = 5 \pm 0.5$ in. and $k = 20 \pm 0.5$ lbf/in., calculate the maximum possible error and the uncertainty of the measured force in absolute (dimensional) and relative (%) terms.

7.5 A mechanical speed control system works on the basis of centrifugal force, which is related to angular velocity through the formula

$$F = mr\omega^2$$

where F is the force, m is the mass of the rotating weights, r is the radius of rotation, and ω is the angular velocity of the system. The following values are measured to determine

ω: $r = 20 \pm 0.02$ mm, $m = 100 \pm 0.5$ g, and $F = 500 \pm 0.1\%$N. Find the rotational speed in rpm and its uncertainty. All measured values have a confidence level of 95%.

7.6 Young's modulus of elasticity, E, relates the strain, $\delta L/L$, in a solid to the applied stress, F/A, through the relationship $F/A = E(\delta L/L)$. To determine E, a tensile machine is used, and F, L, δL, and A are measured. The uncertainties in each of these quantities are 0.5%, 1%, 5%, and 1.5%, respectively, with 95% confidence. Calculate the uncertainty in E in percentage form. Which of these measurements has the greatest effect on the uncertainty of E? How can we reduce that uncertainty by 50%?

7.7 The variation of resistance with temperature is expressed by the relationship $R = R_0[1 + \alpha(T - T_0)]$, where R_0 is the resistance at the reference temperature T_0 and α for the resistor material has been determined to be $0.0048 \pm 0.1\%/°C$. In the range 0 to 100°C, in which we are calibrating this resistor, temperature measurements have shown a standard deviation of 0.1°C. The systematic uncertainty of the temperature-measurement device is known to be 0.1°C. Calculate the percent uncertainty of R at a temperature of 25°C with a 95% confidence level. At 0°C, the resistance is 100.00 ohms.

7.8 One of the parameters that is used to evaluate the performance of an engine is its brake-specific fuel consumption (bsfc), defined as

$$\text{bsfc} = \frac{\dot{m}_{\text{fuel}}}{2\pi N \tau}$$

where, for a given engine, the parameters on the right-hand side of the equation are measured and found to be as follows:

Mass flow rate $\dot{m}_{\text{fuel}} = 5.0 \times 10^{-4}$ kg/s
Rotational speed $N = 50$ rev/s (3000 rpm)
Torque $\tau = 150$ N–m

We can tolerate a total uncertainty of 1% in the measurement of bsfc.

(a) Calculate the rated bsfc.

(b) Determine the maximum tolerable uncertainty in each measurement. This will occur when the uncertainties of the other measurements are negligible.

(c) For a total uncertainty of $\pm 1\%$ in bsfc, calculate the relative (%) uncertainty of the measured parameters if all of them have the same relative uncertainty.

(d) Calculate the absolute value of the uncertainty in each of the measured parameters, based on the results of part (c).

7.9 In using a temperature probe, the following uncertainties were determined:

Hysteresis	± 0.1C
Linearization error	$\pm 0.2\%$ of the reading
Repeatability	$\pm 0.2°C$
Resolution error	$\pm 0.05°C$
Zero offset	$\pm 0.1°C$

Determine the type of these errors (random or systematic) and the total uncertainty due to these effects for a temperature reading of 120°C.

7.10 A digital scale is used to measure the mass of a product on a manufacturing line. The scale has a range of 0–2 kg and uses a 12-bit A/D converter. In addition, it has an accuracy of 2% of its reading. List the uncertainties and categorize them as systematic or random. Calculate the uncertainty in a measurement of 1.25 kg by this instrument.

7.11 A digital-output pressure-measuring system has the following specifications:

Range	0 to 1000 kPa
Accuracy	±0.5% of range
Resolution	±1 kPa
Temperature stability	±2 kPa (0 to 50°C)

Specify the types of these errors and calculate the uncertainty of pressure measurement with this transducer. For a nominal pressure of 500 kPa, calculate the absolute and relative uncertainties.

7.12 In a measurement of temperature (in °C) in a duct, the following readings were recorded:

$$248.0, 248.5, 249.6, 248.6, 248.2, 248.3, 248.2, 248.0, 247.5, 248.1$$

Calculate the average temperature, the standard deviation of the sample, and the random uncertainty of the average value of the measurements. Assume a 95% confidence level.

7.13 A manufacturer of plastic pipes uses a scale with an accuracy of 1.5% of its range of 5 kg to measure the mass of each pipe the company produces in order to calculate the uncertainty in mass of the pipes. In one batch of 10 parts, the measurements are as follows:

$$1.93, 1.95, 1.96, 1.93, 1.95, 1.94, 1.96, 1.97, 1.92, 1.93 \text{ (kg)}$$

Calculate

(a) the mean mass of the sample.

(b) the standard deviation of the sample and the standard deviation of the mean.

(c) the total uncertainty of the mass of a single product at a 95% confidence level.

(d) the total uncertainty of the average mass of the product at a 95% confidence level.

7.14 In Problem 7.13, if the measurement is performed on, say, 50 pieces of pipe, and the same mean and standard deviation are obtained, repeat parts (b), (c), and (d) of that problem.

7.15 In a cheese factory, 4.5-kg blocks of cheese are cut manually. For a large number of blocks, the standard deviation of the cutting process is measured and found to be 0.10 kg. The measurement was done with a scale with an accuracy of 1.5% of the full scale of 12 kg. Calculate the total uncertainty of the weight of the blocks of cheese at a 95% confidence level.

7.16 The managers in the cheese factory of Problem 7.15 have decided to reduce the uncertainty in the mass of cheese and reduce their labor as well by automating the cheese-cutting line. They will use a 12-bit online digital scale with a calibration accuracy of 1% of its reading and a range of 12 kg. Calculate the acceptable standard deviation of the cheese blocks that will reduce the uncertainty in the mass of cheese blocks to 1.5% of the mass of each block.

7.17 In a yogurt-filling line, the containers are filled with 1 kg of yogurt. The mass of the yogurt is measured while the filling nozzle is open. The nozzle flow rate is 0.25 kg per second. Several factors affect the flow rate, including the density and the viscosity of the yogurt, which, combined, can introduce an uncertainty of about 1% of the flow, based on statistical analysis of the data. The dispensing time of the yogurt is controlled mechanically through a cam system. The uncertainty in the time during which the nozzle is open was established to be 0.1 sec in a calibration process.

(a) Calculate the filling time of each container.

(b) Categorize each of the foregoing uncertainties as systematic or random.

(c) Calculate the uncertainty in the mass of the filled yogurt containers.

(d) If the plant managers decide to reduce the uncertainty in the mass of the yogurt, on what do you suggest they concentrate?

7.17E In a yogurt-filling line, the containers are filled with 2 pounds of yogurt. The weight of the yogurt is measured while the filling nozzle is open. The nozzle flow rate is 0.5 pound per second. Several factors affect the flow rate, including the density and the viscosity of the yogurt, which, combined, can introduce an uncertainty of about 1% of the flow, based on statistical analysis of the data. The dispensing time of the yogurt is controlled mechanically through a cam system. The uncertainty in the time during which the nozzle is open was established to be 0.1 sec in a calibration process.

(a) Calculate the filling time of each container.

(b) Categorize each of the foregoing uncertainties as systematic or random.

(c) Calculate the uncertainty in the mass weight of the filled yogurt containers.

(d) If the plant managers decide to reduce the uncertainty in the weight of the yogurt, on what do you suggest they concentrate?

7.18 To improve the accuracy of measuring the heating value of natural gas in Examples 7.4 and 7.5, a new calorimeter with the same range, but with an accuracy of 1% of the full range, is used to make 15 measurements. The average heating value of natural gas is measured to be 49,200 kJ/kg, and the standard deviation is 450 kJ/kg. Calculate the total uncertainties in the estimates of

(a) the mean heating value and

(b) a single measurement of the heating value.

7.19 The following information about a linear submersible depth pressure transmitter is available:

Range	0–20 m H_2O
Output	4–20 mA
Accuracy (including linearity, hysteresis, and repeatability)	0.2% Span
Zero balance	2% Span
Thermal effects	1.5% Span

(a) What will be the nominal output of the device (in mA) at a depth of 15 m of water?

(b) Estimate the uncertainty due to each of the error sources that can be determined from the specifications. Express the uncertainties in mA, in m H_2O, and as a percentage of the reading.

(c) Calculate the total output uncertainty in mA, in m H_2O, and as a percentage of the output.

7.20 The following information is from an instrument manufacturer's catalog for a load cell:

Range (full scale, FS)	0–500 N
Operating temperature	−50 to 120°C
Signal output	3 mV/V excitation voltage, nominal
Excitation voltage	10 volts DC

Linearity	±0.1% Span
Hysteresis	±0.08% Span
Repeatability	±0.03% Span
Temperature effect	0.002% Span/°C (20°C reference)

The device is operated in an environment where the temperature range is 10–40°C.

(a) For an unamplified output of 20 mV, what is the applied load (assuming that the device is linear)?

(b) Estimate the uncertainty due to each of the error sources that can be determined from the specifications. Express the uncertainties in mV, in N, and as percentage of the reading at 20 mV.

(c) Calculate the total output uncertainty in mV, in N, and as a percentage of the output.

7.21 The following information is from an instrument manufacturer's catalog for a positive-displacement flowmeter for high-viscosity fluids:

Maximum capacity	10 lpm (liters per min)
Turndown ratio*	10:1
Accuracy	±6% of flow rate
Repeatability	±1% of flow rate

*The turndown ratio is the ratio of the maximum to the minimum value of the measurand that the instrument can be used to measure.

(a) Estimate the uncertainty due to each of the error sources that can be determined from the specifications for readings of 2 and 10 lpm.

(b) What are the total uncertainties in the measured flow rates of 2 and 10 lpm?

7.22 In a calibration process, it has been determined that a thermocouple is accurate to ±0.2°C with 95% confidence. To determine the total uncertainty of temperature measurement in a vessel, a sample of 15 measurements was taken. The average temperature is 250.0°C, and the standard deviation of the measurements is calculated to be 0.2°C.

(a) Calculate the random uncertainty and the total uncertainty of the average temperature at a 95% confidence level.

(b) Calculate the random uncertainty and the total uncertainty of a single reading of the temperature at a 95% confidence level.

(c) If another thermocouple, which is accurate to 0.1°C is used, will it significantly affect the total uncertainty of temperature measurement?

7.23 In Example 7.6, if the temperature reading is 600°C and the wall temperature is 545°C, calculate the temperature correction and the uncertainty in the correction. All other parameters remain the same.

7.24 Strain is to be measured remotely in a structure by means of a strain gage. To estimate the total uncertainty in the measurement of strain, the strain gage and the transmission line are tested separately. Ten measurements of the strain gage output under the same loading produce a standard deviation of 0.5 mV in an average output of 80 mV. Fifteen measurements of the transmitted voltage produced a standard deviation of 1 mV. Determine the random uncertainty of the voltage measurement of strain produced by the strain gage at a 95% confidence level.

7.25 The pressure-measuring system of Problem 7.11 is used to measure the pressure in a compressed air line. To determine the uncertainty in the nominal pressure of 800 kPa, 15 measurements were made, yielding a standard deviation of 5 kPa. Determine the total uncertainty of each pressure measurement in the line at a 95% confidence level.

7.26 In Problem 7.25, suppose that you make over 30 measurements and still obtain the same standard deviation for the pressure in the line. Will the uncertainty in each measurement of the pressure change? Why? Discuss your results.

7.27 In measuring the power of a three-phase electrical motor running a pump, the following data were obtained (based on one-phase current measurement):

V (volts)	460	459	458	460	461	462	460	459
I (amps)	30.2	31.3	30.4	32.0	31.7	30.7	30.8	31.2
PF	0.78	0.80	0.79	0.82	0.81	0.77	0.78	0.80

For a three-phase motor, the power (P) is related to the voltage (V), current (I), and power factor (PF) by the formula

$$P = V \times I \times PF\sqrt{3}$$

(a) Calculate the mean, the standard deviation, and the random uncertainty of each parameter (V, I, PF, and P) at a 95% confidence level.

(b) Calculate the mean value of the power and the random uncertainty of the mean power at a 95% confidence level.

(c) If all the measured parameters have an accuracy of 1% of the reading of the measuring device, calculate the systematic uncertainty in the power measurement at a 95% confidence level.

(d) Calculate the total uncertainty in the power measurement at a 95% confidence level.

7.28 In Example 7.8, to improve the accuracy of the temperature measurements, a temperature sensor with a better degree of repeatability is to be used. For the new probe, which has an accuracy of ±0.2°C, a standard deviation of 0.5°C was obtained in 15 measurements. Further tests performed on data transmission resulted in a standard deviation of 0.3°C in a total of 30 tests. Calculate the systematic and random uncertainties and the total uncertainty in the temperature measurement. Perform the calculations at a 95% confidence level.

7.29 In measuring the flow rate of diesel fuel into a test engine, a graduated burette is used. The burette has a resolution of ±1 cc. A technician measuring time with a stopwatch produced measurements with an uncertainty of approximately ±0.3 s. Discuss the type of these errors, and determine the uncertainty in the volumetric flow rate of fuel ($Q = \Delta V/\Delta t$, where V = volume and t = time) if the volume measured is 100 cc and the time is 25 s. What other errors may be involved in this process?

7.30 As detailed in an instrument catalog, a power quality analyzer has the following specifications:

Voltage (V):	600 V (rms) range, accuracy ±1% of reading
Current (I):	1–1000A, ±1% of reading
Power factor (PF):	±2% of reading

Using the formula $P = V \times I \times \text{PF}\sqrt{3}$,

(a) Calculate the uncertainty in power measurement as a percentage of reading due to the measurement device for measured values of 460 V, 36 amp, and a power factor of 0.81.

(b) In logging data with a high sampling rate, the calculated standard deviations of the readings are 3.00 V, 2.00 A, and 0.03 for the power factor, while the mean values are 460 V, 35.00 A, and 0.81 for the power factor. Determine the random and systematic uncertainty of the measurements.

(c) What is the total uncertainty in the power measurement at a 95% confidence level?

7.31 A pressure transducer with the following specifications (taken from an instrument catalog) is used for measuring water-pressure discharge from a pump:

Range   0–700 kPa

Accuracy

Linearity	0.5% of span
Hysteresis	0.1% of span
Repeatability	0.1% of span
Stability	0.3% of span

Temperature error (reference temperature, 20°C)

On zero	0.04% of span/°C
On span	0.03% of span/°C

(a) Calculate the total uncertainty of the transducer in measuring the fluid pressure at 15°C.

(b) The same transducer is used to measure water pressure. The standard deviation of the measurement is 3.5 kPa, with a mean value of 550 kPa. Calculate the total uncertainty in measuring pressure with this transducer, assuming 95% confidence. Is the uncertainty due more to the accuracy of the transducer or the fluctuation of the fluid pressure?

(c) The signal from the transducer is transmitted to a 12-bit DAS with a range of 1400 kPa. Does the quantization error have any significant effect on the uncertainty of the final result?

(d) If the noise induced in the transmission signal is estimated to ±2% of the value of the signal, is the signal-transmission error significant in the uncertainty of the final result?

7.31E A pressure transducer with the following specifications (taken from an instrument catalog) is used for measuring water-pressure discharge from a pump:

Range   0–100 psi

Accuracy

Linearity	0.5% of span
Hysteresis	0.1% of span
Repeatability	0.1% of span
Stability	0.3% of span

Temperature error (reference temperature, 70°F)

On zero	0.022% of span/°F
On span	0.016% of span/°F

(a) Calculate the total accuracy of the transducer in measuring the fluid pressure at 60°F.

(b) The same transducer is used to measure water pressure. The standard deviation of the measurement is 0.50 psi, with a mean value of 80.00 psi. Calculate the total uncertainty in measuring pressure with this transducer, assuming 95% confidence. Is the uncertainty due more to the accuracy of the transducer or the fluctuation of the fluid pressure?

(c) The signal from the transducer is transmitted to a 12-bit DAS with a range of 200 psi. Does the quantization error have any significant effect on the uncertainty of the final result?

(d) If the noise induced in the transmission signal is estimated to ±2% of the value of the signal, is the signal-transmission error significant in the uncertainty of the final result?

7.32 One of the methods for measuring the power (P) of rotating machinery is to measure the rotational speed (with a magnetic pickup, for example) and the rotational shaft torque (with a shaft torque meter, for example) and then calculate the power transmitted through the shaft. The formula for power is $P = \tau \times \omega$, where τ is the torque, $\omega (= 2\pi N)$ is the rotational speed in radians per second, and N is the number of revolutions per second. From the measurement of the power of a small engine, the following information is available:

As per the manufacturers' information, the accuracy of the torque meter is ±0.7 N-m and the accuracy of the rotational speed-meter is ±5 rpm. The values of the measured torque and rotational speed are 165 N-m and 3000 rpm, respectively. In repeating the speed and torque measurements, it is found that the standard deviations of these measurements are 4 N-m and 5 rpm, respectively.

(a) Calculate the power of the engine.

(b) If the number of samples used for calculating the standard deviations of the torque and speed are 10 and 20, respectively, calculate the standard deviation of the power.

(c) Calculate the random and systematic uncertainty of the power.

(d) Calculate the total uncertainty of each power measurement at a 95% confidence level.

7.33 In Problem 7.32, if the same standard deviations are obtained for torque and rotational speed through a large number of measurements,

(a) calculate the power of the engine.

(b) calculate the standard deviation of the power.

(c) calculate the random and systematic uncertainty of the power.

(d) calculate the total uncertainty of each power measurement at a 95% confidence level.

7.34 In Problem 7.32, with the same instruments, 10 independent measurements of the engine power are performed. The average and standard deviation of the power are calculated to be 51.5 kW and 0.75 kW, respectively.

(a) Calculate the systematic and random uncertainty of the power measurement at a 95% confidence level.

(b) Calculate the total uncertainty of the power measurement at a 95% confidence level.

7.35 To measure the efficiency of a pump, the parameters in the following formula are usually measured and applied: $\eta = Q\Delta P/W$. In this formula, η is the pump's efficiency, Q is the volumetric flow rate of the pump, ΔP is the pressure differential between the inlet and outlet of the pump, and W is the power input into the pump.

The following equipment is used:

Differential pressure gage	Range	0–1200 kPa
	Accuracy	±0.2% of span (includes linearity, hysteresis, and repeatability)
	Stability	±0.2% of span

Flowmeter	Range	1200 lpm (liters per minute)
	Accuracy	±1.5% of reading

The power is measured through the input of an electrical motor (with specified efficiency), with an expected accuracy of 0.07 kW.

In repeating the measurements, the average values of the pressure differential, flow rate, and power are found to be 700 kPa, 340 lpm, and 5 kW, respectively. The standard deviations of 10 kPa, 5.6 lpm, and 0.15 kW are calculated for the respective parameters. If the number of measurements made to calculate the mean values and standard deviations is 15,

(a) calculate the efficiency of the pump.

(b) calculate the standard deviation of the efficiency of the pump.

(c) calculate the random and systematic uncertainty of the efficiency of the pump.

(d) calculate the total uncertainty of the efficiency of the pump at a 95% confidence level.

7.35E To measure the efficiency of a pump, the parameters in the following formula are usually measured and applied: $\eta = Q\Delta P/W$. In this formula, η is the pump's efficiency, Q is the volumetric flow rate of the pump, ΔP is the pressure differential between the inlet and outlet of the pump, and W is the power input into the pump.

The following equipment is used:

Differential pressure gage	Range	0–170 psi
	Accuracy	±0.2% of span (includes linearity, hysteresis, and repeatability)
	Stability	±0.2% of span
Flowmeter	Range	40 ft³/min
	Accuracy	±1.5% of reading

The power is measured through the input of an electrical motor (with specified efficiency), with an expected accuracy of 0.10 horsepower.

In repeating the measurements, the average values of the pressure differential, flow rate, and power are found to be 100.0 psi, 12.0 ft³/min, and 6.50 horsepower, respectively. The standard deviations of 1.5 psi, 0.2 ft³/min, and 0.20 hp are calculated for the respective parameters. If the number of measurements made to calculate the mean values and standard deviations is 15,

(a) calculate the efficiency of the pump.

(b) calculate the standard deviation of the efficiency of the pump.

(c) calculate the random and systematic uncertainty of the efficiency of the pump.

(d) calculate the total uncertainty of the efficiency of the pump at a 95% confidence level.

7.36 In Problem 7.35, if the same standard deviations and mean values are obtained through a large number of measurements,

(a) calculate the efficiency of the pump.

(b) calculate the standard deviation of the measurements of pump efficiency.

(c) calculate the random and systematic uncertainty of the measurements of pump efficiency at a 95% confidence level.

(d) calculate the total uncertainty of the efficiency of the pump at a 95% confidence level.

7.37 For the pump experiment of Problems 7.35 and 7.36, with the same instruments, a large number of independent measurements of the efficiency of the pump is performed. The mean and standard deviation of the efficiency are determined to be 0.82 and 0.01, respectively.

(a) Calculate the systematic and random uncertainties of the measurements.

(b) Calculate the total uncertainties of the measurements at a 95% confidence level.

7.38 In Example 7.10, 10 engines are tested, and the average value of 30.8% with a standard deviation of 0.25 is obtained for the efficiency of the engine. Calculate the uncertainty of the average efficiency of the engines that are produced.

7.39 The drag coefficient measured in model testing is often used to estimate the drag of actual systems (e.g., airplanes, cars, and ships). The drag force F is related to the drag coefficient C_D, density ρ, velocity V, and frontal area A via the formula

$$C_D = \frac{F}{0.5\rho V^2 A}$$

In a test of a bus model, the following information was obtained:

$A = 3000 \pm 50 \text{ cm}^2$

$F = 1.70 \text{ kN}$, average of 20 tests with a standard deviation of 0.05 kN and a calibration uncertainty of ± 0.05 kN

$V = 30.0 \text{ m/s}$, average of 20 tests with a standard deviation of 0.5 m/s and a calibration uncertainty of ± 0.2 m/s

$\rho = 1.18 \text{ kg/m}^3$, with a spatial variation of $\pm 0.01 \text{ kg/m}^3$

Another factor that has been determined to affect the measurement is the data-transmission error for force, which has been measured to have a standard deviation of 0.01 kN (10 samples). Determine the drag coefficient and the total uncertainty in its measurement at a 95% confidence level. Explain your assumptions.

7.40 In Example 7.12, a new natural-gas supplier is chosen. Fifteen tests of the heating value of the natural gas produced an average value of 47,500 kJ/kg, with a standard deviation of 300 kJ/kg. Twenty new tests of the engine produced an average power of 48.5 kW, with a standard deviation of 0.2 kW. All other parameters of the test remained the same. Calculate the total uncertainty in the measurement of the efficiency of the engine.

7.41 Use the DAS of Example 7.13 for a pressure transducer with the following specifications:

Range	0 to 2000 kPa
Output	0 to 5 V
Linearity and hysteresis (combined)	$\pm 0.25\%$ FS
Repeatability	$\pm 0.03\%$ FS
Thermal span uncertainty	$\pm 0.003\%$ FS/°C

The temperature is uncertain to ± 10°C. Select the DAS input range that gives the best accuracy. Estimate the uncertainty of a pressure measurement made with this system.

7.42 In Example 7.14, a reevaluation of the data and test equipment showed a loading error of 2.5°C (instead of 1.0°C) and steam-line temperature scatter with a standard deviation of 2.0°C. Estimate the zeroth-, first-, and Nth-order uncertainty limits.

7.43 What is the difference between a single-measurement test and a multiple-measurement test? Discuss the advantages and disadvantages of each, together with an example.

Measurement of Solid-Mechanical Quantities

In this chapter we provide the technical basis for common systems to measure the motion of solid-body systems and determine applied forces. Where applicable, appropriate signal conditioners are described. The most common devices for such measurements are included, but it should be noted that many other devices are used in engineering practice.

8.1   MEASURING STRAIN

When a force is applied to a structure, the components of the structure change slightly in their dimensions and are said to be *strained*. Devices to measure these small changes in dimensions are called *strain gages*.

8.1.1   Electrical Resistance Strain Gage

The electrical resistance strain gage is an extremely common device used to measure strain in structures and also as a sensing element in a wide variety of transducers, including those used to measure force, acceleration, and pressure. Electrical-resistance strain gages and associated signal conditioners are simple, inexpensive, and quite reliable. To understand the function of a strain gage, we examine the measurement of strain in a simple structure. Figure 8.1 shows a situation in which a supported beam is bent by applying a lateral force. With this type of loading, the beam will become longer on the bottom surface and shorter on the top surface. Attached to the beam using two standoffs is a wire of length l, which functions as a simple strain gage. When the beam is loaded, this wire is stretched and its length becomes $l + \delta l$. The ratio $\delta l / l$ is known as the strain and is usually given the symbol ϵ. For this simple loading case, the strain in the wire is approximately the same as the strain in the lower surface of the beam. The stretching of the wire will cause its electrical resistance to change so that the wire is a detector of strain.

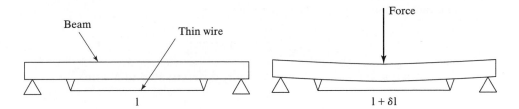

FIGURE 8.1

A simple strain gage.

Strain has units of inches per inch or millimeters per millimeter and hence is dimensionless. In most structures the values of strain are usually very small; for example, a low-strength steel will yield (take a permanent deformation) at a strain of only about 0.0014. As a result it is common to talk about strain in units of *microstrain* (μstrain). Microstrain is the actual strain multiplied by 10^6. Thus, a strain of 1400 μstrain is an actual strain of 0.0014.

In the engineering design process, it is often necessary to determine the stresses in a structure experimentally to demonstrate that the structure is sound. It is difficult to measure the stress directly, but a strain gage can be used to measure the strain, and then the material properties can be used to determine the stress. For elastic materials stressed in a single direction (*uniaxial stress*), the strain is related to the stress by *Hooke's law*,

$$\sigma = E\epsilon \tag{8.1}$$

where σ is the normal stress and E is a property of the material called the *modulus of elasticity* (also called *Young's modulus*). The equation is somewhat more complicated if the material is stressed in more than one direction, and this subject is addressed later.

For a wire to function as a strain gage, we must determine the relationship between the strain and the change in resistance. The resistance of a wire such as that shown in Figure 8.1 is given by

$$R = \frac{\rho L}{A} \tag{8.2}$$

where R is the resistance, ρ is a property of the wire material called the *resistivity*, L is the wire length, and A is the cross-sectional area of the wire. This equation can be logarithmically differentiated to obtain

$$\frac{dR}{R} = \frac{d\rho}{\rho} + \frac{dL}{L} - \frac{dA}{A} \tag{8.3}$$

This result can readily be obtained by taking logarithms of both sides of Eq. (8.2), separating the terms, and then differentiating each term.

Equation (8.3) relates a small change in resistance to changes in resistivity, length, and cross-sectional area. The term dL/L is the *axial strain*, ϵ_a. The term dA/A

can be evaluated further. The area is given by

$$A = \pi \frac{D^2}{4}$$ (8.4)

This equation can also be logarithmically differentiated to obtain

$$\frac{dA}{A} = 2\frac{dD}{D}$$ (8.5)

The change in diameter divided by the diameter, dD/D, is known as the *transverse strain* ϵ_t. Solid mechanics provides the following relationship between the axial and the transverse strain:

$$\epsilon_t = -v\epsilon_a$$ (8.6)

Here, v is a property of the material known as *Poisson's ratio*; the minus sign indicates that as the wire becomes longer, the transverse dimension decreases.

Equations (8.3), (8.5), and (8.6) can be combined to obtain

$$\frac{dR}{R} = \frac{d\rho}{\rho} + \epsilon_a(1 + 2v)$$ (8.7)

Equation (8.7) is a relationship between the change in resistance of the wire, strain, and the change in resistivity of the wire. At this point, it is useful to define the *strain gage factor, S*:

$$S = \frac{dR/R}{\epsilon_a}$$ (8.8)

Combining Eqs. (8.7) and (8.8), we obtain

$$S = 1 + 2v + \frac{d\rho/\rho}{\epsilon_a}$$ (8.9)

If the temperature is held constant, it turns out that the change in resistivity is approximately proportional to the strain. This being true, we can see by examining Eq. (8.9) that the gage factor is approximately a constant. Unfortunately, the gage factor is sensitive to temperature.

Although some strain gages are constructed in the form of straight wires, it is more common to etch them from thin foil metal sheets that are bonded to a plastic backing, as shown in Figure 8.2. This backing can, in turn, be glued to the structure for

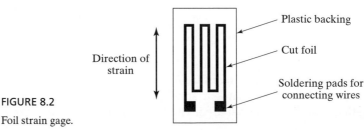

FIGURE 8.2

Foil strain gage.

which it is desired to measure the strain. The dimensions of strain gages vary, but can be as small as 0.2 mm. Gage factors, which depend on the metal, are usually about 2.0 but can be as high as 6.0. Strains as high as 40,000 $\mu\epsilon$ can be measured routinely, and strains as high as 200,000 $\mu\epsilon$ (20% elongation) can be measured with special gages (Norton, 1982).

While metallic foil strain gages are widely used, strain gages can also be constructed from semiconductor materials. Semiconductor strain gages, which are commonly used as sensing elements in pressure and acceleration transducers, have the advantage of a higher gage factor than foil gages, 125 being common. However, they are not as ductile and normally can measure strains only as high as about 3000 $\mu\epsilon$ (Norton, 1982).

In many situations, the surface of a structure is stressed simultaneously in more than one direction, leading to a situation called *biaxial stress*. If a structure is loaded in a single direction, there exists a transverse strain [as predicted by Eq. (8.6)] but no transverse stress. This effect is included when manufacturers determine gage factors. In biaxial stress, however, there is a transverse strain that results from the transverse stress. This transverse strain will affect the strain gage output and can be described with a transverse gage factor, S_t. Similar to Eq. (8.8), which defines the axial gage factor, S_t is defined by

$$S_t = \frac{dR/R}{\epsilon_t} \tag{8.10}$$

Manufacturers measure a factor, K_t, called the *transverse sensitivity*, which is supplied to the user. This is defined as

$$K_t = \frac{S_t}{S_a} \tag{8.11}$$

Values of K_t are normally quite small, less than 0.01 being possible. For a single gage, Budynas (1977) provides the following formula for the error in axial strain due to an applied transverse strain:

$$\frac{\hat{\epsilon}_a - \epsilon_a}{\epsilon_a} = \frac{K_t}{1 - vK_t}\left(v + \frac{\epsilon_t}{\epsilon_a}\right) \tag{8.12}$$

In this equation, ϵ_a is the true axial strain and $\hat{\epsilon}_a$ is the strain that measurement would predict if the transverse strain were neglected. For $K_t = 0.01$, $v = 0.3$, and $\epsilon_t/\epsilon_a = 2$, the error in axial strain is 2.3%. Transverse sensitivity effects are commonly neglected in strain experiments.

Although slightly sensitive to transverse strain, for practical purposes, a single strain gage can measure the strain only in a single direction. To define the state of strain on a surface, it is necessary to specify two orthogonal linear strains ϵ_x and ϵ_y and a third strain called the *shear strain*, γ_{xy}, the change in angle between two originally orthogonal lines when the solid is strained. These strains can be determined by three suitably placed strain gages in an arrangement called a *strain rosette*. Figure 8.3 shows the two most common arrangements of these three gages: the rectangular rosette and equiangular rosette. In the rectangular rosette, the gages are placed at angles of 0°, 45°,

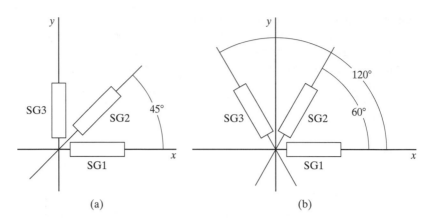

FIGURE 8.3

Orientation of strain gages in common rosettes: (a) rectangular; (b) equiangular.

and 90°. In the equiangular rosette, they are arranged at 0°, 60°, and 120°. Each of these gages measures the linear strain in the direction of the axis of the gage.

According to Popov (1976), if the plane strain field at a point on a solid can be described by values of ϵ_x, ϵ_y, and γ_{xy}, the linear strain in a direction θ to the x-axis can be represented by

$$\epsilon_\theta = \epsilon_x \cos^2\theta + \epsilon_y \sin^2\theta + \gamma_{xy} \sin\theta \cos\theta \tag{8.13}$$

This equation can be applied to each of the strain gages in a rosette, resulting in three simultaneous equations:

$$\begin{aligned}
\epsilon_{\theta_1} &= \epsilon_x \cos^2\theta_1 + \epsilon_y \sin^2\theta_1 + \gamma_{xy} \sin\theta_1 \cos\theta_1 \\
\epsilon_{\theta_2} &= \epsilon_x \cos^2\theta_2 + \epsilon_y \sin^2\theta_2 + \gamma_{xy} \sin\theta_2 \cos\theta_2 \\
\epsilon_{\theta_3} &= \epsilon_x \cos^2\theta_3 + \epsilon_y \sin^2\theta_3 + \gamma_{xy} \sin\theta_3 \cos\theta_3
\end{aligned} \tag{8.14}$$

The rosette provides measurements of ϵ_{θ_1}, ϵ_{θ_2}, and ϵ_{θ_3}, and we seek values of ϵ_x, ϵ_y, and γ_{xy}. For the rectangular rosette, the solution is

$$\begin{aligned}
\epsilon_x &= \epsilon_{0°} \\
\epsilon_y &= \epsilon_{90°} \\
\gamma_{xy} &= 2\epsilon_{45°} - (\epsilon_{0°} + \epsilon_{90°})
\end{aligned} \tag{8.15}$$

For the equiangular rosette, the solution is

$$\begin{aligned}
\epsilon_x &= \epsilon_{0°} \\
\epsilon_y &= \frac{2\epsilon_{60°} + 2\epsilon_{120°} - \epsilon_{0°}}{3} \\
\gamma_{xy} &= \frac{2}{\sqrt{3}}(\epsilon_{60°} + \epsilon_{120°})
\end{aligned} \tag{8.16}$$

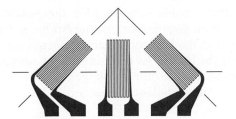

FIGURE 8.4

Strain-gage rosette.

Methods to evaluate the maximum normal and shear stresses from these strain values are included in Hetenyi (1960) and most mechanics of materials textbooks, such as Popov (1976).

Experimenters do not normally construct strain rosettes using three individual strain gages. Instead, manufacturers supply rosettes with three gages already bonded to single plastic backing, as shown in Figure 8.4.

8.1.2 Strain Gage Signal Conditioning

As mentioned, a strain of 1400 μstrain is actually quite large, and for a gage factor of 2.1, the change in resistance, $\delta R/R$, is only about 0.003 (0.3%). In addition, it is often necessary to measure accurately much smaller values of strain. Consequently, simply measuring the gage resistance before and after applying the load will not work very well since the change in resistance may well be on the order of the resolution of the measuring device. What is needed is a device that will measure the change in resistance rather than the resistance itself.

A simple circuit that performs this function, called the *Wheatstone bridge*, is shown in Figure 8.5. If all four resistors in Figure 8.5 have the same value (this restriction will be relaxed shortly), by symmetry, the voltage at points B and D must be the same, so V_0 must be zero. The bridge is then said to be *balanced*. This is the starting condition, before R_3, the strain gage is strained. As R_3 changes from its initial value, the bridge becomes unbalanced and V_0 will change. If another resistor in the bridge is adjusted, the bridge can be brought back into balance. For example, if R_2 is adjusted such that $V_0 = 0$, the value of R_2 will be the same as the strained resistance of the strain gage. In older strain-gage measuring devices, the bridge was balanced by varying a bridge resistor manually. The value of the variable resistor could be determined by the

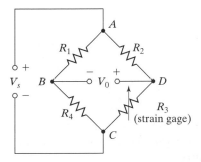

FIGURE 8.5

Constant-voltage Wheatstone bridge.

mechanical position of the variable resistor. (See the potentiometer described in Section 8.2.) In modern strain gage indicators, the bridge is balanced automatically using electronics. For use with modern computer data-acquisition systems, it is usually not convenient to balance the bridge. As demonstrated momentarily, the bridge is allowed to become unbalanced and the output voltage can be used to determine the gage resistance. If the bridge is initially balanced, the output voltage, V_0 is proportional to the *change* in gage resistance.

Using Ohm's law, the current flows through branches ABC and ADC in Figure 8.5 are

$$I_{ABC} = \frac{V_s}{R_1 + R_4} \quad I_{ADC} = \frac{V_s}{R_2 + R_3} \tag{8.17}$$

Using Ohm's law again, the voltage drop across R_4, $V_B - V_C$, is $I_{ABC}R_4$, and the voltage drop across R_3, $V_D - V_C$, is $I_{ADC}R_3$. Then $V_o = V_D - V_B$, or

$$V_o = \frac{R_3 V_s}{R_2 + R_3} - \frac{R_4 V_s}{R_1 + R_4} \tag{8.18}$$

Creating a common denominator, this becomes

$$V_o = V_s \frac{R_3 R_1 + R_3 R_4 - R_4 R_2 - R_3 R_4}{(R_2 + R_3)(R_1 + R_4)} = V_s \frac{R_3 R_1 - R_4 R_2}{(R_2 + R_3)(R_1 + R_4)} \tag{8.19}$$

Before a strain is applied, the initial resistance of the strain gage is R_{3i}. We are free to select values for all other bridge resistors, and we choose these so that the bridge is initially balanced. For the bridge to be initially in balance (i.e., $V_o = 0$), Eq. (8.19) requires that

$$R_{3i} R_1 - R_4 R_2 = 0 \tag{8.20}$$

Now, for the situation where the strain gage is strained, we set $R_3 = R_{3i} + \Delta R_3$ in Eq. (8.19), where ΔR_3 is the resistance change due to strain. Using Eq. (8.20), Eq. (8.19) then becomes

$$V_o = \frac{V_s R_1 \Delta R_3}{(R_2 + R_{3i} + \Delta R_3)(R_1 + R_4)} \tag{8.21}$$

In the denominator, ΔR_3 is usually small compared with R_{3i} and can be neglected. Neglecting the ΔR_3 in the denominator of Eq. (8.21) makes V_o a linear function of ΔR_3. Combining Eq. (8.21) with Eqs. (8.8) and (8.20) (and noting that we make the approximation that $\Delta R = dR$), we obtain the following equation relating the strain to the output voltage:

$$\epsilon_a = \frac{V_o(R_2 + R_{3i})^2}{V_s S R_2 R_{3i}} \tag{8.22}$$

This analysis is what is known as a *quarter-bridge circuit*. This means that there is a single strain gage and three fixed resistors. This arrangement is common when many strain gages are applied to a structure.

For Eq. (8.22) to be valid, it has been assumed that the bridge was initially in balance before a strain was applied to the strain gage. Unfortunately, it is not always practical to obtain resistors and strain gages that have exactly the correct values: they each have uncertainties on the order of 0.1%. This means that V_o may have a nonzero value, V_{oi}, even before the strain is applied. Although in some systems, it is possible to adjust a resistor to bring the bridge into initial balance, not all systems include such an adjustment mechanism. If V_{oi} cannot be adjusted to zero, its value should be subtracted from the actual readings before the voltage is substituted into Eq. (8.22). If V_{oi} is very large, actions should be taken to alter a bridge resistor since assumptions underlying Eq. (8.22) will become invalid, resulting in measurement error.

Sometimes it is possible to use strain gages for the other bridge resistors. The use of more strain gages increases the sensitivity of the circuit—a larger change in voltage output for a given strain. If we assume that each of the bridge resistors shows a variation ΔR, neglecting higher-order terms, Eq. (8.19) can be expanded to obtain

$$V_o = \frac{V_s R_2 R_3}{(R_2 + R_3)^2} \left(\frac{\Delta R_3}{R_3} + \frac{\Delta R_1}{R_1} - \frac{\Delta R_2}{R_2} - \frac{\Delta R_4}{R_4} \right) \qquad (8.23)$$

where R_1, R_2, R_3, and R_4 are unstrained gage resistances. [See Dally (1993).] For example, if R_1 is another strain gage, identical to R_3, which is placed in a location such that it has the same strain as R_3, the output of the bridge will be double that obtained from Eq. (8.21) for a single gage.

Many loading situations have a symmetry such that there are regions with the same strain but opposite sign—symmetrical regions of tension and compression. This is often the case when strain gages are used as sensors in transducers. In such a situation, we can make R_1 and R_3 be compressive gages and R_2 and R_4 be tensile gages. The resulting bridge is now a *full bridge* and has an output that is four times the output of a quarter bridge for the same strain.

Including more than one gage in the bridge provides another benefit: *temperature compensation*. The resistivity of most metals is a fairly strong function of the temperature, and the resistivity affects the gage factor S. Furthermore, there is a differential thermal expansion between the strain gage and the structure to which it is attached. Consequently, a change in temperature of an installed strain gage produces an apparent strain in the absence of any real structural load. Figure 8.6 shows the temperature-induced change in gage factor and the apparent strain for a typical foil strain gage. The total change in resistance is caused by the two effects, the actual strain and the apparent strain caused by temperature effects

$$\Delta R_{\text{total}} = \Delta R_{\text{strain}} + \Delta R_{\text{temperature}} \qquad (8.24)$$

In an experiment, then, a change in temperature might be confused with a change in strain. For example, a strain gage attached to an airplane structure will show major temperature changes when the airplane flies at high altitudes. This problem can be solved if we place two strain gages on the structure of interest and connect them into the bridge as R_2 and R_3 (Figure 8.5). Let the structure remain unstrained but allow the temperature to change. The resistance of R_2 and R_3 will change the same amount in the same direction. As a result, the voltage at point D will not change—the temperature effects cancel out. The same conclusion will be reached using Eq. (8.23). In the simplest

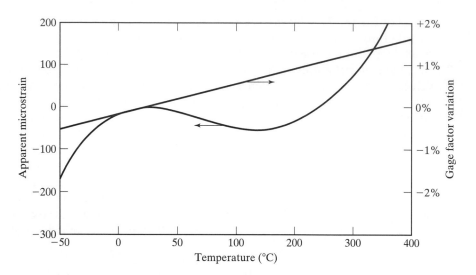

FIGURE 8.6

Typical temperature effects on strain gage.

application of this concept, gage R_3 is the active gage and gage R_2 is attached to the same structure but in a region at the same temperature that will not be subject to strain. In this way the strain can be measured, but any temperature effects will cancel out. The gage at R_2 is known as a *dummy* or *compensating gage*. Gage R_2 can also be an active gage if it can be placed in a location with strain of the same magnitude but opposite sign to that of gage R_3, the situation in many transducers.

The strain gage shown in the Wheatstone bridge schematic (Figure 8.5) is located on the test apparatus and connected to the bridge with lead wires. Lead wires can be a problem since they may have significant resistance compared to the resistance of the gage. Lead-wire resistance may be on the order of 1 to 4 Ω compared to a typical gage resistance of 120 Ω. This lead-wire resistance can initially unbalance the bridge and cause some error if not accounted for. If a compensating gage is used, lead wires of identical length and diameter can be used for the active and compensating gages. The lead wire effect will be minimized. If there is no compensating gage, additional resistance can be included in another leg of the bridge to produce initial balance.

It is common to instrument a structure with a large number of strain gages and to monitor these gages with a multichannel data-acquisition system. At first, it might seem desirable to use a single power supply with three fixed resistors and simply switch the various strain gages into the bridge, as shown in Figure 8.7. Although this arrangement appears to be economical, it has some serious shortcomings. First, there is current flowing through the switches, and the switch contact resistances will cause voltage drops that will be interpreted erroneously as strain. Second, there is some energy dissipated in a strain gage (on the order of 0.02 W), and this heats the strain gage slightly. The temperature will not come to a steady value if the gage is being switched in and out of the circuit. Finally, this arrangement is not compatible with the switching system provided by the multiplexer on microcomputer data-acquisition systems. (See Figure 4.4.)

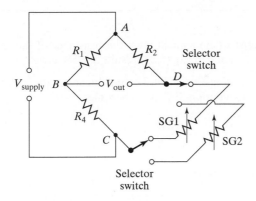

FIGURE 8.7

Multiple strain gages: switching within bridge.

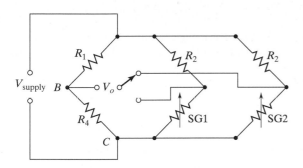

FIGURE 8.8

Multiple strain gages: switching outside of bridge.

One satisfactory approach is to have a separate bridge for each test gage. The output terminals of each bridge are connected to the input channels of the DAS. Only a single power supply is required, so there is some economy for multiple channels. Another switching approach is shown in Figure 8.8. In this arrangement, all the strain gages share a common power supply and two of the fixed resistors. Each active gage is paired with a single fixed resistor (which could also be a compensating or active gage), and this pair is connected permanently to the common fixed resistors and the power supply. In this arrangement the bridge is always complete, and switching is required only to measure the output voltages. The measurement of the output voltage draws negligible current, so the contact resistances have insignificant effects on the measured voltages.

Example 8.1

A single strain gage has a nominal resistance of 120 Ω and a gage factor of 2.06. For a quarter bridge with 120-Ω fixed resistors, what will be the voltage output with a strain of 1000 μstrain for a supply voltage of 3 V?

Solution: Substituting into Eq. (8.22), we obtain

$$1000 \times 10^{-6} = \frac{V_{out}(120 + 120)^2}{3 \times 2.06 \times 120 \times 120}$$

Solving yields

$$V_{out} = 1.544 \text{ mV}$$

Example 8.2

For a quarter bridge using 120-Ω resistors, how much must the resistance of the single strain gage change before the linearization assumed in Eq. (8.22) results in a 5% error?

Solution: We can evaluate this effect using Eq. (8.21). The nonlinearity results from the ΔR_3 term in the denominator. For a linear assumption, Eq. (8.21) becomes

$$V_{\text{out,lin}} = \frac{V_s R_1 \Delta R_3}{(R_2 + R_3)(R_1 + R_4)}$$

Dividing this equation into Eq. (8.21), we obtain

$$\frac{V_{\text{out,nonlin}}}{V_{\text{out,lin}}} = \frac{R_2 + R_{3i}}{R_2 + R_{3i} + \Delta R_3} = \frac{120 + 120}{120 + 120 + \Delta R_3} = 0.95$$

Solving for ΔR_3, the result is 12.6 Ω, or 10% of R_3. For a gage factor of 2, this corresponds to a strain of approximately 50,000 μstrain (5% elongation). This strain would be beyond the yield point of most metals.

8.2 MEASURING DISPLACEMENT

The most common devices for measuring linear and angular displacement are discussed here. Fiber-optic displacement sensors are discussed in Section 9.4.3.

8.2.1 Potentiometer

The linear potentiometer is a device in which the resistance varies as a function of the position of a slider, as shown in Figure 8.9. With the supply voltage as shown, the output voltage will vary between zero and the supply voltage. The potentiometer is a variable voltage divider. Potentiometers are actually very common in everyday experience. Angular potentiometers, as shown in Figure 8.10, are used in common devices such as radios and televisions as volume and tone controls. For the linear potentiometer, the output is a simple linear function of the slider position:

$$V_o = \frac{x}{L} V_s \qquad (8.25)$$

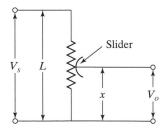

FIGURE 8.9

Linear potentiometer.

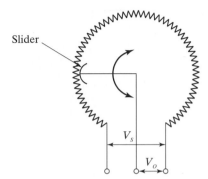

FIGURE 8.10

Angular potentiometer.

It should be noted that the device measuring V_o must have a high impedance to maintain a linear response and avoid a loading error. (See Section 3.2.3.) Linear potentiometers can be used to measure displacements as small as 0.1 to 0.2 in. up to displacements of more than 1 ft. Angular potentiometers can be designed to measure angular displacements of 3500° (multiple rotations). Potentiometers are quite inexpensive, are readily available, and require no special signal conditioning. Both linear and rotary potentiometers can be constructed to have a nonlinear relationship between displacement and output voltage for specialized applications.

The resistive element of a potentiometer can be constructed from materials such as carbon-impregnated plastics, but in instruments it is frequently constructed by tightly winding a fine wire around an insulating form. As a result, the resistance changes in discrete steps as the slider moves from wire to wire, with consequent limits on the resolution of the device. The wire-wound devices with the best resolution use on the order of 2000 wires per inch, to give a resolution of 0.0005 in. Potentiometers do have significant limitations. Because of the sliding contact, they are subject to wear and may have lifetimes of only a few million cycles. Furthermore, the output tends to be somewhat electrically noisy since the slider–resistor contact point has some resistance, and this can affect the output in a somewhat random manner. This effect often becomes worse with the age of the device due to contamination of the contact surface.

8.2.2 Linear and Rotary Variable Differential Transformers

Variable differential transformers are devices used to measure displacement by modifying the spatial distribution of an alternating magnetic field. They are used to measure displacements directly and as the sensing element in a number of transducers, which create a displacement in response to the measurand. A linear variable differential transformer (LVDT), shown schematically in Figure 8.11, has a single primary coil and two secondary coils. An oscillating excitation voltage, generally between 50 Hz and 25 kHz, is applied to the primary coil. The current flowing through the primary coil creates an alternating magnetic field, which induces alternating voltages in coils B and C. The core, made of a ferromagnetic material, tends to concentrate the magnetic field in its vicinity, and as a result, if it is closer to one of the secondary coils, the voltage in that coil will be higher. In Figure 8.11, the voltage generated in coil C will be higher than that generated

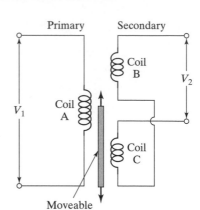

FIGURE 8.11

Schematic LVDT.

in coil B since the core is displaced downward. The two coils are connected electrically as shown in Figure 8.11, and consequently, the two voltages from coils B and C oppose each other. If the core is centered between the two secondary coils and the voltages in coils B and C are the same, the net output voltage V_2 will be zero (or at a minimum). If the core is not centered, there will be a net output voltage, as shown in Figure 8.12. Two points should be noted. First, the output, being an ac rms voltage, is always positive and gives no obvious information about the direction in which the core has been displaced from the centered position. This is a problem that can be solved with appropriate signal conditioners, as discussed shortly. Second, the voltage versus displacement is linear up to a certain point, but beyond that point, the voltage response becomes quite nonlinear. Practical LVDTs operate in the linear region.

   LVDTs are constructed as shown in Figure 8.13. Both primary and secondary coils are wrapped around a hollow, nonmagnetic bobbin, and the core slides inside the bobbin. LVDTs are often supplied with compatible signal conditioners and power supplies. They are reliable and durable displacement-measuring devices. There is no surface contact, so there is negligible wear and they are not affected by moderate levels of surface contamination. They are available in ranges from a few thousandths of an inch

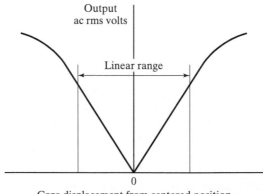

FIGURE 8.12

LVDT sensor output.

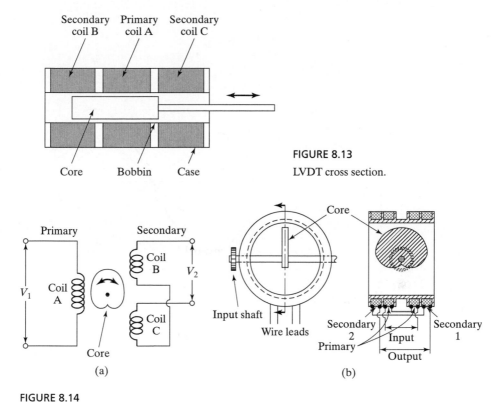

FIGURE 8.13

LVDT cross section.

FIGURE 8.14

Rotary variable differential transformer: (a) schematic; (b) simplified construction. [(b) Based on Herceg, 1976.]

up to several inches. In principle, they can be made to any size, but since the physical length is about three times the range, they are not the method of choice for measuring larger displacements. The resolution can be very good—on the order of microinches in the best cases. Variable differential transformers are also constructed to measure angular displacements, as shown in Figure 8.14. The rotating core is shaped such that the output voltage is linear with angular displacement over a range of up to $\pm 40°$.

LVDTs have some limitations for dynamic measurements. They are not normally suitable for frequencies greater than 1/10 the excitation frequency, and the mass of the core may introduce a significant mechanical loading error in dynamic devices.

Signal Conditioning   The output signal from the LVDT sensor actually does give information about the direction in which the core is displaced. Figure 8.15 shows what the actual LVDT output looks like as a function of time. Figure 8.15(a) shows the primary voltage, the voltages in coils B and C, and the output voltage (the sum of coil B and coil C voltage) for a situation in which the core is closer to coil B (above the centered position in Figure 8.11). Remember that the manner in which the two secondary coils are connected, as shown in Figure 8.11, effectively inverts the voltage of coil C relative to coil B. Figure 8.15(b) is similar to 8.15(a) except that the core is below the centered position, closer to coil C. In Figure 8.15(a), the output voltage has the same phase as the

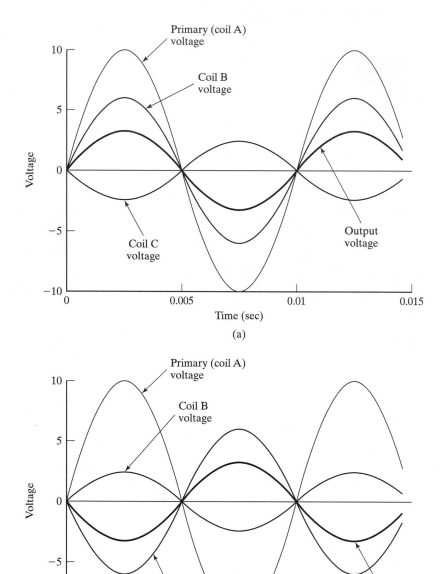

FIGURE 8.15

Output voltage versus time for LVDT: (a) core displaced upward from center; (b) core displaced down from center.

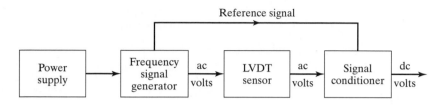

FIGURE 8.16

LVDT system block diagram.

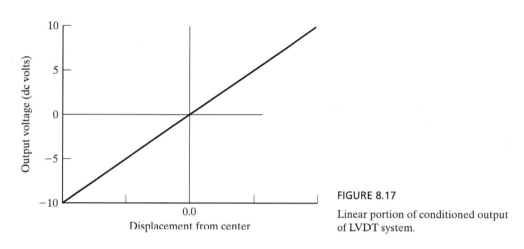

FIGURE 8.17

Linear portion of conditioned output of LVDT system.

primary voltage, but in Figure 8.15(b), the output voltage is 180° out of phase with the primary voltage. By sensing the phase of the output voltage relative to the primary voltage, the signal conditioner can determine the direction of the core displacement. The signal conditioner outputs a dc voltage for which the magnitude is proportional to the displacement and the sign indicates the direction of the displacement. A block diagram for an LVDT measuring system (sensor plus signal conditioner) is shown in Figure 8.16. The linear portion of the signal conditioner output is shown in Figure 8.17.

8.2.3   Capacitive Displacement Sensor

Capacitive displacement sensors are most appropriate for measuring very small displacements (0.00001 to 0.01 in.), although larger ranges are sometimes appropriate. These devices are based on the principle that the capacitance of a capacitor is a function of the distance between the plates and the area of the plates; that is,

$$C = K \in_0 \frac{A}{d} \tag{8.26}$$

where K is the dielectric coefficient of the substance between the plates ($K = 1$ for air), $\in_0$ is the permittivity of a vacuum with a value of 8.85×10^{-12} C^2/N-m^2, A is the overlapping plate area, and d is the distance between the plates. If A has units of m^2 and d is in meters, C will have units of farads.

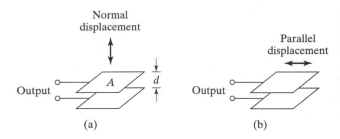

FIGURE 8.18

Capacitive displacement transducer.

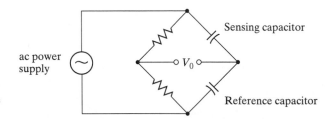

FIGURE 8.19

Wheatstone bridge for capacitance
measurement.

As shown in Figure 8.18, there are two ways of using a capacitive transducer for displacement measurements. In Figure 8.18(a), one plate is moved so that the distance, d, between the plates varies. Alternatively [Figure 8.18(b)], one of the plates can be moved parallel to the other plate so that the overlapping area varies. In the first case, the capacitance is a nonlinear function of the displacement, whereas in the second, the capacitance is approximately a linear function of the displacement. Since the output of the capacitive sensor is not a voltage, signal conditioning is required. An ac-powered Wheatstone bridge as shown in Figure 8.19 can be used for this purpose. If the reference capacitor has the same value as the initial sensor capacitor and the two resistors are the same, the voltage output will be a function of the change in capacitance. Commercial signal conditioners are readily available for capacitance measurement.

Example 8.3

A capacitor for displacement measurement consists of two disks 25 mm in diameter. What will the capacitance be if the distance between them is

    (a) 0.025 mm,

    (b) 0.05 mm, and

    (c) 0.10 mm?

Solution:

    (a)   Using Eq. (8.26), for 0.025-mm separation we obtain

$$C = 8.85 \times 10^{-12} \frac{\pi(0.025)^2/4}{0.000025}$$

    Solving, we get $C = 1.738 \times 10^{-10}$ F or 1.738×10^{-4} μF, a small capacitance.

    (b, c) The values for spacings of 0.05 and 0.10 mm are 0.869×10^{-4} and 0.4345×10^{-4} μF, respectively.

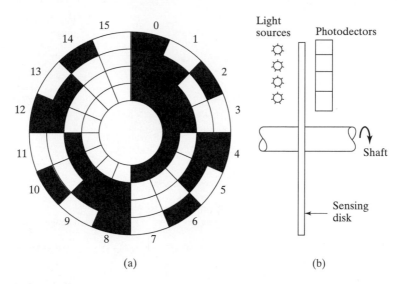

FIGURE 8.20

Angular digital encoder: (a) binary shaft encoder; (b) side view.(Based on Turner, 1988.)

8.2.4 Digital Encoders

Digital encoders are devices that convert a displacement directly into a digital signal. Figure 8.20 is a device for measuring angular position. In this example the disk has 16 sectors and each sector is divided into four bands. Each band in each sector is either opaque or transparent (a window). A radial array of four light sources shines on one side of the disk, and an array of four photodetectors is located on the other side. At each of the 16 sectors, the combination of windows is unique, so the photodetectors respond with a 4-bit digital code indicating the angular position. Similar devices can also be used to measure linear position. Since the output of each of the photodetectors is either on or off, the signal is in digital form. In a typical device, the light sources are light-emitting diodes (LEDs) and the detectors are photoresistive elements. Photoresistive elements are semiconductor devices that show a sharp change in resistance when illuminated. Encoders have a well-defined resolution, which depends on the number of bits used to represent each position. A 4-bit device has only 16 possible positions. A larger number of bits is used in typical commercial devices. A more complete discussion of encoders is given by Woolvert (1977) and Doebelin (1990).

8.3 MEASURING LINEAR VELOCITY

8.3.1 Linear Velocity Transducer

The *linear velocity transducer* (LVT) is an inductive device suitable for measuring the velocity of components in machines. As shown in Figure 8.21, velocity is measured by driving a permanent-magnet core past a fixed coil. As the north pole of the magnet approaches the coil, the magnetic flux lines will cut across the coils and generate a voltage

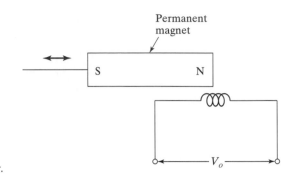

FIGURE 8.21

Linear velocity transducer.

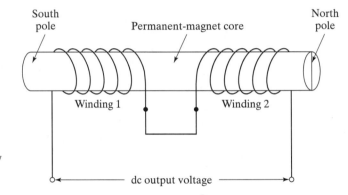

FIGURE 8.22

Two-winding linear velocity transducer (LVT). (Based on Herceg, 1976.).

that is proportional to the velocity. However, the number of flux lines cutting the coil will vary with position, and hence the output voltage is a function of both velocity and position. In fact, when the magnet is centered with respect to the coil, the flux lines due to the negative pole will also induce a voltage in the coil of opposite polarity to that induced by the north pole, and the net output will be zero. To avoid this problem, two coils are used, as shown in Figure 8.22. The south pole induces a voltage primarily in winding 1 and the north pole induces a voltage primarily in winding 2. By connecting the coils with opposing polarity, a voltage proportional to velocity is generated and this voltage is relatively independent of position over a limited range. An LVT is constructed much like an LVDT—the coils are wrapped around a hollow bobbin and the core slides inside the bobbin. For a typical LVT 7 in. long, the output will be proportional to the velocity and independent of position for displacements that are within ±1 in. of the centered position. As with LVDTs, the mass of the core may place a significant load on the system measured. If this becomes a problem, doppler radar, described below, may be more suitable for velocity measurement.

8.3.2   Doppler Radar Velocity Measurement

If a beam of radio waves is directed at a moving object, the frequency of radiation reflected from the object will be altered. A device taking advantage of this effect is shown schematically in Figure 8.23. For the situation shown, the change in frequency of

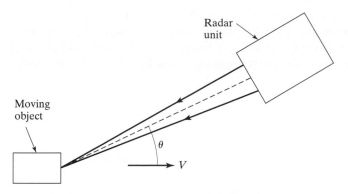

FIGURE 8.23

Doppler velocity measurement.

radiation directed at the moving object is given by

$$f_D = \frac{2V \cos \theta}{\lambda} \tag{8.27}$$

where λ is the wavelength of the incident radio waves. For most engineering applications, the radar beam places essentially no load on the measured system. Doppler radar velocity-measuring devices are readily available commercially. They are used by police to measure vehicle velocities, and they are often used to measure velocities in sports. Doppler velocity measurements can also be made using laser-generated light beams. Devices that use the Doppler effect with laser light, called laser-Doppler velocimeters, arc commonly uscd to mcasurc fluid vclocitics and arc dcscribcd in Chaptcr 10.

Example 8.4

An automobile is moving directly toward a Doppler radar unit that uses a frequency of 10,000 MHz. What will be the change in frequency if the automobile is moving at 30 m/s?

Solution: The frequency and wavelength of radio waves are related by

$$f\lambda = c$$

where c is the velocity of light and has a value of 3×10^8 m/s. Thus the radar wavelength is $3 \times 10^8/10,000 \times 10^6 = 0.03$ m. Using Eq. (8.27), the frequency shift is then

$$f_D = 2 \times 30 \times \frac{1}{0.03} = 2000 \text{ Hz}$$

8.3.3   Velocity Determination Using Displacement and Acceleration Sensors

Displacement and acceleration transducers provide data that can be used to determine velocity. If displacement data are available in the form $x(t)$, the velocity can be obtained by differentiating these data:

$$V = \frac{dx(t)}{dt} \tag{8.28}$$

This approach has some problems, however. The displacement data may have a discontinuous first derivative, particularly if potentiometric sensors are used. This will produce spikes in the velocity, which are artifacts of the differentiation process. If acceleration data are available in the form $a(t)$, then, by integration, the velocity can be estimated as

$$V(t) = V_0 + \int_{t_0}^{t} a(t)\, dt \qquad (8.29)$$

where V_0 is the velocity at an initial time t_0. If the acceleration data are of high quality, this integration process of determining velocity should be quite reliable. Integration and differentiation of signals can be performed mathematically on the test data or by using appropriate signal-conditioning circuits. (See Chapter 3.)

Example 8.5

An object is moving along steadily at a velocity of 88 m/s. A braking system is applied, and the following acceleration data are obtained from an accelerometer:

Time (s)	0	2	4	6	8	10	12
Acceleration (g)	−0.95	−0.95	−0.92	−0.93	−0.92	0	0

Estimate the velocity at 6 s.

Solution: Since the acceleration does not vary much in the interval 0 to 6 s, we can obtain an approximate answer using an average acceleration. This value is 0.9375 g or 0.9375 × 9.8 = 9.19 m/s². Using Eq. (8.29), the velocity is

$$V(6) = 88 + \int_{0}^{6} -9.19 dt = 32.9 \text{ m/s}$$

Alternatively, we could have integrated the data point by point using the trapezoidal rule. Electronic circuits using op-amps can also be used to integrate the data.

8.4   MEASURING ANGULAR VELOCITY

Many common machines have rotating shafts in which it is necessary to measure the angular velocity, commonly referred to as the *shaft speed*. There are several types of devices that perform this function, and they are generically called *tachometers*.

8.4.1   Electric Generator Tachometers

In this approach, a small dc electric generator is attached to the end of the shaft. The output of the generator will be a voltage that is a function of the shaft angular velocity. The voltage can be connected directly to a voltmeter to give an inexpensive direct readout, or the output can be connected to a data-acquisition system. Alternatively, an alternating-current generator can be used. In this case, although it is also possible to measure the voltage, it is generally preferable to measure the frequency of the output voltage using standard frequency measuring instruments.

8.4.2 Magnetic Pickup

A simple and reliable device for measuring the angular velocity is the *magnetic pickup*, sometimes called the *variable reluctance pickup*. This device is sketched in Figure 8.24. A small permanent magnet is wrapped with a coil of insulated wire and then contained in a metal case. The permanent magnet causes a magnetic field to form in the region around the magnet that passes through the coil. If a piece of iron or other magnetic material is brought near one pole of the magnet, it distorts the magnetic field (changes the reluctance), causing a voltage pulse to be generated in the coil. When the magnetic material is removed, there is another voltage pulse of opposite sign. If the pickup is placed on a rotating gear, a continuous string of pulses will be generated such as shown in Figure 8.25. The frequency of these pulses is proportional to angular velocity. With a suitable shape of the gear teeth, the output wave can be made to appear sinusoidal.

The output of the transducer can be processed in two ways. The transducer can be connected to a pulse counter that repeatedly counts the number of pulses in a fixed period of time (1 s, for example) and provides a digital result, which can be input to a

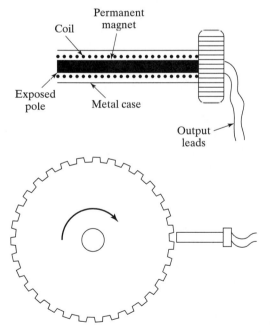

Coil

Permanent magnet

Exposed pole

Metal case

Output leads

FIGURE 8.24

Magnetic pickup for angular velocity measurement.

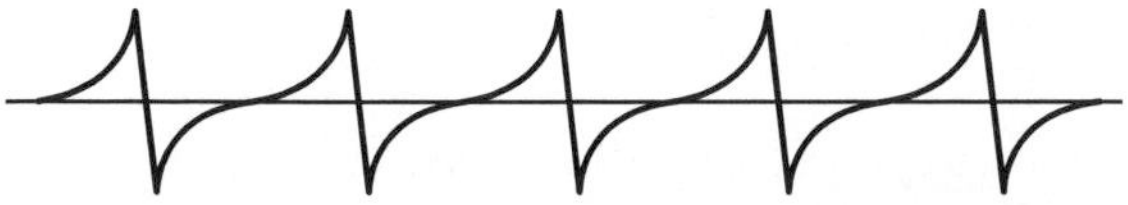

FIGURE 8.25

Pulsed output of magnetic-pickup transducer.

computer directly. Alternatively, the transducer can be connected to a frequency-to-voltage converter circuit, which results in a dc voltage that is proportional to the angular velocity. In most applications, a special gear is attached to the shaft. However, it is possible to obtain satisfactory performance with common gears or even hexagonal nuts on the shaft. Since magnetic pickups are relatively inexpensive and very reliable, they are an excellent choice for permanently installed systems. The magnetic pickup is used in the distributor of many modern automobiles with electronic ignition to measure engine shaft speed and to time the spark. Magnetic pickups are also used as wheel angular speed sensors for automotive antilock braking systems.

Example 8.6

Each time a tooth of a toothed wheel passes a magnetic pickup, a single positive pulse will be generated. If a counter indicates the number of pulses for 1 s, how many teeth must the toothed wheel have to give a readout in revolutions per minute?

Solution: Consider the toothed wheel with n teeth to be turning at N revolutions per minute or $N/60$ revolutions per second. We want the pickup to generate N counts per second. The counts per second will then be

$$\text{count} = N = (\text{pulses/rev}) \times (\text{rev/s}) = n \times N/60$$

Solving for n, we obtain 60 teeth. The counter will give a readout every second but will read directly in rpm.

8.4.3   Stroboscopic Tachometer

A common laboratory device for measuring shaft speed is the *stroboscopic tachometer*, shown schematically in Figure 8.26. This device produces flashing light at a user-controlled frequency. This light is shined on the shaft or pulley, which has a mark at one circumferential position. The flashing frequency of the light is adjusted so that the mark on the shaft appears stationary. These devices normally read out directly in revolutions per minute and are very accurate since the flashing frequency is quartz-crystal controlled

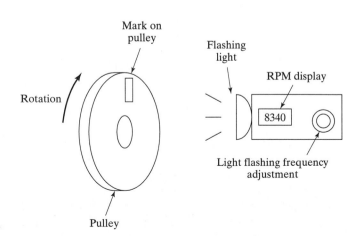

FIGURE 8.26

Stroboscopic tachometer.

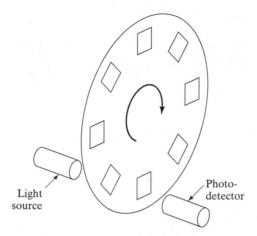

FIGURE 8.27

Photoelectric tachometer.

on modern units. One problem is that the mark will appear to stand still not only when the shaft is moving at the strobe frequency but also at integer fractions of the strobe frequencies. With proper procedure, it is possible to match the frequencies correctly. These devices depend on operator adjustment and are not well suited for remote operation.

8.4.4 Photoelectric Tachometer

There are a number of systems that use photodetectors to measure angular velocity; one such system is shown in Figure 8.27. A light beam shines through a perforated disk onto a photodetector. The detector produces a pulsed output as the disk rotates. An alternative design has reflective tape attached to the disk, with the light sources and the photodetector being combined into a single unit. In either case the pulsed signal is input into a counter that counts pulses over a fixed period of time to produce a digital value for angular velocity. One problem with photodetectors of this type is that over time, oil and other contamination from the machine may cover the various optical surfaces, leading to inaccurate output or no output at all.

8.5 MEASURING ACCELERATION AND VIBRATION

Acceleration measurement is very common in engineering experimentation—vibration and impact testing being major applications.

8.5.1 Piezoelectric Accelerometers

Piezoelectric Sensing Element Certain materials, when deformed, are capable of generating an electric charge. The best known of these is the quartz crystal. This property of generating an electric charge when deformed makes piezoelectric materials useful sensors in several common types of transducers. Figure 8.28(a) shows a piezoelectric material under compression. The faces have been coated with a conducting material such as silver. When the load is applied, electrons move to one of the conducting faces and away from the other, which results in a charge being stored by the inherent capacitance of the

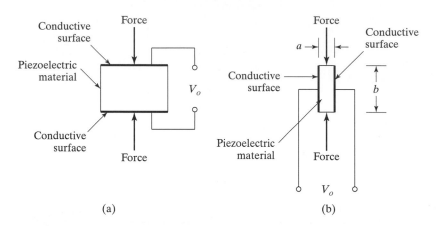

FIGURE 8.28

Piezoelectric sensors: (a) longitudinal effect; (b) transverse effect.

piezoelectric material itself. For the arrangement shown in Figure 8.28(a), called the *longitudinal effect*, the charge generated is given by Kail and Mahr (1984) as

$$Q = F \times d \tag{8.30}$$

where F is the applied force and d is the piezoelectric coefficient of the material. d depends on the piezoelectric material and its crystal orientation relative to the force, F. For a typical quartz element, d has a value of 2.3×10^{-12} C/N. Although Eq. (8.30) shows the output to be proportional to the applied force, it can also be viewed as being proportional to the displacement. As with other materials, the piezoelectric element is slightly flexible, and the imposition of a force will produce a small, proportional displacement.

The configuration shown in Figure 8.28(b) is called the *transverse effect*, and the charge generated is again given by Kail and Mahr (1984):

$$Q = F \times d \times \frac{b}{a} \tag{8.31}$$

If the ratio of the dimensions, b/a, is greater than 1 (the usual case), the transverse effect will produce a greater charge than the longitudinal effect. With either loading, the charge is proportional to the applied force. The charge will result in a voltage; however, this voltage depends not only on the capacitance of the piezoelectric element but also on the capacitances of the lead wires and the signal-conditioner input.

Since the piezoelectric element generates a charge when loaded, this charge must be sensed in a manner that does not dissipate the charge. This is normally performed with a device called a *charge amplifier*, a typical configuration being shown in Figure 8.29. R_1 is very high, so this circuit draws very low current and produces a voltage output that is proportional to the charge. It should also be noted that the capacitance of the lead wire from the transducer to the amplifier is important and affects calibration. Normally, the charge amplifier will be located close to the transducer. Charge amplifiers are normally supplied by the manufacturer of the piezoelectric transducer, and in some cases the amplifier is incorporated into the body of the transducer.

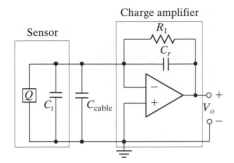

FIGURE 8.29

Charge amplifier for piezoelectric sensor.

   Although piezoelectric sensors are capable of measuring quasi-static forces, they cannot measure completely static forces since the charge will decay with time. This decay is often accelerated by the signal conditioner. Consequently, in many common applications, transducers using piezoelectric sensors have a lower limit on frequency response and are not suitable for steady or quasi-steady measurands. The primary advantage of piezoelectric sensors is their ability to respond to high-frequency measurands. As discussed in Chapter 11, many transducers behave like second-order spring–mass systems and have natural frequencies, as do other second-order systems. When making time-varying measurements, it is normally necessary to have a transducer that has a natural frequency much higher than the highest frequency in the measurand. Most piezoelectric materials are very stiff and have a high modulus of elasticity. Quartz has a modulus of elasticity of about 85 GPa, which can be compared to concrete (35 GPa) and steel (200 GPa). Since a high spring constant (relative to mass) leads to a high natural frequency, transducers using piezoelectric sensors usually have very high natural frequencies[†] compared to sensors using strain gages or LVDTs.

Accelerometers Using Piezoelectric Sensing Elements   An accelerometer using a piezoelectric material as the sensing element is shown schematically in Figure 8.30. It consists of a housing, a mass called the *seismic mass*, and a piezoelectric sensing element, which typically uses the longitudinal piezoelectric effect. An initial force between the mass and sensor is obtained with a preloading spring sleeve. As the housing of the accelerometer is subject to an acceleration, the force exerted by the mass on the quartz crystal is altered. This generates a charge on the crystal, which can be sensed with a charge amplifier. Piezoelectric accelerometers are available in many ranges up to ±1000g, (g is the acceleration of gravity, 9.8 m/s^2). Quartz crystal accelerometers can have very high values of natural frequency, 125 kHz being possible, which results in the ability to measure frequencies as high as 25 kHz. In most applications the charge amplifier causes a lower-frequency limit, on the order of 100 Hz. Piezoelectric accelerometers (when combined with the appropriate signal conditioner) are relatively expensive.

[†]It is this resonant frequency characteristic that is utilized when quartz crystals are used in clocks. An oscillating voltage is applied to the crystal, causing it to vibrate at its natural frequency. Since these crystals can be made to very accurate dimensions, the natural frequency is very well defined and hence provides excellent control of the clock speed.

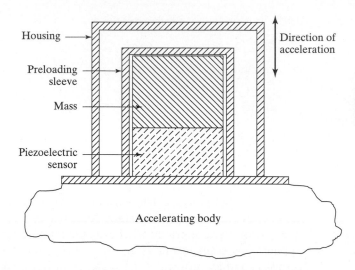

FIGURE 8.30

Schematic piezoelectric accelerometer.

8.5.2 Strain-Gage Accelerometers

An alternative type of accelerometer uses strain gages as the sensing elements. One type is shown in Figure 8.31. As the base is accelerated, the force accelerating the mass is transmitted through the cantilever beam. The bending of this beam is sensed with strain gages, which are normally of the semiconductor type, to maximize sensitivity. A Wheatstone bridge is formed with four active gages, two on the top of the beam and two on the bottom. A damping fluid may fill the housing to damp oscillations. They are available in ranges up to ±1000g. Strain gage accelerometers will respond properly to a steady acceleration. However, they normally have a low natural frequency and usually have usable upper frequency limits of a few hundred hertz. Some have upper frequency

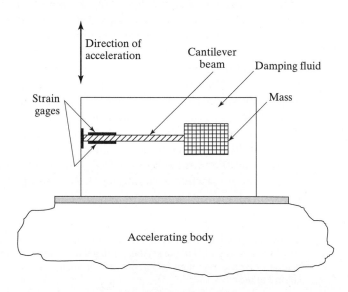

FIGURE 8.31

Strain-gage accelerometer.

limits as high as 1 to 2 kHz, but this is considerably lower than the limit for piezoelectric transducers. Strain-gage accelerometers are significantly less expensive than piezoelectric accelerometers. There are other sensors that can be used for accelerometers—the motion of the mass can be sensed with potentiometers, LVDTs, or capacitive sensors. These devices are usually used for fairly low frequencies.

8.5.3 Servo Accelerometer

Another type of accelerometer is the *servo accelerometer*, shown schematically in Figure 8.32. In this device acceleration causes the pendulous mass to move slightly. This motion is sensed, and, using a feedback network, a voltage is generated that drives the torque motor, causing the pendulous mass to move back close to its initial position. The required torque is proportional to the acceleration. The voltage used to drive the torque motor is thus a measure of the acceleration. Servo accelerometers are very accurate and are frequently used in aircraft navigation systems and satellite control systems. They can be obtained for linear acceleration ranges up to ±50g and can also be used to measure angular acceleration. They are used primarily for steady and low frequencies since they have low natural frequencies, usually less than 200 Hz.

8.5.4 Vibrometer

An instrument that is used to measure ground motion in earthquakes and sometimes to measure vibration in machines is the *vibrometer*, shown in Figure 8.33. Although the basic components are the same as in piezoelectric and strain-gage accelerometers, the mode of operation is different. The spring is actually quite soft, and as the housing

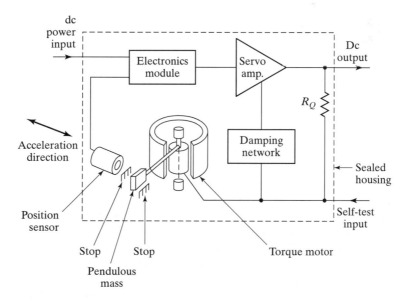

FIGURE 8.32

Servo accelerometer. (Based on Lucas Schaevitz literature.)

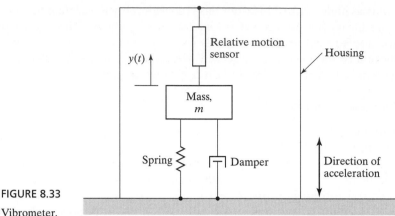

FIGURE 8.33

Vibrometer.

moves, the mass remains approximately stationary. The relative motion, y, is large and is sensed with a potentiometer or LVDT. These devices are used to measure vibrations with frequencies that are high relative to the natural frequency of the spring–mass system, which is often less than 1 Hz. Although acceleration and displacement are related mathematically, the vibrometer effectively measures the displacement of the base rather than the acceleration. Thus, these devices are most sensitive to vibrations with moderate frequencies and fairly large displacement amplitudes. High-frequency vibrations usually have relatively small values of displacement amplitude and are better measured with accelerometers.

8.6 MEASURING FORCE

8.6.1 Load Cells

Virtually any simple metal structure will deform when subjected to a force, and as long as the resulting stresses are below the material yield stress, the deflection (δ) and resulting strains (ϵ) will be linear functions of the applied force:

$$F = C_1\delta \tag{8.32}$$

$$F = C_2\epsilon \tag{8.33}$$

In these equations, the C's are constants determined from analysis or calibration. The most common force-measuring devices are *strain-gage load cells*. A wide variety of simple structures are used for this purpose; two common designs are shown in Figure 8.34. Figure 8.34(a) shows a cantilever beam instrumented with four strain gages—two on the top and two on the bottom. These four gages form the Wheatstone bridge and offer effective temperature compensation. The output of the bridge will be four times the output of an individual gage. (See Section 8.1.2.)

The hollow-cylinder load cell [Figure 8.34(b)] also uses four strain gages and is also temperature compensated. As the cylinder is compressed, it not only becomes slightly shorter but also becomes slightly larger in diameter. As a result, two of the gages

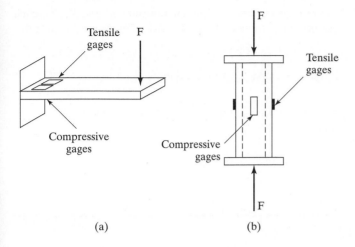

Tensile F
gages

Compressive
gages

F

Tensile
gages

Compressive
gages

F

(a)   (b)

FIGURE 8.34

Common types of load cells: (a) cantilever beam; (b) hollow cylinder.

measure the axial compression; the other two are mounted transversely and measure the tensile diametral strain. Since the transverse strain is only Poisson's ratio times the axial strain, the output is less than four times the output of a single axial strain gage. For a value of Poisson's ratio of 0.3, the output will be about 2.6 times the output of a single axial strain gage. Commercial load cells are available with ranges from ounces up to several hundred thousand pounds. Due to their simple design, any range can be readily manufactured.

Unlike accelerometers, it is not useful to specify the frequency response of commercial load cells. This is because the mass and flexibility of the instrumented system control the dynamic response. Furthermore, an installed load cell will add flexibility to the system and also affect the dynamic response. If the flexibility of strain-gage load cells is too high, load cells using piezoelectric sensors, which are much stiffer (see Section 8.5), are commercially available.

Example 8.7

A hollow-cylinder load cell of the type shown in Figure 8.34(b) has an outside diameter of 1 in. and a wall thickness of 0.040 in. It is made of aluminum with a modulus of elasticity of 10×10^6 lbf/in^2 and Poisson's ratio of 0.3. What axial force is necessary to produce an axial stress of 20,000 psi? What will be the electrical output with this force if we have four active gages, each with a nominal 120-Ω resistance and a gage factor of 2.06, and a 10-V power supply?

Solution: The axial stress is simply the force over the cross-sectional area:

$$\sigma = \frac{F}{A} = \frac{F}{\pi(D_0^2 - D_i^2)/4}$$

$$F = 20,000 \times \frac{\pi[1^2 - (1 - 2 \times 0.004)^2]}{4} = 2412 \text{ lb}$$

This is a uniaxial stress situation, so we can use Eq. (8.1) to compute the axial strain. The axial strain is then $\epsilon_a = \sigma/E = -20,000/10 \times 10^6 = -2000\ \mu$strain. Based on Eq. (8.6), the transverse strain is then $-0.3 \times -2000 = 600\ \mu$strain. Referring to Figure 8.5, we will use the two axial gages in positions R_2 and R_4 and the two transverse gages in positions R_1 and R_3. For each axial strain gage, the change in resistance can be obtained from Eq. (8.8). For gage R_4, the change in resistance is

$$\frac{\Delta R_4}{R_4} = S\epsilon_a = 2.06 \times (-2000 \times 10^{-6}) = -4120 \times 10^{-6}$$

Using the same method, $\Delta R_2/R_2 = -4120 \times 10^{-6}$ and $\Delta R_1/R_1 = \Delta R_3/R_3 = 1236 \times 10^{-6}$. These values can be substituted in Eq. (8.23):

$$V_o = \frac{10 \times 120 \times 120}{(120 \times 120)^2}(1236 \times 10^{-6} + 1236 \times 10^{-6} + 4120 \times 10^{-6} + 4120 \times 10^{-6})$$

$$= 26.8 \text{ mV}$$

8.6.2   Proving Rings

When calibrating small force-measuring devices, it is usually possible to use a set of ac-curate weights. This is not practical for larger forces, 50,000 lb for example. A useful de-vice for these larger forces is the *proving ring* (Figure 8.35), which is a simple metal ring distorted by a transverse load. Due to the simple geometry, proving rings can be manu-factured accurately and the force–deflection relationship determined from analysis. Based on Roark and Young (1989), for a thin ring loaded as shown in Figure 8.35, the relationship between the vertical deflection δ and the force F is given by

$$F = \frac{\delta E I}{R^3(\pi/4 - 2/\pi)} \tag{8.34}$$

where E is the modulus of elasticity, I is the moment of inertia of the cross section of the ring, and R is the radius from the ring center to the centroid of the ring cross sec-tion. Thick rings require a more complicated formulation.

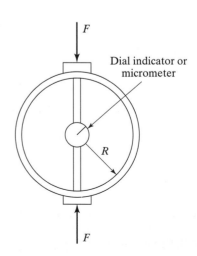

FIGURE 8.35

Proving ring for force measurement.

8.7   MEASURING ROTATING SHAFT TORQUE

It is frequently necessary to measure the torque and power associated with rotating machines, such as pumps, turbines, electric motors, electric generators, and internal-combustion engines. To obtain power, it is necessary to measure shaft speed and shaft torque. Shaft speed measurement was discussed earlier in this chapter. In this section we discuss the measurement of the torque applied to rotating shafts.

The most versatile device for determining the shaft torque of rotating machines is the electric-motoring dynamometer, shown in Figure 8.36. This dynamometer can be used either to power an energy-absorbing device such as a pump (motoring mode) or to absorb power from an energy-producing device such as an engine. In the absorption mode, the dynamometer functions as an electric generator, and the absorbed energy is dissipated in a connected resistor bank. In the powering mode, the dynamometer functions as an electric motor, and the supplied energy is obtained from a dc power supply.

As shown in Figure 8.36, a dc electric motor/generator is mounted in bearings into a support cradle. The motor would be free to turn relative to the cradle if it were not for the torque arm, which attaches to the cradle through a force-measuring device. Hence any torque applied to the motor/generator shaft produces a reaction in the motor/generator body that can be sensed by multiplying the force in the force-sensing device by the moment arm to the motor/generator axis. The motor/generator itself is constructed in the same manner as other dc machines.

The main disadvantage of electric-motoring dynamometers is that they are physically large in size and quite expensive when the rated power is high. When only energy absorption is required, the eddy-current dynamometer is often preferable. It is relatively compact even for power levels of 1000 hp or more. As shown in Figure 8.37, the eddy-current dynamometer physically resembles the electric-motoring dynamometer. Instead of generating an electric current that is transferred to a resistor bank, the electric current

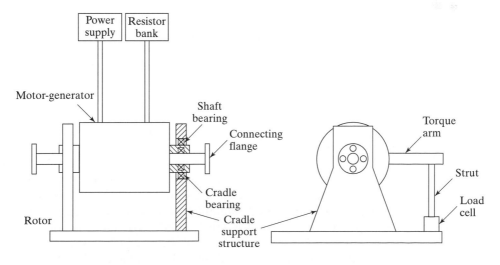

FIGURE 8.36

Electric-motoring dynamometer.

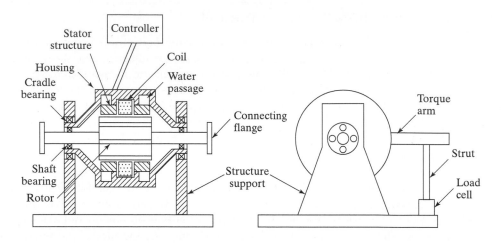

FIGURE 8.37

Eddy-current dynamometer.

circulates in the iron in the rotor and the internal magnetic structure, where it is dissipated. These internal currents are known as *eddy currents*. The rotor has a set of radial lobes (somewhat like a gear). In the stationary portion of the device a dc-powered coil creates a magnetic field that passes through the stationary stator structure and through the lobes of the rotor. As the rotor turns, the reluctance of the magnetic circuit changes, causing the magnetic field to fluctuate. These fluctuating fields generate currents within the rotor and stator, which dissipate due to electrical resistance. The dissipated energy is carried away with cooling water. In smaller units, only the strator is cooled. In larger units, water also flows over the rotor.

The magnetic field produces a drag on the rotor, which results in a torque being applied to the dynamometer body, which is resisted by the torque arm. The magnitude of the torque can be controlled by varying the dc current flow through the electric coil. The fact that eddy-current dynamometers cannot supply power to the tested device has become a major disadvantage in recent years as exhaust emissions have become a major concern. In parts of the engine emissions testing process, the vehicle is coasting and the engine is actually absorbing power. If this mode is not of concern, eddy-current dynamometers are probably the best choice for testing moderate and high-horsepower internal-combustion engines.

Two other types of absorption dynamometers are frequently used. One is the *prony brake*, which dissipates energy with mechanical friction. One design resembles an automotive disk brake. Prony brakes are usually limited to a few horsepower and are not as stable as eddy-current dynamometers in controlling shaft speed. The other common absorption dynamometer is the *hydraulic dynamometer* or *water brake*. In this device a rotor with vanes is used to churn water, which flows through a chamber. The water is heated in the process. Hydraulic dynamometers are less expensive than eddy-current dynamometers and can absorb substantial power. They are, however, less capable than eddy-current dynamometers in providing stable speed control.

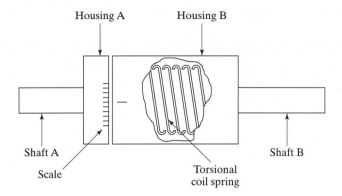

FIGURE 8.38

Shaft torque meter.

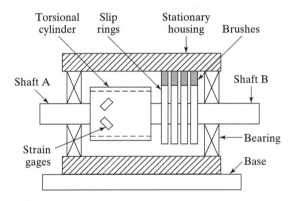

FIGURE 8.39

Shaft torque meter with electric output signal.

If other devices can be used to supply or absorb power, shaft torque can be measured with shaft torque meters. One relatively simple device is shown in Figure 8.38. The input and output shafts are connected with a torsional spring such that there is a relative angular displacement between the input and output shafts. This displacement is displayed on a scale on two parts of the housing, which are connected to the input and output shafts, respectively. Since the entire device is rotating, it is necessary to use a stroboscope to read the scale. The main disadvantage of this device is that it cannot be read remotely.

A shaft torquemeter that can be read remotely is shown in Figure 8.39. In this device the input and output shafts are connected with a cylinder in torsion. Strain gages that produce output in response to the torque are attached to the cylinder. The signal is transmitted to the stationary housing through slip rings and brushes.

REFERENCES

[1]   BUDYNAS, R.G. (1977). *Advanced Strength and Applied Stress Analysis*, McGraw-Hill, New York.

[2]   DALLY, J., RILEY, W., AND McCONNELL, K. (1993). *Instrumentation for Engineering Measurements*, 2d ed., Wiley, New York.

[3] DOEBELIN, E. O. (1990). *Measurement Systems: Application and Design*, 4th ed., McGraw-Hill, New York.

[4] HERCEG, E. (1976). *Handbook of Measurement and Control*, Schaevitz Engineering, Pennsauken, NJ.

[5] HETENYI, M. (1960). *Handbook of Experimental Stress Analysis*, Wiley, New York.

[6] KAIL, R., AND MAHR, W. (1984). *Piezoelectric Measuring Instruments and Their Applications*, Kistler Instrument Corp., Amherst, NY.

[7] NORTON, H. N. (1982). *Sensor and Analyzer Handbook*, Prentice Hall, Englewood Cliffs, NJ.

[8] POPOV, E. P. (1976). *Mechanics of Materials*, Prentice Hall, Englewood Cliffs, NJ.

[9] ROARK, R., AND YOUNG, W. (1989). *Formulas for Stress and Strain*, 6th ed., McGraw-Hill, New York.

[10] TURNER, J. D. (1988). *Instrumentation for Engineers*, Springer-Verlag, New York.

[11] WOOLVERT, G. A. (1977). *Transducers in Digital Systems*, Peter Peregrinus and the Institution of Electrical Engineers (UK), Stevenage, Herts, England.

PROBLEMS

8.1 A strip of cast iron is 8 in. long and has a cross section of 1.0 by 1/16 in. It is subjected to an axial load of 1000 lb. Two strain gages are attached to the surface, one in the axial direction and one in the transverse direction. The axial gage indicates a strain of 1143 μstrain and the transverse gage indicates a strain of -286 μstrain (compression). Estimate the modulus of elasticity and Poisson's ratio for this material.

8.2 A strain gage has a nominal resistance of 120 Ω and a gage factor of 2.09. If it is used in a quarter bridge with the other resistors being 120 Ω and with a supply voltage of 2 V, what will be the strain if V_o is 12.5 mV?

8.3 A strain gage has a nominal resistance of 120 Ω and a gage factor of 2.00. It is used in a quarter bridge, with the other resistors being 120 Ω. The power supply voltage is 3 V. If the active gage is subjected to a strain of 2500 μstrain, what will be the bridge output voltage?

8.4 Two strain gages with a nominal resistance of 120 Ω (each) and a gage factor of 2.10 are used in a half-bridge (resistors R_1 and R_3 in Figure 8.5), with other unstrained gage resistors being 120 Ω. The supply voltage is 5 V. Calculate the microstrain if the output voltage is 20 mV.

8.5 In Problem 8.4, the bridge is balanced at 20°C. Calculate the error in the reading of the output (both voltage and strain) if the temperature of the strained gages decreases to -40°C while the temperature of the unstrained resistors remains unchanged. The apparent resistance of these strain gages changes 0.5%/°C. How can we eliminate this error? Neglect the change in S.

8.6 In Problem 8.4, suppose the four lead wires connecting the strained gages to the bridge have a resistance of 2 Ω each. Calculate the error in reading the output (both voltage and strain) due to the connecting wires. How can we eliminate this error?

8.7 Three legs of a Wheatstone bridge have resistances of 120.00 Ω, and the fourth leg has a resistance of 119.11 Ω. The power supply is 3 V. Determine the output voltage and the loading error produced by measuring the output with a voltmeter having an input impedance of 50 kΩ.

8.8 An electrical resistance strain gage has a gage factor of 2 and a resistance of 120 Ω. If this gage is connected to lead wires with a combined resistance of 10 Ω, what will be the effective gage factor for the combined system?

8.9 A strip of high-strength steel has a length of 30 cm and a cross section of 1 mm by 20 mm. The modulus of elasticity is 200 GPa, and Poisson's ratio is 0.27. It is subjected to an axial load of 15,000 N, and it is instrumented with two axial strain gages, with $R = 120 \ \Omega$ and a gage factor of 2.10. These two gages are connected into opposite legs of a Wheatstone bridge (R_1 and R_3 in Figure 8.5). The two fixed resistors are also 120 Ω, and the supply voltage is 2.5 V. The bridge is adjusted to zero-voltage output before the load is applied. Find the output of the bridge with the load applied.

8.10 In measuring the strains applied to a symmetrical I-beam, a full bridge is used. Two strain gages are attached to the upper flange and two to the lower flange at the same axial section. The nominal resistance of each gage is 120 Ω, the gage factor of each is 2.05, and the supply voltage is 5 V.

(a) Calculate the strains in the upper and lower flanges if the beam is subjected to a bending load and the bridge output voltage is 40 mV.

(b) If the temperature of the gages change (by the same amount), will the temperature change introduce any error in the results? Why?

(c) Can the same configuration of the strain gages be used to measure strain due to an axial force on the beam? Why?

8.11 Four strain gages are attached to a cantilevered beam and are used to measure a load **F** as shown in Figure P8.11. Sketch a Wheatstone bridge, and show how the four gages are located to give maximum output.

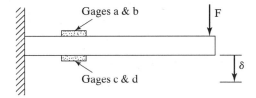

FIGURE P8.11

8.12 The metal strip in Problem 8.9 is instrumented with four of the same types of strain gages. Two gages are located axially and correspond to R_1 and R_3 in Figure 8.5. The other two gages are located transversely and correspond to R_2 and R_4 in Figure 8.5. What will be the bridge output with the same axial load as in Problem 8.4?

8.13 In Problem 8.2, the systematic uncertainty in the variables are

Strain-gage resistance	±0.15%
Strain-gage factor	±0.5%
Bridge resistors	±0.1%
Supply voltage	±0.2%
Output voltage	±0.3%

Estimate the systematic uncertainty in the measured strain assuming that the bridge is balanced before taking the reading.

8.14 In Figure 8.5, assume that R_2 is an unstrained temperature-compensating gage that shows a variation $R_2 = R_{2i} + \Delta R_{2t}$ and that R_3 is an active gage of the same type that shows a variation $R_3 = R_{3i} + \Delta R_{3t} + \Delta R_{3s}$. The subscript i indicates the unstrained value at the reference temperature, and the subscripts t and s indicate the temperature and strain effects, respectively. Assuming that the temperature variations are the same, $\Delta R_{3t} = \Delta R_{2t}$, show that Eq. (8.22) is still valid.

8.15 An installed strain gage has the temperature characteristics shown in Figure 8.6, a nominal resistance of 360 Ω, and a nominal gage factor of 2.05. It is installed in a quarter-bridge arrangement with a 3-V power supply and the bridge is adjusted to provide a zero output. The system to which the gage is attached is heated to 300°C without any stress applied at 20°C.

(a) What will be the bridge output?

(b) If the structure is now strained to a value of 500 μstrain with the temperature at 300°C, what will be the bridge output?

8.16 A strain gage is carefully matched with three fixed-bridge resistors, each having a resistance of 120 Ω. The gage factor is 2.00, and the bridge supply voltage is 3.00 V. The strain gage is installed on a structure and connected to the bridge with a pair of 40-ft leads with a resistance of 0.026 Ω/ft.

(a) What will be the apparent strain before application of the load?

(b) If the structure is then strained to 1000 μstrain, what will be the bridge output voltage?

8.17 For the system described in Problem 8.2, what amount of strain will cause the nonlinearity error to be 1%?

8.18 What are the potential sources of error in measuring strain with strain gages? How can these errors be minimized?

8.19 If the random uncertainties in the supply and output voltages in a quarter-bridge measurement of strain are 0.1% and 0.3%, respectively, determine the random uncertainty of the measured strain. Assume that the resistors are measured accurately.

8.20 Consider a linear potentiometer in which the resistor $(R_1 + R_2)$ has a value of 1000 Ω and the input voltage is 90 V (independent of any loading). If the output is connected to a load of 5000 Ω, prepare a plot of V_o versus x/L. Is it a straight line?

8.21 In Problem 8.20, what will be the power dissipated in the potentiometer when the slider is at the center of its travel? What might you do to reduce the power dissipation, and what might be the consequences?

8.22 An angular potentiometer has a resistance range of 0–500 Ω for an input rotation of 300 degrees. The output-voltage measuring device has an input impedance (take it as resistance) of 10 kΩ. The supply voltage is 5 V.

(a) Derive the equation for calculating the systematic error (i.e., drop in voltage) due to low impedance of the output measurement device.

(b) At what angle, does the maximum error occur?

8.23 The following data have been taken for the calibration of an LVDT. Using a least-squares fit, calculate the calibration constant (V/in.) for a linear calibration curve, the maximum error, and the mean deviation of the measurements.

Displacement (in.)	0.00	0.50	1.00	1.50	2.00
Output (V)	0.15	0.63	1.30	1.92	2.65

8.24 A capacitive sensor consists of two parallel 1-cm-square plates separated by a distance of 0.2 mm.

(a) Find the capacitance in pF.

(b) If the plates are displaced in a normal direction, find the sensitivity in pF/mm.

(c) If the plates are displaced in a parallel direction, estimate the sensitivity in pF/mm.

8.25 Consider the two types of capacitive displacement sensors discussed in Section 8.2.3.

(a) Which of the two configurations provides a linear relation between the input and the output of the sensor?

(b) Derive the formula for sensitivity of output (capacitance, in farads) with respect to distance between the capacitor plates [Fig. 8.18(a)]. How can this sensitivity be increased?

(c) Derive the formula for sensitivity of output (capacitance, farads) with respect to overlapping area of the capacitor plates [Fig. 8.18(b)]. How can this sensitivity be increased?

8.26 Repeat Problem 8.24 if the plates are 0.5 cm square and are separated by 0.1 mm.

8.27 An angular digital encoder has 8 bits in each sector. What is the angular resolution?

8.28 An accelerometer has produced the following results for the acceleration of an object. Find the velocity of the object at 30 s if its initial velocity is zero. What distance will it travel in the period shown?

Time (s)	0	5	10	15	20	25	30
Acceleration (m/s^2)	5	5.5	6.1	6.7	7.5	8.5	9.6

8.29 A policeman standing on the side of a road measures the speed of a car moving at 120 km/hr with a Doppler radar gun. The Doppler frequency is 10,000 MHz. The measurement of speed is based on a straight-line assumption, while the actual angle of measurement is close to 10 degrees. Calculate the error in the apparent speed of the car.

8.30 A magnetic pickup is used to detect the shaft speed of a turbocharger. A standard hexagonal nut is used to excite the pickup. If the shaft is turning 50,000 rpm, how many positive pulses will be created per second? By what number should the number of counts in a second be multiplied to produce a result in rpm?

8.31 To measure the speed of a variable-speed motor, a hexagonal sleeve is fitted on its shaft to drive a magnetic pickup. The output on an oscilloscope screen shows 20 positive peaks in four divisions. The scope setting is 0.5 s/div. Calculate the shaft speed in rpm.

8.32 A magnetic pickup is excited by the rotation of a gear with 12 teeth on the shaft. The shaft rotates at 3000 rpm. How many cycles per second do you expect to be displayed by an oscilloscope that is used to indicate the signal?

8.33 A magnetic pickup is excited by the rotation of a gear with 12 teeth on the shaft. The position of the teeth can be off by ±0.5 degrees. Calculate the uncertainty in the speed of the shaft at 300 rpm if we measure the period of individual cycles.

8.34 A strobotachometer (strobe) is shined at a disk rotating at 3000 rpm. The disk has a single circumferential mark for speed measurements.

(a) How many marks will you see if the strobe flashing rate is 3000 per minute?

(b) Repeat part (a) for flashing rates of 1000, 1500, 6000, and 9000 flashes/min.

8.35 Suggest a procedure for obtaining the correct rpm value when using a stroboscopic tachometer.

8.36 In a photoelectric tachometer, 24 equally spaced reflectors have been placed on a disk. Calculate the rpm of the disk if 600 cycles per second are detected.

8.37 An aluminum cylindrical strain-gage load cell [Figure 8.34(b)] is to be specified to measure a force of 100,000 lb. The maximum stress in the metal is to be 20,000 psi, the modulus of elasticity is 9.5×10^6 psi, and Poisson's ratio is 0.33. What must be the wall thickness of a 6-in.-OD cylinder. If it is instrumented with four active strain gages (two axial, two transverse) with $S = 2.1$, $R = 350 \ \Omega$, and the power supply is 10 V, what will be the output at maximum load?

8.38 If the strain gages in Problem 8.37 have the temperature characteristics shown in Figure 8.6, what will the change in output be for a 180°C increase in the temperature of the load cell (initially at 20°C)?

8.39 The following data are available from the calibration of a strain-gage load cell:

Load (lb)	0.00	2.00	5.00	10.00	20.00
V (mV)	−0.02	0.10	0.26	0.53	1.05

If the power supply is 6 V, determine the calibration constant in mV/lb-V (i.e., mV/lb per power-supply volt). If the voltage of the power supply is doubled, what is the input load for an output of 1.50 mV?

8.40 A steel proving ring has a radius R of 4 in. and a moment of inertia I of 0.0045 in^4. The modulus of elasticity of the steel is 29×10^6 psi. Find the deflection δ for a load of 300 lb.

8.41 An internal-combustion engine is driving an electric dynamometer. The torque arm length is 0.5 m (from the axis) and the force is 200 N. A tachometer indicates that the shaft speed is 3490 rpm. What are the engine torque and power output?

8.42 A hydraulic dynamometer is used to measure the power of a small motor. The speed and torque are measured to be 2000 rpm and 20 N-m, respectively. Calculate the amount of mechanical energy that dissipates into the water in one hour.

8.43 An automotive engine is estimated to produce a maximum of 600 N-m of torque and can operate at up to 5500 rpm. Select a suitable power specification for a dynamometer to test this engine.

8.44 Why are eddy-current dynamometers not used for accurate simulation of driving cycles of automobile internal-combustion engines? What type of dynamometer would you recommend for this type of application?

CHAPTER 9

Measuring Pressure, Temperature, and Humidity

In this chapter, we provide the technical basis for common systems used to measure pressure, temperature, and humidity. The most common devices for measurements are included, but it should be noted that other devices are also used in engineering practice. Also included in this chapter is an introduction to the technology of fiber-optic measurement systems, which includes sensors for pressure and temperature as well as other measurands.

9.1   MEASURING PRESSURE

Pressure is measured in three different forms: absolute pressure, gage pressure, and differential pressure. Absolute pressure is that used in thermodynamics to determine the state of a substance. Gage pressure is the pressure relative to the local ambient air pressure. Differential pressure is simply the difference in pressure at two points in a system. Devices are available to measure pressures directly in each form. Although some devices measure absolute pressure directly, it is common to make two measurements with two devices—one to determine the ambient absolute pressure and a second to determine the gage pressure. The absolute pressure is then

$$p_{abs} = p_{gage} + p_{ambient} \qquad (9.1)$$

In the English unit system, absolute, gage, and differential pressures are normally stated in units of pounds per square inch in the form psia, psig, and psid, respectively. In the SI unit system, pressure is expressed in pascals (or kilopascals) with the word *absolute*, *gage*, or *differential* addended. One pascal is a pressure of 1 newton per square meter.

9.1.1   Traditional Pressure-Measuring Devices

Three traditional pressure-measuring devices that have no electrical output are still widely used and deserve mention. The *manometer* and the *bourdon gage* are used because they enable a pressure to be read very quickly (and reliably), and the third

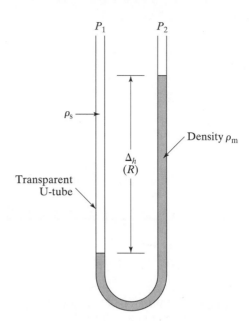

FIGURE 9.1

U-tube manometer.

device, the *dead-weight tester*, is valuable for use in the calibration of other pressure-measuring devices. None is suitable for dynamic measurements.

Manometers   The simplest manometer is the *U-tube manometer*, shown in Figure 9.1. It consists of a U-shaped transparent glass or plastic tube partially filled with a liquid. The sensing lines of the pressures to be sensed are connected to the tops of the tubes. The device is used readily for measuring either differential or gage pressure for either liquids or gases. If the fluid being sensed is a liquid, then the manometer fluid must be both denser than the sensed fluid and be immiscible with it. The fluids should also have different colors so the interface (meniscus) is readily visible. The pressure difference at the top of the manometer can be computed from

$$\Delta P = P_1 - P_2 = \Delta h g(\rho_m - \rho_s) = Rg(\rho_m - \rho_s) \tag{9.2}$$

where Δh is the difference in the levels of the two interfaces (given by the symbol R), ρ_m is the manometer liquid density, ρ_s is the density of the sensed fluid, and g is the acceleration of gravity. For gases, ρ_s is very small relative to ρ_m and can be neglected, giving $\Delta P = Rg\rho_m$.

While pressure has units of psi or Pa, it is common to express pressure as a height of a fluid column. If a pressure is divided by ρg, the result has the unit of height. For example, in the English unit system, if we use the density of water, the pressure can be expressed as feet of water or inches of water. Atmospheric absolute pressures are usually expressed this way—30 in. or 760 mm of mercury, for example. When a pressure is expressed as the height of a fluid column, it is also necessary to know the temperature of the fluid since this affects the density. For example, the density of water varies 0.75% between 10 and 40°C. It is common to use the density of water at 4°C, 1000 kg/m³ or 62.43 lbm/ft³. It is also common to specify the density of a fluid using the term *specific gravity*, S, which is the ratio of the fluid density to the density of water at a specified temperature (usually 4°C).

Manometers are normally accurate devices even without calibration. The main factors affecting accuracy are the scale and density of the manometer fluid. Scales can be constructed accurately, and they maintain their accuracy with time. Fluid densities are also well known and can be readily checked. Thermal expansion will affect both the scale and the fluid density, but analytical corrections can be made to eliminate errors.

U-tube manometers are rather inconvenient since it is necessary to read the locations of the two interfaces. A common variation, a *well-type manometer*, is shown in Figure 9.2. In this configuration, the cross-sectional area of the well is very large relative to the area of the transparent tube, and when a pressure is applied, the change in surface elevation of the well is extremely small compared to the elevation change in the tube. As a result, only one reading is required. These devices have an adjustment, so the reading is zero when there is no pressure differential applied. In using well manometers for gases, Eq. (9.2) can be used directly since $\rho_s \approx 0$. For liquids, the applicable formula is more complicated since the well port and the column port are not at the same elevation. For liquid applications, the user should follow manometer analysis as given in White (1999).

Manometers can be used for a rather limited range of pressures. The most dense fluid normally available is mercury, with a density 13.6 times that of water. For a device height of 10 ft (a ladder is required to read it), the maximum pressure difference is about 4 atm. With other fluids, such as light oils, differential pressures as low as 0.5 in. of water can be measured with a resolution of the order of ±0.02 in.

For somewhat lower pressure differentials, a manometer called an *inclined manometer* (Figure 9.3) has increased sensitivity and will improve accuracy due to improved resolution. These can be used to measure small differential pressures as low as

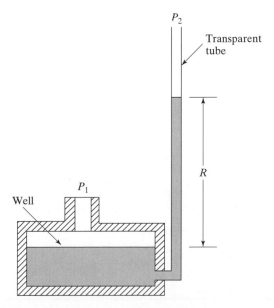

FIGURE 9.2

Well-type manometer.

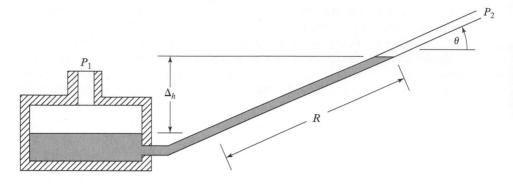

FIGURE 9.3

Inclined manometer.

0.1 in. of water column. The slanting tube makes a small change in fluid height cause a large displacement in the direction of the transparent tube. For gases,

$$\Delta P = \Delta h \, \rho g = R \sin \theta \, \rho g \qquad (9.3)$$

where R is the reading and θ is the angle between the manometer tube and the horizontal direction. An oil with density lower than water is generally used. In most cases, the scale on the manometer is stretched so that readings are in the appropriate pressure units. As for the vertical well manometer, the formula is more complicated if the sensed fluid is a liquid. The reader may refer to White (1999) for details.

Most laboratories also contain a barometer for determining absolute atmospheric pressure (Figure 9.4). This device is essentially a well-type manometer in which one leg is evacuated so that the pressure on that leg is mercury vapor pressure. While barometers read directly in inches (or mm) of mercury, a small temperature correction has to be made to account for the vapor pressure of mercury and differential thermal expansion of the mercury and the measuring scale.

There are a number of other variations of manometers, which can be used for higher or lower pressures than discussed here. For higher pressures, nonmanometric devices such as pressure transducers (discussed later) are usually preferable to manometric devices. A variety of devices called *micromanometers* are used for low pressures. For vacuum (very low) pressures, a manometric device called a *McLeod gage* is often used. (See Section 9.1.3.)

Example 9.1

A gas-pressure difference of 125 kPa is applied to the legs of a U-tube manometer. The manometer contains mercury with a specific gravity of 13.6. Determine the reading of the manometer.

Solution: Using Eq. (9.2), with $\rho_s = 0$, we obtain

$$R = \frac{\Delta P}{\rho_m g} = \frac{\Delta P}{S \rho_{water} g} = \frac{125,000}{13.6 \times 1000 \times 9.8}$$

$$= 0.938 \text{ m}$$

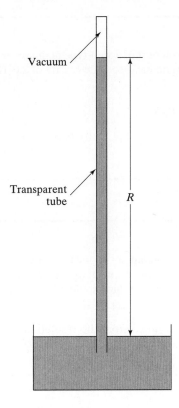

FIGURE 9.4

Mercury barometer.

Example 9.2

A pressure is given as 58 psi. What is this pressure expressed as inches of mercury ($S = 13.6$) and feet of water?

Solution: Using Eq. (9.2) with $\rho_s = 0$ yields

$$\Delta h = \frac{\Delta P}{\rho g}$$

For inches of mercury, this becomes

$$\Delta h = \frac{58\dfrac{\text{lbf}}{\text{in.}^2}144\dfrac{\text{in}^2}{\text{ft}^2}32.17\dfrac{\text{lbm} - \text{ft}}{\text{lbf} - \text{sec}^2}}{13.6 \times 62.43\dfrac{\text{lbm}}{\text{ft}^3}32.17\dfrac{\text{ft}}{\text{sec}^2}} = 9.84 \text{ ft Hg} = 118.0 \text{ in Hg}$$

Similarly, for feet of water, we have

$$\Delta h = \frac{58\dfrac{\text{lbf}}{\text{in}^2}144\dfrac{\text{in}^2}{\text{ft}^2}32.17\dfrac{\text{lbm} - \text{ft}}{\text{lbf} - \text{sec}^2}}{62.43\dfrac{\text{lbf}}{\text{ft}^3}32.17\dfrac{\text{ft}}{\text{sec}^2}} = 133.8 \text{ ft H}_2\text{O}$$

Comment: Note that we had to make two unit conversions in this problem. First, we converted psi to lbf/ft^2 and then we had to introduce g_c to convert the lbm in the density.

Example 9.3

A gas pressure is applied to the well port of a well-type manometer. The column port is open to atmosphere and the manometer fluid has a specific gravity of 2.0. If the reading is 47.5 cm, find the applied pressure.

Solution: Using Eq. (9.2), we obtain

$$P_1 - P_2 = P_1 - 0 = Rg\rho_m = 0.475 \times 9.8 \times 2.0 \times 1000$$
$$P_1 = 9.31 \text{ kPa}$$

Example 9.4

It is desired to design an inclined manometer to measure a gas pressure between 0 to 3 in. of water column with a resolution of 0.01 in. The manometer fluid is water, and it is possible to read the sloping scale to a resolution of 0.05 in. What should be the angle θ, and how long must the inclined tube be?

Solution: The angle can be determined from the resolution requirement; 0.01 in. vertically corresponds to 0.05 on the slanting tube. Hence $\sin \theta = 0.01/0.05$, $\theta = 11.5°$. Since the total rise is 3 in. in the length of the tube, $\sin \theta = 3/L$, $L = 15$ in.

Bourdon Gage A very common pressure-measuring device, the *Bourdon gage*, is shown in Figure 9.5. It is a simple device for obtaining rapid readings of fluid pressures. The basic principle of operation is that a curved, flattened tube (Bourdon tube) will attempt to straighten out when subjected to internal pressure. The end of the tube is connected with a linkage to a rotary dial indicator. Relatively inexpensive Bourdon gages can be obtained to measure a wide range of pressures from low vacuums up to 1500 atmospheres or more. The less expensive devices are not normally highly accurate—uncertainties of up to 5% of full scale are common. The accuracy of the more expensive gages can be quite good—on the order of 0.5% of full scale. Bourdon tubes are sometimes used as remote pressure-sensing devices. The deflection of the tube is sensed with an LVDT or potentiometer, which transmits an electrical signal to the data-acquisition location.

Dead-Weight Tester The dead-weight tester, shown schematically in Figure 9.6, is a device that is often used to calibrate other pressure-measuring devices at moderate to high pressures. The pressure-measuring device to be calibrated senses the pressure of oil contained in a chamber. A piston–cylinder arrangement is attached to the top of the chamber, and weights can be placed on the piston. A separate screw and piston can be used to adjust the volume of the chamber so that the weighted piston is in the middle of its possible movement range. The fluid pressure is then the weight of the piston–weight assembly divided by the piston area. The device is very accurate since the piston area

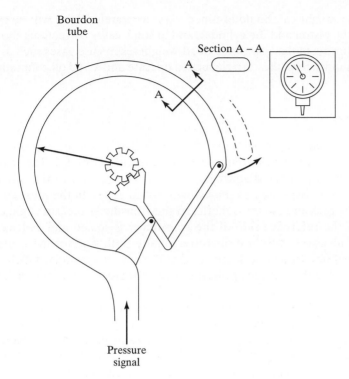

FIGURE 9.5

Bourdon gage.

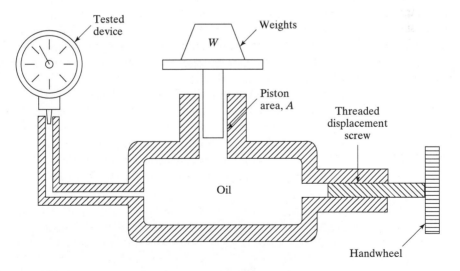

FIGURE 9.6

Dead-weight tester.

and the value of the weight can be determined very accurately. There will be a slight clearance between the piston and the cylinder, and in some cases corrections should be made for the fluid flowing in this gap. Some dead-weight testers use gases such as nitrogen as the working fluid; hence, they are cleaner to operate and avoid oil contamination of the tested device.

9.1.2   Pressure Transducers

A very common and relatively inexpensive device used to measure fluid pressure is the diaphragm strain-gage *pressure transducer*, sketched in Figure 9.7. The test pressure is applied to one side of the diaphragm, a reference pressure to the other side, and the deflection of the diaphragm is sensed with strain gages. In the most common design, the reference pressure is atmospheric, so the transducer measures gage pressure. In some cases the reference side of the transducer is sealed and evacuated so that the transducer measures absolute pressure. Finally, both sides of the transducer can be connected to different test pressures so that the measurement is of differential pressure. Transducers for these three applications have slightly different construction details.

   In the past, the diaphragm was usually made of metal, and foil strain gages were used. More recently, it has become common to make the diaphragm of a semiconductor material (usually silicon) with semiconductor strain gages formed into the diaphragm. This is a less expensive construction technique, and since semiconductor gages have high gage factors, the sensitivity is improved. The silicon is subject to corrosion by some fluids, and as a result it is common to include an additional corrosion-resistant metal diaphragm with the regions between the two diaphragms filled with a fluid. Normally, the Wheatstone-bridge signal conditioner is built into the transducer (all branches of the bridge are active gages), and the strain gages are connected to give temperature compensation, as discussed in Section 8.1.2. Most strain-gage pressure transducers produce a dc output in the millivolt range, but some include internal amplifiers and have outputs in the range 0 to 5 or 0 to 10 V. The higher-voltage output units are less susceptible to environmental electrical noise.

   Pressures can also be sensed with LVDT devices. Figure 9.8 shows an arrangement with a flexible chamber called a capsule and an LVDT to sense the displacement. This design is more expensive than those using strain-gage sensors but may be more durable in an application requiring a long lifetime. Many heavy-duty pressure transducers used in the process control industry use LVDT sensors. As with LVDTs, the output is a dc voltage with a range on the order of 0 to 5 or 0 to 10 V. In the process

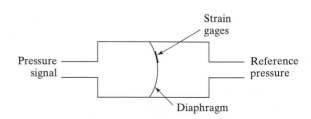

FIGURE 9.7

Strain-gage pressure transducer.

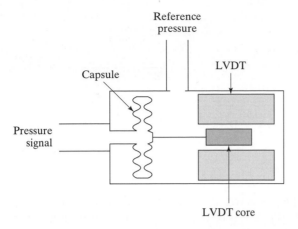

FIGURE 9.8

LVDT pressure transducer.

industries, the voltage output will usually be converted to a 4- to 20-mA current for signal transmission.

Capacitive sensors are sometimes used in pressure transducers and are particularly useful for very low pressures (as low as 0.1 Pa) since capacitive sensors can detect extremely small deflections. A sketch of a capacitance pressure transducer is shown in Figure 9.9.

Measuring pressures that are varying rapidly in time presents a number of technical problems. The fluid and the diaphragm (or other displacing element) form a second-order dynamic system. As discussed in Chapter 11, if the diaphragm (or other displacing element) is very flexible, the natural frequency will be low and the transducer output will be misleading for high-frequency pressure variations. Transducers used for high-frequency pressure measurements, such as the combustion process in an internal combustion engine, usually use a piezoelectric sensing element, as discussed in Chapter 8.

A schematic of a piezoelectric transducer is shown in Figure 9.10. These transducers generally use transverse-effect piezoelectric sensing elements. The piezoelectric material is very stiff, and the transducers have a high natural frequency in many applications. Piezoelectric pressure transducers can have natural frequencies up to 150 kHz and are usable up to about 30 kHz.

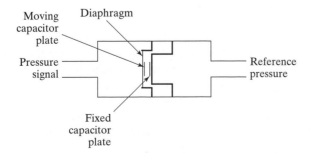

FIGURE 9.9

Capacitive pressure transducer.

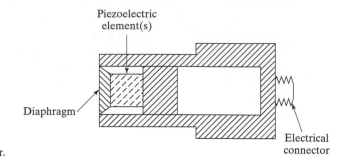

FIGURE 9.10

Piezoelectric pressure transducer.

The geometry piezoelectric transducers is different from the transducers discussed earlier—the diaphragm is of the flush-mounted type, and when the transducer is installed, it comes into direct contact with the fluid in the pipe or chamber. The reasons for this are twofold. If a cavity were included as in the other transducers, it might significantly alter the measurand, due to loading. Furthermore, the natural frequency would be reduced and the ability to respond to transients would be impaired. As discussed in Chapter 11, sensing lines affect natural frequency, making the determination of natural frequency application dependent. Other types of pressure transducers are also available with flush mounting, but this is often for use in dirty fluids, in which the cavity might become plugged or difficult to clean.

9.1.3   Measuring a Vacuum

The need to measure very low absolute pressures (vacuums) exists both in the laboratory and in manufacturing. The freeze-drying of food is performed in a vacuum environment, and many operations in the manufacture of computer chips are completed in a vacuum. Vacuum absolute pressures are measured in units of torr. This unit is defined as 1/760 of the standard atmosphere. Since the standard atmosphere is 760 mm of mercury, 1 torr is 1 mmHg. Norton (1982) gives the following definitions for ranges of vacuum pressures:

Low vacuum	760 to 25 torr
Medium vacuum	25 to 10^{-3} torr
High vacuum	10^{-3} to 10^{-6} torr
Very high vacuum	10^{-6} to 10^{-9} torr
Ultrahigh vacuum	Below 10^{-9} torr

Pressure-measuring devices described previously can be used for low and medium vacuum measurement. Manometers, Bourdon gages, and similar gages that use a bellows instead of a bourdon tube and capacitive diaphragm transducers can measure vacuums to 10^{-3} torr. Specialized capacitive diaphragm transducers can measure vacuums as low as 10^{-5} torr [Norton (1982)]. In this section, three specialized vacuum-measuring devices are discussed: the McLeod gage, thermal-conductivity gages, and ionization gages. The first is a mechanical gage useful for calibration, and the last two provide electrical outputs.

McLeod Gage   The principle of operation is to compress a large volume of low-pressure gas into a much smaller volume and then measure that pressure. A sketch of one variation of McLeod gage is shown in Figure 9.11. Initially, the gage is connected to the vacuum source with the mercury below point A, as shown in Figure 9.11(a). The large chamber with volume V is filled entirely with low-pressure gas. Next, the plunger is pushed downward until the mercury rises to level h_2 in capillary tube 2 [Figure 9.11(b)]. In the mode of operation shown here, the level of h_2 is the same as the top of capillary tube 1. The gas originally contained in volume V has been compressed into the capillary tube 1 and has a volume and pressure given by

$$V' = a(h_2 - h_1) \tag{9.4}$$

and

$$P' = P_{vac} + (h_2 - h_1) \tag{9.5}$$

where a is the cross-sectional area of the capillary tubes. Since we will determine the pressure in torr, the units of the h's are mmHg. In the range that McLeod gages are used, the vacuum pressure, P_{vac}, is usually negligible compared with $(h_2 - h_1)$.

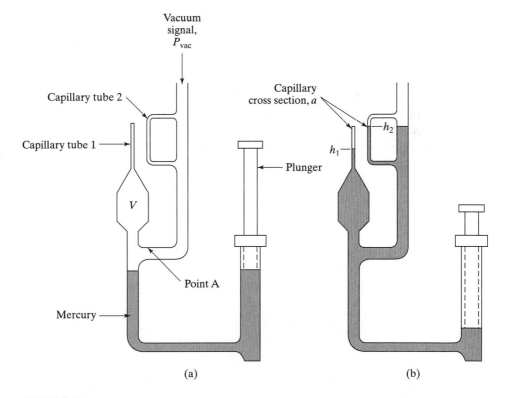

FIGURE 9.11

McLeod vacuum gage. (Based on Van Atta, 1965.)

The *ideal gas law* relates the conditions before and after compression:

$$\frac{P_{vac}V}{T} = \frac{P'V'}{T}$$

(9.6)

If the system comes into thermal equilibrium after compression, the initial and final temperatures will be the same. Combining Eqs. (9.4) through (9.6), we obtain

$$P_{vac} = \frac{(h_2 - h_1)a(h_2 - h_1)}{V} = k(h_2 - h_1)^2$$

(9.7)

That is, the sensed pressure is equal to the height difference squared times a constant k. The scale can be marked to read directly in units of torr.

McLeod gages are available to measure vacuums in the range 10^{-3} to 10^{-6} torr. They must only be used with dry gases that will not condense when compressed into the capillary tube. McLeod gages are somewhat inconvenient to use and are useful primarily for calibrating other vacuum-measuring devices.

Thermal-Conductivity Vacuum Sensors   These devices are based on the fact that the thermal conductivity of gases at low pressures is a function of the pressure. Although these devices do not normally sense vacuums as low as the McLeod gage, they provide an electrical output and are simple to use. A thermal-conductivity sensor called a *Pirani gage* is sketched in Figure 9.12(a). A heated filament is located at the center of a chamber connected to the vacuum source. The heat transfer from the filament to the wall is given by

$$q = C(T_f - T_w)P_{vac}$$

(9.8)

where T_f is the filament temperature, T_w is the chamber wall temperature, and C is a coefficient that depends on the gas in the chamber, the wall temperature, the geometry of the chamber, and the filament surface area. The vacuum pressure must be low enough

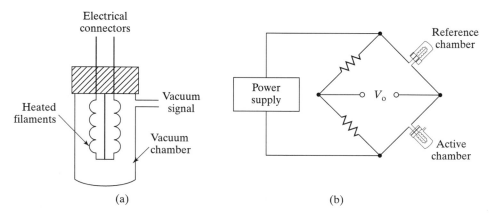

(a)   (b)

FIGURE 9.12

Pirani thermal-conductivity vacuum gage: (a) sensing chamber; (b) bridge circuit.

for the mean free path of the gas to be large compared to the dimensions of the chamber. Details on the theoretical basis for Eq. (9.8) are given in Van Atta (1965).

One method of using a Pirani gage is shown in Figure 9.12(b), where it is placed into a Wheatstone bridge. As the pressure decreases, the temperature difference between the filament and the wall will increase, increasing the resistance of the filament. The output of the bridge, which is a function of the sensor resistance, is thus a measure of the gas pressure. Since the heat transfer is also a function of the ambient temperature, a sealed reference chamber is included in the bridge for compensation. There are several designs of thermal-conductivity gages that are usable to pressures as low as 10^{-3} torr.

Ionization Vacuum Gages This sensor is based on the principle that as energetic electrons pass through a gas, they will ionize some of the gas molecules. The number of ions generated depends on the density of the gas and hence its pressure. An ionization gage is shown schematically in Figure 9.13. The sensor physically resembles the vacuum tube known as a triode, although the mode of operation is different. The cathode is a heated filament, and the circuit creates an electron current between the cathode and the grid. The electrons will ionize some of the gas molecules, creating positive ions and more electrons. The electrons will be attracted to the grid but the ions will be attracted to the plate, which is maintained at a negative voltage (unlike the triode, where the plate is maintained at a positive voltage). The ion current and the plate current are measured separately. The pressure can then be obtained from

$$P_{\text{vac}} = \frac{i_+}{si_-} \tag{9.9}$$

where i_+ is the plate (ion) current, i_- is the grid (electron) current, and s is a constant for a given circuit. Ionization gages cannot be used at pressures greater than 10^{-3} torr since the filament will deteriorate. However, they can measure pressures as low as 10^{-7} torr. A variation of the gage described here, the Bayard–Alpert gage, can measure pressures as low as 10^{-12} torr.

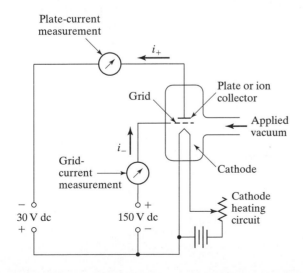

FIGURE 9.13

Ionization vacuum-gage circuit. (Based on Van Atta, 1965.)

9.2 MEASURING TEMPERATURE

Two traditional electric output devices for measuring temperature—thermocouples and resistance temperature detectors (RTDs)—are still widely used, but a number of semiconductor devices are finding applications at moderate temperatures. Traditional mechanical measurement devices still have some applications. Noncontact devices called *infrared thermometers* (for lower temperatures) and *pyrometers* (for higher temperatures) are also commonly used.

9.2.1 Thermocouples

If any two metals are connected together, as shown in Figure 9.14, a voltage is developed that is a function of the temperature of the junction. This junction of two metals used as a temperature sensor is called a *thermocouple*. The voltage is generated by a thermoelectric phenomenon called the *Seebeck effect* (named after Thomas Seebeck who discovered it in 1821). It was later found that the Seebeck voltage is the sum of two voltage effects: the *Peltier effect*, generated at the junction, and the *Thompson effect*, which results from the temperature gradient in the wires. The general simplicity of thermocouples has led to their very wide use as temperature-measuring sensors.

There are, however, a number of complications in their use:

1. The voltage measurement must be made with no current flow.
2. Connections to voltage-measuring devices result in additional junctions.
3. Voltage depends on the composition of metals used in the wires.

To be used for temperature measurement, there must be no current flow through the wires and the junction. This is because current flow will not only result in resistive losses but will also affect the thermoelectric voltages. Meeting this requirement is not a problem today since electronic voltmeters and data-acquisition systems with very high input impedance are readily available. Input impedance is usually greater than 1 MΩ, and the current draw is negligible. In the very recent past, it was necessary to use delicate and expensive devices called *balancing potentiometers* for the thermocouple voltage-measuring function.

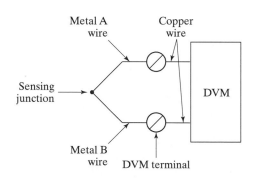

FIGURE 9.14

Thermocouple.

The second complication concerns the fact that there are actually three junctions shown in Figure 9.14. Besides the sensing junction, there are two more junctions where the thermocouple wires connect to the DVM. The voltage reading is thus a function of three temperatures (the sensing junction and the junctions at the DVM terminals), two of which are of no interest at all. The solution to this problem is shown in Figure 9.15(a). Two thermocouple junctions are used, the second being called a *reference junction*. The reference junction is held at a fixed, known temperature, most commonly the temperature of a mixture of pure water and pure water ice at 1 atm (0°C). Electronic devices are available that electrically simulate an ice reference junction without the actual need for ice. There are still two junctions at the DVM terminals, but each of these junctions consists of the same two materials, and if the two terminals can be held at the same temperature, the terminal voltages will cancel out. The two terminals can be held at the same temperature by placing them both in the same thermally insulated box or by mechanically connecting them with a thermally conducting but electrically insulating structure. With the known reference junction temperature, the measured voltage is a unique function of the materials of the thermocouple wires and the temperature of the sensing junction. The circuit shown in Figure 9.15(b) is electrically equivalent to Figure 9.15(a) and will produce the same voltage at the DVM.

Finally, the voltage generated depends strongly on the composition of the wires used to form the thermocouple. This problem has been solved by restricting the materials used to construct thermocouples. When wires and wire pairs are manufactured according to standards established by the National Institute of Standards and Technology (NIST) (formerly called the National Bureau of Standards), standard calibration curves can be used to determine the temperature based on the measured voltage. Since the voltage output is generally a nonlinear function of temperature, tables, graphs, or polynomial curve fits are required to interpret voltage data. Table 9.1 lists common thermocouple pairs. Chromel is an alloy of nickel and chromium, alumel is an alloy of

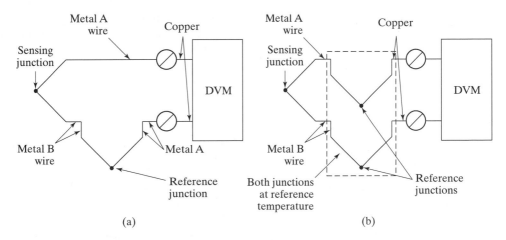

FIGURE 9.15

Thermocouple with reference junction.

TABLE 9.1   Characteristics of Standard Thermocouples

Type	Materials	Lead-wire color	Operating range (°C)	Approximate sensitivity (mV/°C)
T	Copper/constantan	Blue	−250 to 400	0.052
E	Chromel/constantan	Purple	−270 to 1000	0.076
J	Iron/constantan	Black	−210 to 760	0.050
K	Chromel/alumel	Yellow	−270 to 1372	0.039
R	Platinum/platinum–13% rhodium	Green	−50 to 1768	0.011
S	Platinum/platinum–10% rhodium	Green	−50 to 1768	0.012
C	Tungsten, 5% rhenium/tungsten, 26% rhenium	White, red trace	0 to 2320	0.020

nickel and aluminum, and constantan is an alloy of nickel and copper. Each pair has been assigned a letter, used by all manufacturers, designating the type. There is also a standard color for the lead-wire outer insulation and all the connectors for each type. For example, the color for type K thermocouples is yellow. The final column lists the approximate sensitivity of the pair in millivolts of output per Celsius degree of sensed temperature.

Figure 9.16 shows the calibration curves for several thermocouples when the reference junction is held at 0°C. Table 9.2 lists voltages (in mV) as a function of temperature for the most common thermocouple pairs. This table is quite abbreviated—tables in typical use list the temperature for increments of 1 degree. Complete tables for most thermocouples can be found in a publication of a major thermocouple supplier, *The Temperature Handbook*, by Omega Engineering.

Manufacturers carefully control the composition of their wires and test them to make sure that they have the correct electrical properties so that the standard calibration tables can be used. Generally, it is acceptable to mix components of thermocouple systems of the same type purchased at different times from different sources. Determining junction temperature can in some cases be quite accurate, uncertain to less than ±0.2°C. Accuracy of thermocouples depends on the type of the thermocouple, the quality grade of the thermocouple (manufacturers supply different quality grades), and the temperature range. The actual measurement process may introduce a number of other errors (discussed later in this chapter).

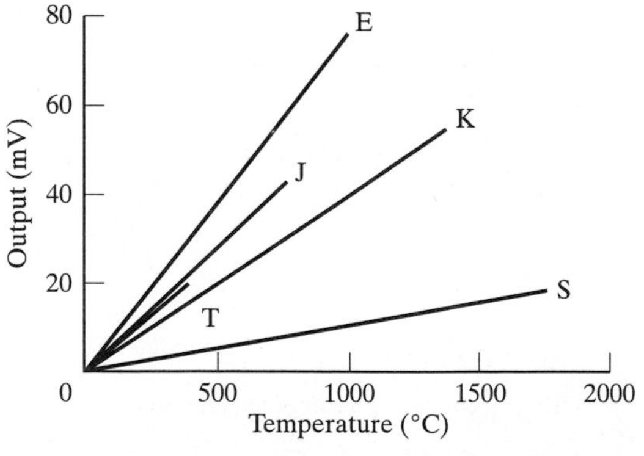

FIGURE 9.16

Output of some common thermocouples.

TABLE 9.2   Millivolt Output of Common Thermocouples (Reference Junction at 0°C)

Temperature (°C)	Thermocouple type					
	T	E	J	K	R	S
−250	−6.181	−9.719		−6.404		
−200	−5.603	−8.824	−7.890	−5.891		
−150	−4.648	−7.279	−6.499	−4.912		
−100	−3.378	−5.237	−4.632	−3.553		
−50	−1.819	−2.787	−2.431	−1.889		
0	0.000	0.000	0.000	0.000	0.000	0.000
20	0.789	1.192	1.019	0.798	0.111	0.113
40	1.611	2.419	2.058	1.611	0.232	0.235
60	2.467	3.683	3.115	2.436	0.363	0.365
80	3.357	4.983	4.186	3.266	0.501	0.502
100	4.277	6.317	5.268	4.095	0.647	0.645
120	5.227	7.683	6.359	4.919	0.800	0.795
140	6.204	9.078	7.457	5.733	0.959	0.950
160	7.207	10.501	8.560	6.539	1.124	1.109
180	8.235	11.949	9.667	7.338	1.294	1.273
200	9.286	13.419	10.777	8.137	1.468	1.440
220	10.360	14.909	11.887	8.938	1.647	1.611
240	11.456	16.417	12.998	9.745	1.830	1.785
260	12.572	17.942	14.108	10.560	2.017	1.962
280	13.707	19.481	15.217	11.381	2.207	2.141
300	14.860	21.033	16.325	12.207	2.400	2.323
350	17.816	24.961	19.089	14.292	2.896	2.786
400	20.869	28.943	21.846	16.395	3.407	3.260
450		32.960	24.607	18.513	3.933	3.743
500		36.999	27.388	20.640	4.471	4.234
600		45.085	33.096	24.902	5.582	5.237
700		53.110	39.130	29.218	6.741	6.274
800		61.022		33.277	7.949	7.345
900		68.873		37.325	9.203	8.448
1000		76.358		41.269	10.503	9.585
1100				45.108	11.846	10.754
1200				48.828	13.224	11.947
1300				52.398	14.624	13.155
1400					16.035	14.368
1500					17.445	15.576
1600					18.842	16.771
1700					20.215	17.942

There are a number of factors that must be considered in selecting thermocouples for a given application:

Sensitivity (voltage change per degree temperature change)
Linearity of output
Stability and corrosion resistance
Temperature range
Cost

Type R and type S thermocouples are very expensive and are not very sensitive; however, they are satisfactory at high temperatures (up to 1768°C) and are resistant to a number of corrosive chemicals. Type C thermocouples are usable to very high temperatures but are relatively expensive and cannot be used in an oxidizing environment. Type T thermocouples are inexpensive and very sensitive but corrode rapidly at temperatures over 400°C. Type K thermocouples are popular for general use since they are moderately priced, reasonably corrosion resistant, and usable at temperatures up to 1372°C. They also have a relatively linear output, which means that for applications in which accuracy requirements are not too severe, the temperature can be computed by assuming a linear relationship between temperature and voltage.

Thermocouples can be purchased in a number of forms. It is possible to purchase wires and form a thermocouple junction by welding together (in some cases, soldering or mechanically connecting) the ends of the wires. Wire can also be purchased with the junction already formed by the manufacturer. A wide variety of fabricated thermocouple probes can be purchased for various applications. One of the most common configurations is to place the wires and junction and ceramic insulation (generally, MgO) inside a stainless steel or inconel sheath as shown in Figure 9.17. The sheath protects the thermocouple wires from damage or chemical contamination.

Manufacturers also supply lead wire to connect the sensing probe to the voltage-measuring device. This lead wire usually consists of two conductors contained in a plastic sheath. In most cases, the lead wires are made of the same materials as the thermocouples themselves (e.g., the lead wire for an iron–constantan (type J) thermocouple will contain one iron conductor and one constantan conductor). If the lead wire is constructed of the same materials as those used for junctions, it is possible to form junctions from lead wires. However, the composition of lead wire is often not as carefully controlled in quality as the composition of wires sold for junctions; hence, it may be less accurate. For most thermocouple pairs made of expensive metals such as platinum (e.g., type S) and for certain other types, the lead wires are made of lower-cost but electrically compatible materials. These are called *compensating lead wires.*

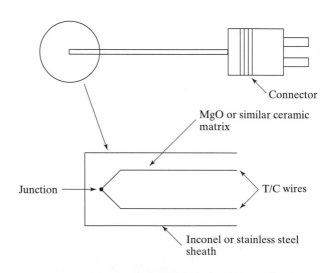

FIGURE 9.17

Common sheathed thermocouple probe.

When compensating lead wires are used, it is not possible to construct sensing junctions from lead wire.

Junctions made directly from wire can be very small. Wires can be purchased in diameters as small as 0.0005 in., and the welded junctions, which are spherical in shape, have a diameter only about three times the wire diameter. A small junction has excellent spatial resolution (i.e., the location of the measured temperature can be known accurately). These small junctions also have excellent transient response (the ability to measure rapidly changing temperatures accurately). When good spatial resolution or good transient response is required, a thermocouple is to be preferred over other common temperature sensors.

In some cases several thermocouples are connected in series in a device called a *thermopile*, as shown in Figure 9.18. When arranged in this manner, the voltage output to the display device is *n* times the voltage of a single junction, where *n* is the number of thermocouples in series. This increases the sensitivity of the system. In addition, it provides a method to average several thermocouples, which are distributed in a spatial region. Thermopiles are also used in some applications as a power source. An example exists in some gas furnaces where a thermopile is used as the power source to control a solenoid-operated gas-shutoff valve.

When a single thermocouple is used, one of the circuits shown in Figure 9.15 is generally used. However, for data-acquisition systems, a simpler arrangement is normally used as shown in Figure 9.19. Each of the thermocouples is connected to a channel of the DAS in an insulated, constant-temperature connection box. The thermocouples then effectively measure the temperature difference between the measurands and the box. In Figure 9.19 one channel is reserved for the reference junction. Software is then used to evaluate the correct voltage by subtracting the reference voltage from the voltage read from the individual channels. (*Note:* the reference voltage will normally be negative.) Another alternative is to use a different kind of temperature sensor, such as a semiconductor device (discussed later) to measure the temperature of the connection box. A voltage for the same thermocouple material corresponding to the temperature of the junction box relative to 0°C is computed and added to the readings for each channel. Further details on thermocouples and their application can be found in Benedict (1984).

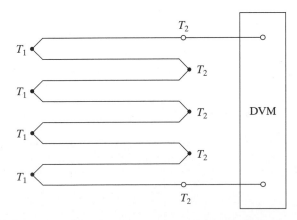

FIGURE 9.18

Four series thermocouples forming a thermopile.

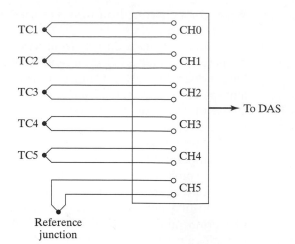

FIGURE 9.19

Thermocouple connections with data-acquisition systems.

Example 9.5

A type R thermocouple system with an ice reference has an output of 9.1 mV. What is the temperature of the sensing junction?

Solution: In Table 9.2, 7.949 mV corresponds to 800°C and 9.203 mV corresponds to 900°C. Linear interpolation gives a temperature of 891.8°C for 9.1 mV. Using a more complete set of tables would give a temperature of 892.1°C.

Example 9.6

A set of two type K thermocouple junctions [Figure E9.6(a)] measures a voltage difference of 30.1 mV. The cooler junction is independently known to have a temperature of 300°C. What is the temperature difference between the two junctions?

Solution: This setup uses a circuit that does not include a standard reference junction. If we enter Table 9.2 with 30.100 mV, we will obtain a temperature difference of 721.7. However, this is not quite correct since the thermocouple calibration is nonlinear. Instead, we should note that the circuit in Figure E9.6(a) is equivalent to the circuit shown in Figure E9.6(b). Two reference junctions have been inserted, but each contributes the same voltage with opposite sign. The complete circuit is then two separate standard circuits connected so as to measure the difference of the output voltages. That is,

$$V_{\text{out}} = V_{T1} - V_{T2} = 30.1 \text{ mV}$$

From Table 9.2, for a standard circuit at 300°C, $V_{T2} = 12.207$ mV. V_{T1}, for the thermocouple at the higher temperature, is then $12.207 + 30.100 = 42.307$ mV. From the table we find that the higher temperature is 1027.0°C. The difference is then $1027 - 300 = 727$°C. Incorrect use of the tables leads to a 5.3°C error in this case.

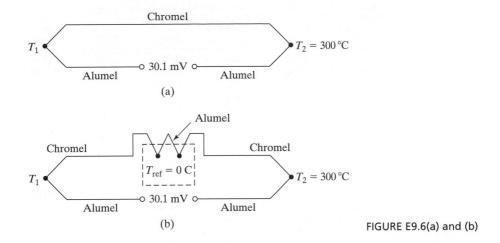

FIGURE E9.6(a) and (b)

Comment: It should be noted that due to nonlinearity in the tables, we cannot accurately evaluate a temperature difference from a voltage difference without independently knowing the temperature of one of the junctions.

9.2.2   Resistance-Temperature Detectors

A normal characteristic of metals is that the electrical resistance is a function of the metal temperature. Thus a length of metal wire combined with a resistance-measuring device is a temperature-measuring system. Temperature sensors based on the temperature effect on metal resistance are known as resistance-temperature detectors (RTDs). RTDs are used by the International Temperature Scale for interpolation purposes. (See Chapter 2.) Compared to thermocouples, platinum RTDs are more accurate (to ±0.001°C in some laboratory measurements) and have an output response that is more linear. They can be used to measure temperature directly, not relatively, as with thermocouples. They also tend to be more stable; that is, the characteristics are less likely to change over time due to chemical or other effects. On the other hand, RTD probes are generally physically larger than thermocouples, resulting in poorer spatial resolution and slower transient response.

The most common RTD sensors are constructed from platinum, although other metals, including nickel or nickel alloys, can be used. For platinum, the resistance temperature relationship is given by the *Calendar–Van Dusen equation*, namely,

$$R_T = R_0\{1 + \alpha[T - \delta(0.01T - 1)(0.01T) - \beta(0.01T - 1)(0.01T)^3]\} \qquad (9.10)$$

where α, β, and δ are constants that are dependent on the purity of the platinum and are determined by calibration. T is the temperature in Celsius degrees. The dominant constant is α, which has a value of either 0.003921/°C for the so-called U.S. calibration curve, or 0.003851/°C for the "European" calibration curve. For the U.S. calibration curve, $\delta = 1.49$ and $\beta = 0$ for $T > 0$ and $\beta = 0.11$ for $T < 0$. Sensors are readily

TABLE 9.3   Platinum RTD: R Versus T (U.S. Calibration)[a]

T (°C)	$R(\Omega)$	T (°C)	$R(\Omega)$	T (°C)	$R(\Omega)$
−100	59.57	100	139.16	300	213.92
−90	63.68	110	143.01	310	217.54
−80	67.78	120	146.85	320	221.14
−70	71.85	130	150.68	330	224.74
−60	75.91	140	154.49	340	228.32
−50	79.96	150	158.29	350	231.89
−40	83.99	160	162.08	360	235.44
−30	88.01	170	165.86	370	238.99
−20	92.02	180	169.63	380	242.52
−10	96.01	190	173.39	390	246.05
0	100.00	200	177.13	400	249.56
10	103.97	210	180.86		
20	107.93	220	184.58		
30	111.87	230	188.29		
40	115.81	240	191.99		
50	119.73	250	195.67		
60	123.64	260	199.35		
70	127.54	270	203.01		
80	131.42	280	206.66		
90	135.30	290	210.30		

[a] $R = 100 \ \Omega$ at 0°C.

available corresponding to either curve. Table 9.3, which is based on a more elaborate higher-order polynomial [slight improvement over Eq. (9.10)], presents an abbreviated list of resistance values versus temperature for the U.S. calibration curve.

There are a large number of configurations of RTD sensing elements. Figure 9.20 shows a coiled platinum wire sensor and a thin-film sensor. In the coiled wire sensor, the platinum is wound around a bobbin and the entire assembly is then coated in a ceramic or sealed in a glass envelope. The outer coating prevents damage or contamination. In the thin-film design, platinum is plated on a ceramic substrate and then coated with ceramic or glass. The thin-film design is a newer technology and is gaining favor due to its lower cost. It is important in the design of RTD probes to minimize strain on the platinum due to thermal expansion since strain also causes changes in resistance. Design details minimize strain.

As with strain gages, the Wheatstone bridge is an appropriate circuit to measure the resistance change for an RTD. Figure 9.21(a) shows a Wheatstone bridge as it might be used to measure the resistance of an RTD. Unlike Figure 8.5 for a strain gage, Figure 9.21(a) includes resistance for the lead wires. For strain gages, the lead wire resistance is usually constant during an experiment and has little effect on strain measurement. For RTDs, at least a portion of each lead wire will be close to the region where the temperature is being measured, so the lead wire resistance will also change. If V_{out} is measured as shown, the lead-wire resistances are in the same arm of the bridge as the RTD and the change in lead-wire resistance will simply add to the resistance change of the RTD. The circuit of Figure 9.21(a) will be adequate if the resistance of the lead wires is low and great accuracy is not required. Neglecting the

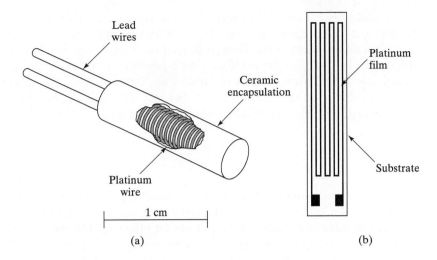

FIGURE 9.20

Resistance temperature detectors: (a) platinum wire; (b) thin film.

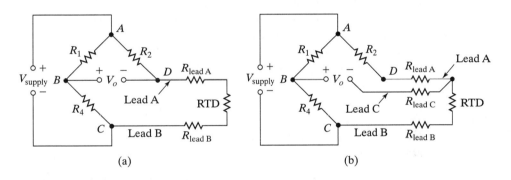

FIGURE 9.21

Wheatstone bridge circuits for RTD: (a) two-wire; (b) three-wire.

lead-wire resistances and assuming that $R_1 = R_4$, circuit analysis yields the following formula for the RTD resistance:

$$R_{RTD} = R_2 \frac{V_{supply} - 2V_o}{V_{supply} + 2V_o} \qquad (9.11)$$

It must be noted that the change in the resistance of RTDs is very large compared to strain gages, and the linearization possible for strain gages is not possible for RTD circuits. As a result, Eq. (9.11) shows a nonlinear relationship between the measured voltage and the RTD resistance.

An alternative circuit called a *three-wire RTD bridge* is shown in Figure 9.21(b), where an additional wire C has been added. With this circuit, $R_{lead\ A}$ is in the same arm of the bridge as R_2, and $R_{\text{lead } B}$ is in the same arm as the RTD. If the lead wires are of the same material, have the same diameter and length, and follow the same routing (not difficult to achieve), the *changes* in lead wire resistance will have a very small effect on V_o. This is the same principle that is involved in using a compensating gage to cancel out temperature effects in strain-gage circuits. There is no current through $R_{\text{lead } C}$, so this resistance has no effect. For this circuit, including the lead wire resistances (with $R_1 = R_4$), the RTD resistance is given by

$$R_{\text{RTD}} = R_2 \frac{V_{\text{supply}} - 2V_o}{V_{\text{supply}} + 2V_o} - R_{\text{lead}} \frac{4V_o}{V_{\text{supply}} + 2V_o} \qquad (9.12)$$

The second term in this equation is usually small, but to obtain the best results, the initial value of the lead resistances should be determined. The fact that R_{lead} (all leads are assumed to have the same resistance) has an effect on the measurement is a consequence of operating the bridge in the unbalanced mode. It is possible to operate the bridge in a balanced mode in which the resistor R_2 is adjusted such that V_o is zero. In this case, $R_{\text{RTD}} = R_2$ and the lead resistances will not affect the result. Unfortunately, it is difficult to use data acquisition systems with the balanced mode. For very high accuracy measurements, however, the balancing mode is preferable.

Figure 9.22 presents two more circuits used to determine the resistance of an RTD. In Figure 9.22(a), the voltage drop across the RTD is sensed with two leads that carry no current and hence have no voltage drop. For this circuit the resistance is a linear function of the measured voltage and is given by

$$R_{\text{RTD}} = V_o I \qquad (9.13)$$

In this circuit V_o is proportional to the resistance of the RTD rather than the change in resistance as is the case with Wheatstone-bridge circuits. Figure 9.22(b) uses four current-carrying leads following the same path to the RTD. Two of the leads plus the RTD are in leg A–D, and the other two leads plus R_3 are in the D–C leg. As with the three-wire bridge, the changes in lead resistance compensate and have a negligible

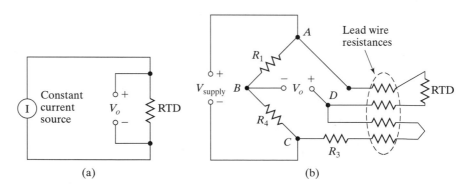

(a)   (b)

FIGURE 9.22

Four-wire RTD circuits.

effect on V_o. The formula to evaluate the RTD resistance is

$$R_{\text{RTD}} = R_3 \frac{V_{\text{supply}} - 2V_o}{V_{\text{supply}} + 2V_o} - R_{\text{lead}} \frac{8V_o}{V_{\text{supply}} + 2V_o} \qquad (9.14)$$

As with the three-wire bridge, the nominal lead resistances should be known for accurate measurements in the unbalanced mode of operation.

Since a current flows through RTDs when they are placed in a resistance-measurement circuit, power dissipates; hence, the RTD is self-heating. This is not normally a problem when measuring liquid temperatures, but it may produce an error in measuring gas temperatures. This self-heating effect can be estimated by using two different supply voltages while measuring a static temperature. Any difference in resistance indicates a potential self-heating problem. Self-heating can be minimized by using low-power supply voltages; however, the output of the sensing circuit will be reduced.

As mentioned, RTD probes have the potential of very high accuracy ($\pm 0.001°C$), but in actual engineering measurements, accuracy will usually be significantly degraded. Uncertainties in bridge resistors and voltage-measuring devices will limit accuracy. As discussed later in this chapter, there are a number of significant errors that affect temperature measurements, particularly for hot gases. Additional information on RTD measurement systems can be found in Benedict (1984).

Example 9.7

An RTD probe has a resistance of 100 Ω at 0°C. The Callendar–Van Dusen constants are $\alpha = 0.00392$, $\delta = 1.49$, and $\beta = 0$ for $T > 0°C$. What will be the resistance at 350°C?

Solution: Substituting into Eq. (9.10) yields

$$R_T = 100\{1 + 0.00392[350 - 1.49(0.01 \times 350 - 1)(0.01 \times 350)$$
$$-0.0(0.01 \times 350 - 1)(0.01 \times 350)^3]\}$$
$$= 232.08 \ \Omega$$

This is a substantial change in resistance—much larger than observed in strain measurements. Alternatively, we could use Table 9.3 and obtain $R = 231.89 \ \Omega$.

9.2.3 Thermistor and Integrated-Circuit Temperature Sensors

As with the RTD, the thermistor is a device that has a temperature-dependent resistance. However, the thermistor, a semiconductor device, shows a much larger change in resistance with respect to temperature than the RTD. The change in resistance with temperature of the thermistor is very large, on the order of 4% per degree Celsius. It is possible to construct thermistors that have a resistance-versus-temperature characteristic with either a positive or a negative slope. However, the most common thermistor devices have a negative slope; that is, increasing temperature causes a decrease in resistance, the opposite of RTDs. They are highly nonlinear, showing a logarithmic relationship between resistance and temperature:

$$\frac{1}{T} = A + B \ln R + C(\ln R)^3 \qquad (9.15)$$

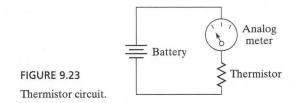

FIGURE 9.23

Thermistor circuit.

As semiconductor devices, thermistors are restricted to relatively low temperatures—many are restricted to 100°C and are generally not available to measure temperatures over 300°C. While probes can be as small as 0.10 in. in diameter, they are still large compared to the smallest thermocouples and have inferior spatial resolution and transient response. Thermistor sensors can be quite accurate, on the order of ±0.1°C, but many are much less accurate.

Thermistors are often used in commercial moderate temperature-measuring devices. Thermistor-based temperature-measuring systems can be very simple, as in Figure 9.23, which shows a circuit frequently used in automobiles to measure engine-water temperature. The current flow through the thermistor is measured with a simple analog mechanical meter. Nonlinearity can be handled by having a nonlinear scale on the meter dial. Thermistors are used in electronic circuits, where they are used to compensate for circuit temperature dependency. Another thermistor application is in simple temperature controllers. For example, thermistor-based circuits can be used to activate relays to prevent overheating in many common devices, such as VCRs. Thermistors are not widely used by either the process industry or in normal engineering experiments since RTDs and thermocouples usually have significant advantages.

As mentioned, the resistance of a thermistor is highly nonlinear. When inserted into a circuit, however, a fairly linear response can readily be achieved. For example, a typical thermistor shows a highly nonlinear resistance variation with temperature, as in Figure 9.24(b). However, the voltage-divider circuit shown in Figure 9.24(a), which includes the thermistor, has an approximately linear output, as shown in Figure 9.24(b). In Figure 9.24(b) both resistance and voltage are normalized by the values at 25°C.

Integrated-circuit temperature transducers, which combine several components on a single chip, have a variety of special characteristics that make them useful for some applications. Some circuits provide a high-level (range 0 to 5 V) voltage output that is a linear function of temperature. Others create a current that is a linear function of temperature. These chips generally have the same temperature-range limitations as thermistors and are larger, giving them poor transient response and poor spatial resolution. One common application is measuring temperature in the connection boxes used when thermocouples are connected to a data-acquisition system. (See Section 9.2.1.)

9.2.4   Mechanical Temperature-Sensing Devices

Liquid-in-Glass Thermometer   Probably the best known of temperature-measuring devices is the liquid-in-glass thermometer as shown in Figure 9.25. The most common liquid is mercury, but alcohol and other organic liquids are also used, depending on the temperature range. Liquid-in-glass thermometers are useful when a ready indication of

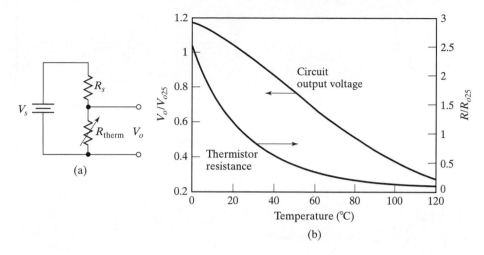

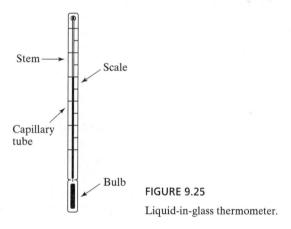

FIGURE 9.24

Linearization circuit for thermistor.

FIGURE 9.25

Liquid-in-glass thermometer.

temperature is required. They are also rather simple devices and are likely to maintain accuracy over long periods of time (many years). Consequently, they are useful for calibrating other temperature-measuring devices. The user should be cautious that the liquid column is continuous (no gaps in the column) and that the glass envelope is free of cracks. The accuracy of liquid-in-glass thermometers can be quite good. Measurement uncertainty depends on the range, but uncertainties in the ±0.2C are quite possible. High-accuracy thermometers are of the total-immersion type. This means that the thermometer is immersed into the fluid from the bulb to the upper end of the liquid column in the stem.

Bimetallic-Strip Temperature Sensors These devices are based on the differential thermal expansion of two different metals that have been bonded together as shown in

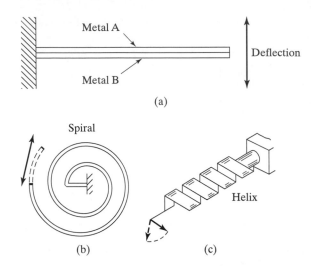

FIGURE 9.26

Bimetallic strip devices. [(b) and (c) Based on
Doebelin, 1990.]

Figure 9.26(a). As the device is heated or cooled, it will bend, producing a deflection of
the end. Figure 9.26(b) and (c) show alternative geometries of these devices. Bimetal
devices have been widely used as the sensing element in simple temperature-control
systems. They have the advantage that they can do sufficient work to perform mechanical functions, such as operating a switch or controlling a valve. Household furnaces are
frequently controlled with bimetallic sensors. They are sometimes used for temperature measurement; for example, domestic oven thermometers are often of this type.
They are not normally used for accurate temperature measurement.

Pressure Thermometers A pressure thermometer, such as the one shown in Figure 9.27,
consists of a bulb, a capillary tube, and a pressure-sensing device such as a bourdon
gage. The capillary is of variable length and serves to locate the pressure indicator in a
suitable location. The system may be filled with a liquid, a gas, or a combination of vapor
and liquid. When filled with a gas, the gage is measuring the pressure of an effectively
constant-volume gas, so pressure is proportional to bulb temperature. With a liquid,
sensing is due to the differential thermal expansion between the liquid and the bulb. In

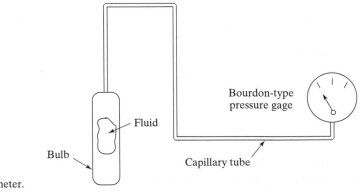

FIGURE 9.27

Pressure thermometer.

both cases the temperature of the capillary might have a slight effect on the reading. In a vapor–liquid system, the gage is reading the vapor pressure of the liquid. These liquid–vapor devices are nonlinear since vapor pressure is usually a very nonlinear function of temperature, but they are insensitive to the temperature of the capillary. Pressure thermometers are used in simple temperature-control systems in such devices as ovens. At one time they were widely used as engine-water temperature-measuring systems in automobiles, but this function has been taken over by thermistor systems.

9.2.5   Pyrometers and Infrared Thermometers

Contact temperature measurements are very difficult at high temperatures since the measurement device will either melt or oxidize. As a result, noncontact devices called *pyrometers* have been developed. These devices measure temperature by sensing the thermally generated electromagnetic radiation emitted from a body. Pyrometers can also be used at lower temperatures as a nonintrusive alternative to contact methods. Any body emits electromagnetic radiation continuously, and the power and wavelength distribution of this radiation are functions of the temperature of the body. An ideal radiating body is called a *blackbody*, and no body at a given temperature can thermally radiate more. The total power radiated by a blackbody, E_b, in watts per square meter of surface, is given by the *Stefan–Boltzmann law*:

$$E_b = \sigma T^4 \tag{9.16}$$

where T is the absolute temperature in kelvin and σ is the Stefan–Boltzmann constant, 5.669×10^{-8} W/m^2-K^4. The distribution of this radiation with wavelength is described with the variable $E_{b\lambda}$, called the *monochromatic emissive power*. The monochromatic emissive power represents the power in a narrow band of wavelengths, $\Delta\lambda$, and is a function of wavelength. The expression for $E_{b\lambda}$ is

$$E_{b\lambda} = \frac{C_1\lambda^{-5}}{e^{C_2/\lambda T} - 1} \tag{9.17}$$

where λ is the wavelength, $C_1 = 3.743 \times 10^8$ W-μm^4/m^2 and $C_2 = 1.4387 \times 10^4 \mu$m-K. Figure 9.28, which plots Eq. (9.17), not only shows that the radiated power increases with increasing temperature but that the wavelength of maximum power decreases with increasing temperature. Thus we have two possible approaches to determining temperature: measurement of total emitted power or determination of the spectral distribution.

One characteristic of the variation in spectral distribution is that the power emitted in the visible spectrum (0.35 to 0.75 μm) depends strongly on the temperature. This fact forms the basis for one of the older designs of pyrometers, the disappearing-filament optical pyrometer, shown schematically in Figure 9.29. This device resembles a telescope but has a red filter in the optics and an incandescent light bulb. The user points the pyrometer at a hot source and adjusts the current flow through the light bulb until the filament appears to disappear, indicating that the filament is at the same temperature as the source. By measuring the current and using a calibration that gives filament temperature versus current, the source temperature can be estimated.

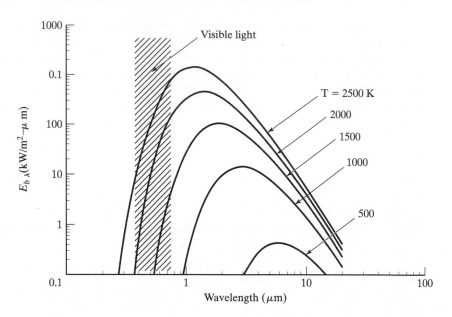

FIGURE 9.28

Blackbody monochromatic emissive power.

Most modern pyrometers giving electrical output signals operate in the infrared region of the spectrum. They are classified as wideband pyrometers, narrowband pyrometers, and ratio pyrometers. The *wideband pyrometer* measures radiation over a large range of wavelengths and effectively determines the temperature using Eq. (9.16). Although the details of the designs vary considerably, a typical optical configuration is shown in Figure 9.30. The detector is usually a thermopile or a thermistor radiation-sensing device. The change in thermopile or thermistor temperature is a function of the radiated energy. The viewing port is used to align the device since the hot object must completely fill the field of view. A block diagram of this type of device is shown in Figure 9.31(a).

A similar kind of pyrometer measures the radiation in a narrow band of wavelengths (the disappearing-filament pyrometer is a form of a *narrowband pyrometer*, in the visible range). For example, it may measure energy only in a band of wavelengths around 0.8 μm. The wavelength used will be different for different temperature ranges. This type of device [Figure 9.31(b)] normally uses a photoconductive type of detector. Photoconductive devices show a variation in resistance when light impinges on them. They are widely used in devices such as cameras.

Unfortunately, most bodies do not radiate as blackbodies. Real bodies only radiate a fraction, called the *emissivity*, of the radiation of a blackbody. The emissivity is defined as

$$\epsilon = \frac{E}{E_b} \tag{9.18}$$

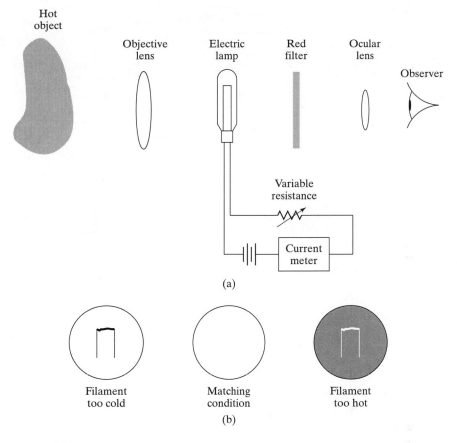

FIGURE 9.29

(a) Disappearing-filament optical pyrometer; (b) appearance to observer.

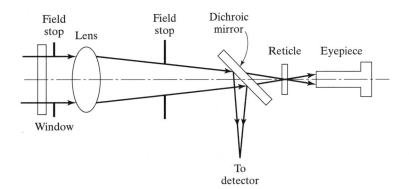

FIGURE 9.30

Optics of wideband pyrometer. (Based on Norton, H. *Sensor and Analyzer Handbook*, Prentice Hall, Englewood Cliffs, NJ, 1982. By permission.)

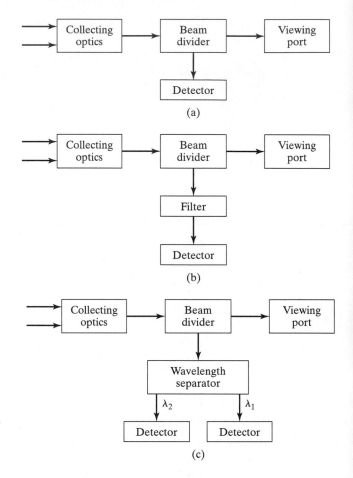

FIGURE 9.31

Pyrometer block diagrams: (a) wideband; (b) narrowband; (c) ratio. (Based on Norton, H. *Sensor and Analyzer Handbook*, Prentice Hall, Englewood Cliffs, NJ, 1982. By permission.)

where E is the emissive power per square meter of the real body. ϵ can vary from low values such as 0.018 for some shiny metals (e.g., gold) to 0.98 for some heavily oxidized surfaces. Variations in emissivity will affect the results obtained from wideband and narrowband pyrometers. An alternative is to measure the radiant energy at two different wavelengths and take the ratio, which is also a function of temperature and less dependent on emissivity. A block diagram for such a *ratio pyrometer* is shown in Figure 9.31(c). Even this type of device will be inaccurate in some cases, since for most substances, the emissivity varies with wavelength.

Pyrometers are available to measure temperatures up to 4000°C and sometimes higher. The dynamic response depends on the sensor with response times (usually the time to achieve 98% of final reading) in the range of a fraction of a second to about 2 s. According to Benedict (1984), optical pyrometers can achieve uncertainties of ±4°C at 1064.43°C (the gold point), ±6°C at 2000°C, and ±40°C at 4000°C.

Infrared measurement is also used in infrared thermometers, which are in widespread use. Like pyrometers, these devices measure the infrared radiation from a surface to determine the surface temperature. Unlike pyrometers, however, they are designed for lower temperatures—some operate from −30°C to 850 °C. Rather than using a telescope type of alignment method, infrared thermometers usually incorporate a laser that points at the location for which the temperature is being determined. Commonly, the sensor is a thermopile, as described in Section 9.2.1. However, the thermopile is constructed using semiconductor chip manufacturing technology and is very small and low in mass, enabling it to respond to changes in temperature very quickly—on the order of 10 ms. [See Schieferdecker, et al. (1995).]

As with pyrometers, infrared thermometers depend on the emissivity of the surface. Common surfaces (e.g., painted) have a high emissivity (greater than 0.8), and the infrared thermometer will give a reasonably accurate reading. On the other hand, shiny metals can have very low emissivity values (often less than 0.1), and the reading of the infrared thermometer will be in serious error. The more expensive infrared thermometers have an adjustment for emissivity, but this is not as effective as one might expect. Shiny metal surfaces reflect infrared radiation from other sources, and this reflected radiation confounds the measurement.

Typical infrared thermometers have an accuracy of 1% of the reading. A common application is the medical ear thermometer, which, due to its narrow range of operation, can be very accurate.

9.2.6   Common Temperature-Measurement Errors

There are three important systematic temperature-measurement errors that are generic and deserve special mention: conduction errors, radiation errors, and recovery errors. It is important to note that a temperature sensor responds to the temperature of the sensor itself, which is not necessarily the temperature of interest. In the cases of the errors discussed earlier, the sensor itself can show excellent accuracy when calibrated but will show major errors when used to make a measurement.

Surface-Temperature Conduction Errors   Figure 9.32 shows a surface to which a thermocouple has been attached. Assuming that the surface is warmer than the surrounding air, heat will be conducted from the surface along the wires. The wires will, in turn, conduct heat to the surrounding air; the thermocouple lead wires act as cooling fins. This heat-conduction process cools the point on the surface where the thermocouple is attached and produces a significant measurement error. The temperature at the measurement point is that which results when a balance exists between the heat transferred from the fin to the surrounding fluid (usually air) and the heat conducted to the

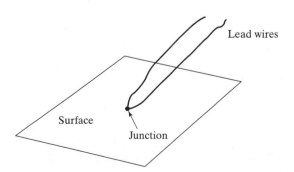

FIGURE 9.32

Surface-temperature measurement.

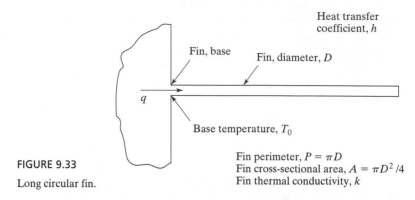

FIGURE 9.33

Long circular fin.

fin through the surface. To understand this surface-cooling process, it is helpful to examine the equations used to describe the heat transfer from fins. Using the nomenclature shown in Figure 9.33, a standard formula for computing the heat loss from a long fin is (Holman, 2002)

$$q = \sqrt{hPkA}\,(T_0 - T_\infty) \qquad (9.19)$$

Substituting for the perimeter P and the cross-sectional area A for a circular fin, this becomes

$$q = \sqrt{\frac{hk\pi^2 D^3}{4}}\,(T_0 - T_\infty) \qquad (9.20)$$

To minimize heat transfer along the fin and hence the error in temperature measurement, we should keep the thermocouple wire small in diameter and use a material with a low thermal conductivity. It should be noted that as the diameter of the thermocouple wire decreases, the heat transfer coefficient increases moderately but not sufficiently to counteract the advantage of the smaller diameter.

The analysis required to estimate the actual temperature error is beyond the scope of this book. Hennecke and Sparrow (1970) present a method to estimate the conduction error in surface measurements. An alternative approach is to set up a three-dimensional heat-transfer model using a finite-element conduction heat-transfer program. In either case, estimates of conduction errors will be approximate at best. Heat-transfer coefficients can only be determined approximately, and insulated sensor wiring is difficult to model.

Conduction Errors in Gas-Temperature Probes   Another kind of conduction error occurs when a thermocouple or similar temperature probe is used to measure a gas temperature, as shown in Figure 9.34. In the situation shown, the sensor approaches an equilibrium temperature based on the various modes of heat transfer to it. These heat transfer modes are convection from the gas, radiation to the duct walls, and conduction along the sensor support to the duct wall. Radiation effects are discussed separately in the

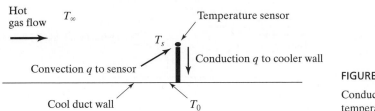

Hot gas flow T_∞

Temperature sensor

T_s

Convection q to sensor

Conduction q to cooler wall

Cool duct wall

T_0

next section. At least approximately, the probe including the sensor can be approximated as a fin, as shown in Figure 9.33. For surface temperature measurement, we were concerned with how the fin affected the attachment point temperature. Here we are interested in how the base affects the temperature at the other end of the fin. Holman (2002) gives the following formula for the temperature distribution along a finite-length fin:

$$\frac{T - T_\infty}{T_o - T_\infty} = \frac{\cosh[m(L - x)]}{\cosh mL} \qquad m = \sqrt{\frac{hP}{kA}} \tag{9.21}$$

where x is the distance from the fin base and L is the fin length. At the end, $x = L$ and $T = $ the sensor temperature T_s, so

$$\frac{T_s - T_\infty}{T_o - T_\infty} = \frac{1}{\cosh mL} \tag{9.22}$$

To minimize the conduction error, we want T_s to approach T_∞. This will occur if $\cosh mL$ (and hence mL) is a maximum. For a fin of circular cross section, mL becomes

$$mL = \sqrt{\frac{4h}{kD}}\, L \tag{9.23}$$

To maximize mL, we want large h, large L, small D, and small k. Convective heat-transfer theory shows that h increases as diameter decreases, so decreasing the diameter has the additional advantage of increasing h. Because values of h are much larger in liquids, this type of conduction error is not likely to be a problem in liquid temperature measurement.

The actual structure of thermocouple probes is often complicated (see Figure 9.17, for example), and the analysis is more complicated than the simple fin theory above. (Finite-element thermal analysis is appropriate.) However, fin theory can be used for a simple probe geometry and as an approximation for more complex situations. The following example demonstrates the application of this approach.

Example 9.8

A thermocouple used to measure the temperature of a gas flow is constructed of two bare wires, as shown in Figure E9.8. The wires are 1.3 mm in diameter and 2 cm long. One wire has a thermal conductivity of 19.2 W/m-K and the other a conductivity of 29.8 W/m-K. The heat-transfer coefficient between the gas and each wire is 200 W/m²-K. The thermocouple reads 375 K

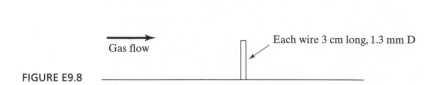

Gas flow

Each wire 3 cm long, 1.3 mm D

FIGURE E9.8

and the wall of the duct is at 320 K. Estimate the conduction error in the measurement of the gas temperature.

Solution: We will treat the probe as two wires with the higher thermal conductivity. (This will slightly overestimate the error.) Using Eq. (9.23), we obtain

$$mL = \sqrt{\frac{4 \times 200}{29.8 \times 0.0013}}\, 0.03 = 4.311$$

Substituting into Eq. (9.21), we obtain

$$\frac{T - T_\infty}{T_o - T_\infty} = \frac{1}{\cosh 4.311} = 0.0268 = \frac{375 - T_\infty}{320 - T_\infty}$$

Solving, $T_\infty = 376.5$ K. The sensor, which reads 275 K, has an error of 1.5 K.

Comment: We could significantly reduce this error by making the wires longer and/or smaller in diameter.

Radiation Errors Radiation errors are normally a problem in measuring hot gas temperatures as shown in Figure 9.35. In this case heat flows into the thermocouple junction by convective heat transfer from the gas but flows out due to radiation heat transfer to the cooler duct walls. The convective heat transfer is given by

$$q = hA(T_{gas} - T_s) \tag{9.24}$$

where h is the heat transfer coefficient and A is the junction surface area. The radiation heat transfer from the junction to the duct wall is given by

$$q = \epsilon A \sigma (T_s^4 - T_w^4) \tag{9.25}$$

where ϵ is a property of the junction surface called the emissivity, T_w is the duct wall temperature and σ is the Stefan–Boltzmann constant, 5.669×10^{-8} W/m²-°K⁴. Combining

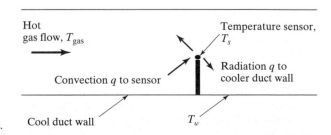

Hot gas flow, T_{gas}

Temperature sensor, T_s

Radiation q to cooler duct wall

Convection q to sensor

FIGURE 9.35
Radiation error source.

Cool duct wall

T_w

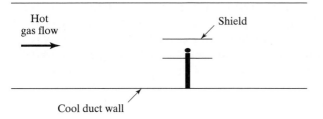

Hot
gas flow

Shield

Cool duct wall

FIGURE 9.36

Reducing radiation error using a radiation
shield.

these equations, we obtain

$$T_{gas} - T_s = \frac{\epsilon}{h}\sigma(T_s^4 - T_w^4) \tag{9.26}$$

Ideally, T_{gas} should equal T_s, so $T_{gas} - T_s$ represents the measurement error. To minimize $T_{gas} - T_s$ it is necessary to minimize ϵ and maximize h. Shiny surfaces on the junction will minimize ϵ. The heat-transfer coefficient for the junction will increase if the
diameter is decreased. Unfortunately, in hot gas environments, most thermocouple materials corrode fairly rapidly, and with small diameters, have a short lifetime. Hence
there is a trade-off between radiation error and durability.

A technique to reduce radiation errors is to install one or more tubular radiation
shields as shown in Figure 9.36. The shield will reach a temperature intermediate between the gas temperature and the wall temperature. The thermocouple effectively ra-
diates to a hotter surface and hence has a smaller radiation error. Commercial probes
using this principle are available.

Example 9.9

A thermocouple is used to measure the temperature of a gas flowing through a duct. The thermocouple gives a reading of 900°C and the wall is found to have a temperature 600°C. The emissivity of the thermocouple sheath is estimated to be 0.1, and the heat-transfer coefficient is
estimated to be 80 W/m²-K. Estimate the radiation error in the thermocouple reading.

Solution: Substituting into Eq. (9.26) yields

$$T_{gas} - T_s = \frac{\epsilon}{h}\sigma(T_s^4 - T_w^4)$$

$$= \frac{0.1}{80}5.669 \times 10^{-8}[(900 + 273)^4 - (600 + 273)^4]$$

$$= 93°C$$

This error, 93°C, indicates how difficult it is to obtain an accurate measurement of hot gases with
insertion probes. If we insert a radiation shield and that shield has a temperature of 800°C, the
error would be reduced to 40°C, still a substantial error.

Recovery Errors Another temperature-measurement error is known as a *recovery*
error. This is most commonly associated with gas flows at high velocities and hence high

Mach numbers (the Mach number is the ratio of the gas velocity to the speed of sound in that same gas). The thermodynamic temperature or static temperature of a gas is that which would be measured by an instrument moving with the velocity of the gas. However, a high-velocity gas has considerable kinetic energy, and if the gas is slowed down, its temperature will increase. Due to viscous effects, the fluid at the surface of a sensor in a gas stream will be slowed to zero velocity. This will increase the temperature of the junction to a temperature higher than the static temperature of the flowing fluid. If there is no other heat transfer to the junction surface, it will reach a temperature called the *adiabatic wall temperature*. This temperature is given by (Shapiro, 1953),

$$T_{aw} = T_{static} + R\frac{V^2}{2c_p} \qquad (9.27)$$

where R is the recovery factor, V is the gas velocity, and c_p is the constant-pressure specific heat of the gas. The recovery factor, which depends on the fluid, the Mach number, and the orientation and shape of the probe, varies between 0.68 and 0.86 (Moffat, 1962). Further information on temperature-measurement errors can be found in Benedict (1984).

Example 9.10

In a centrifugal air compressor test, a thermocouple probe reads 325°C and the air velocity is 160 m/s. The recovery factor is 0.86 and the gas specific heat is 1055 J/kg-K. Estimate the static temperature of the air.

Solution: Solving Eq. (9.58) for T_s, we obtain

$$T_s = T_{aw} - R\frac{V^2}{2c_p} = 325 - 0.86\frac{160^2}{2 \times 1055} = 325 - 10.4 = 314.6°C$$

Comment: This example estimates the static temperature based on a measurement of T_{aw}. This effect is large in this example but will be even greater in higher-velocity situations (actually, higher-Mach-number situations). In many cases the gas is also very hot and there will be significant radiation errors as well. Measuring gas temperature in high-velocity hot gases is often rather difficult.

9.3   MEASURING HUMIDITY

Humidity is a measure of the amount of water vapor in air. Water vapor in air affects the density, and humidity measurement is necessary to determine the performance of many systems. The characteristics of air–water vapor mixtures are discussed in most thermodynamics texts, such as that of Van Wylen, et al. (1994). Humidity can be specified by a number of parameters, the two most common being the humidity ratio and the relative humidity. The *humidity ratio* is defined as the ratio of the water vapor mass to the mass of dry air in the mixture:

$$\omega = \frac{\text{mass water vapor}}{\text{mass of dry air}} \qquad (9.28)$$

The *relative humidity* is defined as the ratio of the mass of water vapor in a volume at a given temperature and pressure to the mass of water vapor in the same volume at the same temperature and total pressure if the water vapor is saturated (i.e., the maximum possible water vapor is present):

$$\phi = \left(\frac{\text{actual water vapor mass}}{\text{saturated water vapor mass}} \right)_{\text{constant } T \text{ and } P} \tag{9.29}$$

The vapor content of the air–water vapor mixture is completely defined by determining either the humidity ratio or the relative humidity. There are a number of devices that can be used to determine humidity.

9.3.1   Hygrometric Devices

Certain hygroscopic materials change properties depending on the moisture content, which in turn depends on the humidity. Human hair, some animal membranes, and certain plastics show a dimensional change in response to humidity. The humidity can be determined by sensing this dimensional change. This principle is used for the humidity gage in inexpensive mechanical home weather stations, although these devices are rarely very accurate. Nevertheless, hygrometric devices suitable for laboratory use can be constructed. Some sense the displacement with strain gages or other electrical output devices and can be used for remote sensing. Certain materials show a change in resistance in response to changes in humidity and can be used for remote-sensing applications. A common probe that is sensitive to a wide range of humidities is based on a moisture-sensitive polymer film, which forms the dielectric of a capacitor. Measuring the capacitance is a function of humidity. At present, capacitance humidity sensors are probably the most widely used hygrometric devices.

9.3.2   Dew-Point Devices

If a sample of air is cooled at constant pressure to a temperature called the *dew-point temperature*, the moisture in the air will start to condense. This temperature, plus the initial temperature, can be used to determine the amount of vapor in the air.

9.3.3   Psychrometric Devices

If the bulb of a thermometer measuring an air temperature is surrounded with a wick wetted with water, it will measure a temperature lower than the true air temperature. Evaporation of the water in the wick will cool the bulb. This is the basis of the sling psychrometer, shown in Figure 9.37. One thermometer measures the true air temperature called the dry-bulb temperature, and the thermometer with the wet wick measures the wet-bulb temperature. If the device is used properly, these two temperatures can be used to determine the air humidity.

The temperature of the wet bulb is dependent on several heat-transfer processes. The sought-after effect is convective heat and mass transfer between the wet wick and the surrounding air. Unfortunately, there is also conduction heat transfer along the stem and radiation heat transfer to the surroundings. If the air velocity relative to the wick is on the order of 5 m/s, it has been observed that the wet-bulb temperature will be very

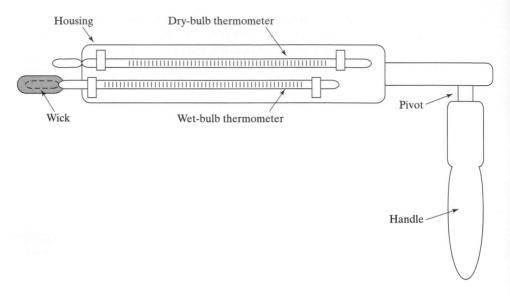

FIGURE 9.37

Sling psychrometer.

close to a temperature called the *adiabatic saturation temperature*. This air velocity is achieved with the sling psychrometer by causing the housing to swing around the handle in a circular motion. It should be noted that the adiabatic saturation temperature is not the same as the dew-point temperature. The dry-bulb temperature can be used, together with the adiabatic saturation temperature, to determine the air humidity. A convenient way to do this is to use a graph called a psychrometric chart, shown in Figure 9.38. Figure 9.38 is valid for an atmospheric pressure of 1 atm. More detailed psychrometric charts for both British and SI units are contained in Appendix B. A complete set of

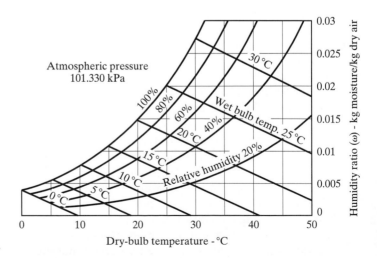

FIGURE 9.38

Psychrometric chart.

charts that includes the effects of altitude, is available from the American Society of Heating, Refrigerating, and Air Conditioning Engineers (ASHRAE). Wet- and dry-bulb temperatures can also be sensed with stationary thermocouples or other temperature sensors. In this case, a fan or blower is required to produce the required air velocity.

Example 9.11

A sling psychrometer measures a dry-bulb temperature of 33°C and a wet-bulb temperature of 21°C. What are the relative humidity and the humidity ratio?

Solution: Using Figure B.1, we find that the relative humidity is approximately 34% and the humidity ratio is 0.0108 kg water/kg air.

9.4   FIBER-OPTIC DEVICES

Fiber-optic sensing is a new and emerging technology that is expected to find widespread use in engineering applications. Fiber-optics systems are more a method of signal transmission than a sensing technology, but two of the more important sensing applications are temperature and pressure measurement. Fundamentals of operating fiber optics and fiber-optic sensors are introduced briefly in this section. For more detail the reader may refer to detailed references, including Allan (1973), Krohn (2000), Hentschel (1989), Grattan and Meggit (2000), and Lopez–Higuera (2002).

9.4.1   Optical Fiber

An optical fiber is basically a guidance system for light and is usually cylindrical in shape. If a light beam enters from one end face of the cylinder, a significant portion of energy of the beam is trapped within the cylinder and is guided through it and emerges from the other end. Guidance is achieved through multiple reflections at the cylinder walls. Internal reflection of a light ray is based on Snell's law in optics. If a light beam in a transparent medium strikes the surface of another transparent medium, a portion of the light will be reflected and the remainder may be transmitted (refracted) into the second medium. Referring to Figure 9.39, the direction of the refracted wave is governed by *Snell's law*,

$$n_1 \sin \theta_1 = n_2 \sin \theta_2 \tag{9.30}$$

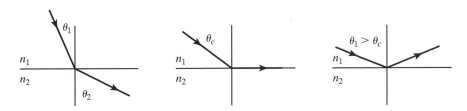

FIGURE 9.39

Refraction and reflection of light rays at a dielectric interface defined by refractive indices n_1 and n_2 ($n_1 > n_2$).

where n is the refractive index of the medium, the ratio of the speed of light in a vacuum to the speed of light in a transparent material, and θ is the angle of the light beam with the normal to the deflection surface. For the case that light is passed from a higher n medium into a medium with a lower n (such as the water–air interface), beyond a certain angle of incidence the light ray will not penetrate the medium with lower n and will be entirely reflected into the higher n medium. The minimum angle resulting in this total reflection is called the *critical angle of reflection* and is derived from Eq. (9.30) by setting θ_2 equal to 90°:

$$\sin \theta_c = \frac{n_2}{n_1} \tag{9.31}$$

A light ray will be completely reflected from the interface at all angles of incidence greater than θ_c.

   Basically, the optical fiber is a cylinder of transparent dielectric material surrounded by another dielectric material, called cladding, with a lower refractive index. In practice, a third protective layer is also required. (See Figure 9.40.) Figure 9.41, which shows the ray diagram of an optical fiber, indicates that the rays that enter the fiber beyond the acceptance angle may not be fully transmitted through the fiber and will eventually be lost.

   Optical fibers have extensive application in telecommunication and computer networking, but their application as sensing devices is not that widespread yet. Applying fiber-optic–based sensors is an emerging technology and is expected to grow in the near

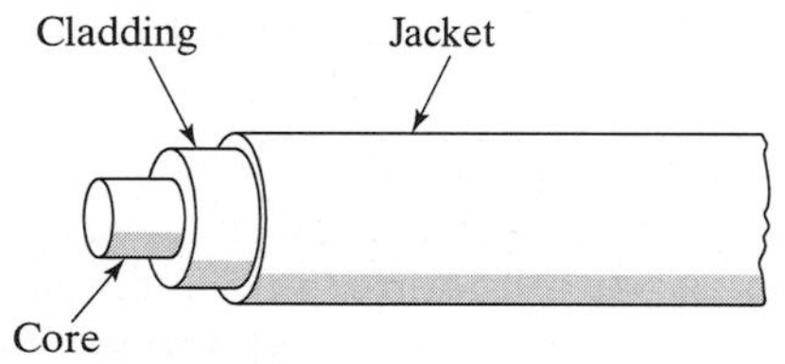

FIGURE 9.40

Typical optical fiber.

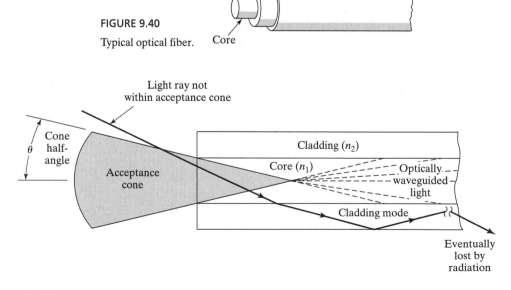

FIGURE 9.41

Ray transmission through an optical fiber.

future. Optical sensing and signal transmission have several potential advantages over conventional electric output transducers and electric signal transmission. Major advantages are as follows:

1. Nonelectric (optical fibers are immune to electromagnetic and radio-frequency interference).
2. Explosion proof.
3. High accuracy.
4. Small size (both fibers and the attached sensors can be very small in size, applicable to small spaces with minimum loading and interference effect).
5. High capacity and signal purity.
6. Can be easily interfaced with data-communication systems.
7. Multiplexing capability (numerous signals can be carried simultaneously, allowing a single fiber to monitor multiple points along its length or to monitor several different parameters).

According to Krohn (2000), most physical properties can be sensed with fiber-optic sensors. Light intensity, displacement (position), pressure, temperature, strain, flow, magnetic and electrical fields, chemical composition, and vibration are among the measurands for which fiber-optic sensors have been developed. In the sections that follow, basic principles and some typical fiber-optic sensors are introduced.

9.4.2 General Characteristics of Fiber-Optic Sensors

Fiber-optic sensors can be divided into two general categories, intrinsic and extrinsic. In *intrinsic sensors*, the fiber itself performs the measurement, while in *extrinsic sensors* a coating or a device at the fiber tip performs the measurement. Figure 9.42 shows schematic

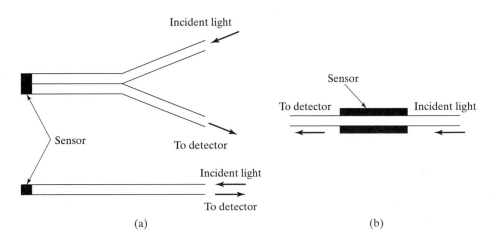

FIGURE 9.42

Two types of fiber-optic sensors: (a) extrinsic; (b) intrinsic.

diagrams of these two types of sensors. Depending on the sensed property of light, fiber-optic sensors are also divided into phase-modulated sensors and intensity-modulated sensors. *Phase-modulated sensors* compare the phase of light in a sensing fiber to a reference fiber in a device called an *interferometer*. Phase difference can be measured with extreme sensitivity. In *intensity-modulated sensors* the perturbation causes a change in received light intensity, which is a function of the phenomenon being measured (measurand). Intensity-modulated sensors are simpler, more economical, and more widespread in application, so the discussion here is limited to this type. What will follow here is a brief introduction of these sensors for some mechanical measurements.

9.4.3 Fiber-Optic Displacement Sensors

Two concepts that are widely used in fiber-optic sensors are reflective and microbending concepts. Both concepts sense displacement but can be used for other measurements if the measurand can be made to produce a displacement. Figure 9.43 shows the basic concept of a reflective-displacement sensor. In a *reflective sensor* a pair or two bundles of fibers are used. One bundle is used for transmitting the light to a reflecting target while the other collects (traps) the reflected light and transmits it to a detector. Any motion or displacement of the reflecting target can affect the reflected light that is being transmitted to the detector. The intensity of the reflected light captured depends on the distance of the reflecting target from the optic probe. A typical response curve is also shown. Following the basics of geometric optics, the behavior of this response curve can be interpreted and used in measuring displacement. Plain reflective displacement sensors have a limited dynamic range of about 0.2 in. This can be improved by using a lens system (shown schematically in Figure 9.44) to 5 or more inches (Krohn, 2000). Disadvantages of this type of sensor are that they are sensitive to the orientation of the reflective surface and to the contamination of the reflective surface.

Microbending is another attractive and widely used technology in fiber-optic sensors. If a fiber is bent as shown in Figure 9.45(a), a portion of the trapped light is lost through the wall of the fiber. The amount of the received light at the detector compared to the light source is a measure of the physical property influencing the bend. As Figure 9.45(b) shows, microbending can be used to measure displacement (or strain).

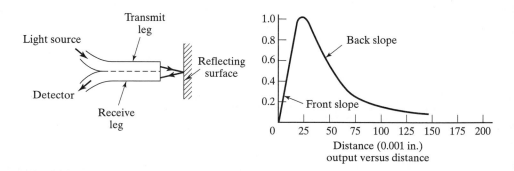

FIGURE 9.43

Reflective fiber-optic response curve for displacement measurement. (After Krohn, 2000.)

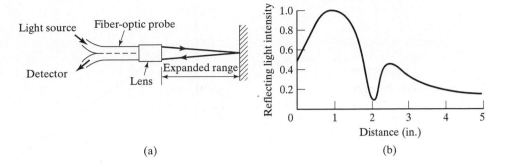

(a)

(b)

FIGURE 9.44

Fiber-optic displacement transducer with lens: (a) configuration; (b) response curve. (After Krohn, 1992.)

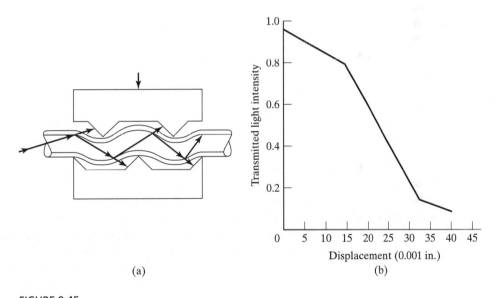

(a)

(b)

FIGURE 9.45

(a) Microbending fiber-optic sensor; (b) output. (Based on Krohn, 2000.)

9.4.4 Fiber-Optic Temperature Sensors

Several fiber-optic sensing concepts have been used in measuring temperature. These include reflective, microbending, intrinsic, and other intensity- and phase-modulated concepts. Some of the more common sensors are described briefly here. In one type of reflective sensor shown in Figure 9.46, the displacement of a bimetallic element is used as an indication of temperature variation. The response curve of such a sensor, taken from Krohn (2000), is also shown. These types of sensors are used both as thermal switches and analog sensors.

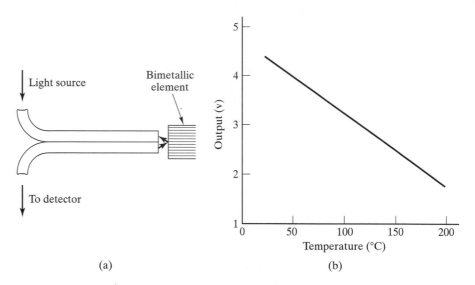

(a) (b)

FIGURE 9.46

(a) Reflective fiber-optic temperature transducer using a bimetallic sensor; (b) typical output curve for such a device. (Based on Krohn, 2000.)

According to Krohn (2000), active sensing materials such as liquid crystals, semi-conductor materials, materials that fluoresce, and other materials that can change spectral response can be placed in the optical path of a temperature probe to enhance the sensing effect. A fiber-optic sensor using fluorescent material to emit light as a function of temperature is shown in Figure 9.47. The fiber tip is coated with a phosphor layer and encapsulated. An incoming ultraviolet (UV) light excites the phosphor, which emits light in the visible spectrum. The visible light is carried back to the electro-optic interface via the same fiber. The incoming and outgoing light beams are separated by a beam splitter, and their intensity is measured. The intensity ratio of the two beams is a direct function of the temperature of the phosphor. According to Krohn (2000), the operating range of this type of sensor is −50°C to above 250°C with accuracy of better than 1°C. Because of the small size of these sensors, they can have superior response time to many conventional methods of temperature measurement. Small fluorescent temperature-measurement devices can have an accuracy of ±0.1°C and a response

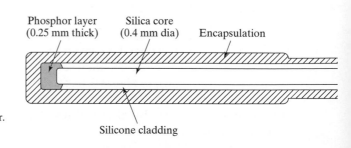

FIGURE 9.47

Sensor tip of a fluorescent sensor.
(From Krohn, 2000.)

time as low as 25 ms. The concept of microbending is also used for temperature measurement. Using thermal expansion of a component structure, the sensor can measure the temperature by altering the fiber bend radius with temperature. This affects the amplitude of the transmitted light.

As discussed in Chapter 8, all materials emit thermal radiation as a function of their temperature and surface properties. The radiated light from a surface (which represents the surface temperature) can be collected and measured by a fiber-optic sensor called a blackbody fiber-optic sensor. Blackbody fiber-optic sensors use silica or sapphire fibers, with the fiber tip coated with precious metal for light collection. These sensors can have a range of 500 to 2000°C. Fiber-optics–based sensors that operate on an interferometric principle, using mirrors on the end of a single-mode fiber, have been reported to operate in the range of −200°C to 1050°C [Lee, et al. (1988)].

9.4.5 Fiber Optic Pressure Sensors

Several fiber optic concepts have been used in design of fiber optic pressure sensors which have demonstrated high accuracy. Figure 9.48 shows a fiber optic pressure sensor that modulates the light intensity transmitted through a fiber by movement of a shutter. The shutter is moved by a diaphragm which senses the pressure. Other forms of transmissive pressure sensors are discussed by Krohn (2000). The reflective concept has also been used in design of fiber optic pressure sensors. Figure 9.49 shows the application of the concept in a diaphragm pressure transducer. The change in the position of the diaphragm changes the amount of light that is reflected into the receiving fiber as an indication of the sensed pressure.

9.4.6 Other Fiber-Optic Sensors

Fiber-optic-type sensors have found applications in practically every field [see LopezHiguera, (2002)], including measuring strain and temperature in complex structures [Measures, (2001)]; biomedical, chemical, and environmental sensing [Gratten and Meggitt, (1999)]; and measuring electrical current and voltage. Details of the principles and technology behind these various sensors are beyond the scope of this book.

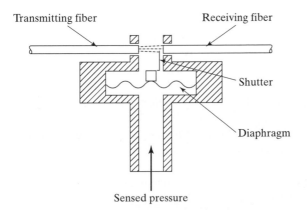

FIGURE 9.48

Transmissive fiber optic pressure sensor. (Based on Krohn, 2000.)

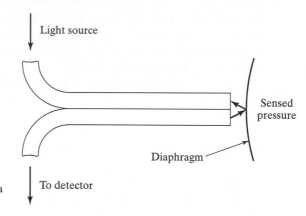

FIGURE 9.49

Reflective fiber optic pressure sensor using a diaphragm. (Based on Krohn, 2000.)

REFERENCES

[1]   ALLAN, W. B. (1973). *Fibre Optics: Theory and Practice*, Plenum Press, New York.

[2]   BENEDICT, R. (1984). *Fundamentals of Temperature, Pressure and Flow Measurements*, Wiley, New York.

[3]   DALLY, J., RILEY, W., AND MCCONNELL, K. (1993). *Instrumentation for Engineering Measurements*, 2d ed., Wiley, New York.

[4]   DOEBELIN, E. O. (1990). *Measurement Systems: Application and Design*, 4th ed., McGraw-Hill, New York.

[5]   GRATTEN, K. T. V. AND MEGGITT, B. T. (1999). Optical Fiber Sensing Technology, *Chemical and Environmental Sensing*, Vol. 4, Kluwer Academic Publishers, Norwell, MA.

[6]   GRATTEN, K. T. V. AND MEGGITT, B. T. (2000). *Optical Fiber Sensor Technology*, Kluwer Academic Publishers, Norwell, MA.

[7]   HENNECKE, D. K., AND SPARROW, E. M. (1970). Local heat sink on a convectively cooled surface: application to temperature measurement error, *International Journal of Heat and Mass Transfer*, Vol. 13, Feb.

[8]   HENTSCHEL, C. (1989). *Fiber Optics Handbook*, Hewlett-Packard, Dallas, TX.

[9]   HOLMAN, J. P. (2002). *Heat Transfer*, 7th ed., McGraw-Hill, New York.

[10] HOTTEL, H. C., AND SAROFIM, A. F. (1967). *Radiative Transfer*, McGraw-Hill, New York.

[11] KAIL, R., AND MAHR, W. (1984). *Piezoelectric Measuring Instruments and Their Applications*, Kistler Instrument Corp., Amherst, NY.

[12] KROHN, D. A. (2000). *Fiber Optic Sensors: Fundamental and Applications*, 2d ed., Instrument Society of America, Research Triangle Park, NC.

[13] LEE, C. E., ATKINS, R. A., AND TAYLOR, H. F. (1988). Performance of a Fiber Optic Temperature Sensor from −200 to 1050 C, *Optics Letters*, Vol. 13, pp. 1038–1040, Nov. 1988.

[14] LOPEZ–HIGUERA, J. M. (2002). *Handbook of Optical Fiber Sensing Technology*, Wiley, New York.

[15] MCGEE, T. D. (1988). *Principles and Methods of Temperature Measurement*, Wiley, New York.

[16] MEASURES, R. M. (2001). *Structural Monitoring with Fiber Optic Technology*, Academic Press, San Diego.

[17] MOFFAT, R. J. (1962). Gas temperature measurements, in *Temperature: Its Measurement and Control*, Vol. 3, Part 2, Reinhold, New York.

[18] NORTON, H. N. (1982). *Sensor and Analyzer Handbook*, Prentice-Hall, Englewood Cliffs, NJ.

[19] OMEGA ENGINEERING (regularly updated). *The Temperature Handbook*, Omega Engineering, Inc., Stamford, CT.

[20] POPOV, E. P. (1976). *Mechanics of Materials*, Prentice-Hall, Englewood Cliffs, NJ.

[21] RESNICK, R. AND HALLIDAY, D. (1966). *Physics*, Part I, Wiley, New York.

[22] SHAPIRO, A. (1953). *The Dynamics and Thermodynamics of Compressible Fluid Flow*, Ronald Press, New York.

[23] SCHIEFERDECKER, J., QUAD, R., HOLZENKAMPFER, E., AND SCHULZE, M. (1995). Infrared thermopile sensors with high sensitivy and very low temperature coefficient, *Sensors and Actuators* A 46–47, pp. 422–427.

[24] STREETER, V. L., AND WYLIE, E. B. (1985). *Fluid Mechanics*, McGraw-Hill, New York.

[25] VAN ATTA, C. (1965). *Vacuum Science and Engineering*, McGraw-Hill, New York.

[26] VAN WYLEN, G., SONNTAG, R., AND BORGNAKKE, C. (1994). *Fundamentals of Classical Thermodynamics*, 4th ed., Wiley, New York.

[27] WHITE, F.M. (1999). *Fluid Mechanics*, 4th ed., McGraw-Hill, New York.

PROBLEMS

9.1 A gas pressure of 150,000 Pa (gage) is applied to one leg of a U-tube manometer, which contains a fluid with density of 2000 kg/m^3. The other leg is open to atmosphere. What is the manometer reading?

9.2 A gas pressure difference is applied to the legs of a mercury-filled ($S = 13.6$) U-tube manometer. The manometer reading is 45.3 in. Find the pressure difference in psid.

9.3 Convert the pressure 5.1 psi into units of inches of mercury and feet of water.

9.4 Convert the pressure 65 kPa into units of millimeters of mercury and meters of water.

9.5 Determine the following conversion factors:

    (a) The factor to multiply mm of Hg to get Pa ($S_{Hg} = 13.6$).

    (b) The factor to multiply inches of H$_2$O to get psi.

    (c) The factor to multiply Pa to get m of H$_2$O.

9.6 A vacuum pressure (below atmospheric) is applied to the column port of a well-type manometer. The well port is open to the atmosphere. The fluid is mercury ($S = 13.6$). If the reading R is 130 mm, what is the applied pressure relative to atmospheric pressure?

9.7 A well-type inclined manometer has a gas pressure difference of 3 in. of water column applied to its ports (higher pressure to well). If the manometer is at an angle of 7° and contains a fluid with a density of 50 lbm/ft^3, what is the reading R?

9.8 An inclined manometer is used to measure the differential gas pressure across an orifice. The angle is 30° and the fluid used in the manometer has a specific gravity of 0.8. Calculate the sensitivity (cm length/Pa) and the resolution of the manometer (in Pa) if 0.5 mm (length) can be resolved with a naked eye.

9.9 In Problem 9.8, calculate the length of the tube for measuring a maximum pressure of 10 cm of water.

9.10 Determine the gas pressure exerted on the reservoir of an inclined manometer (P_1 in Fig. 9.3) if it has a 15-degree angle, uses a fluid with specific gravity of 0.7, and reads 10.2 cm.

9.11 Determine the systematic error as a percentage of the reading if the actual angle of an inclined manometer is 15.5 degrees while it is considered 15.0 degrees.

9.12 The scale of a mercury barometer is made of aluminum, which has a linear coefficient of thermal expansion of 23×10^{-6} m/m-K. Mercury has a volume coefficient of expansion of 1.82×10^{-4} m³/m³ -K. If the reading of 760 mm is correct at 20°C, what will be the reading at 35°C if no correction is made for temperature? Assume that the zero is properly adjusted.

9.13 An R-type thermocouple with a 0°C reference junction reads 2.5 mV. What is the temperature?

9.14 A type S thermocouple with an ice reference has an output of 4.005 mV. What is the temperature of the hot junction?

9.15 Consider the standard chromel–alumel thermocouple circuit shown in the figure.

    (a) Find the voltage measured by the DVM.

    (b) If the DVM measures 17.51 mV, what would the temperature T_1 be?

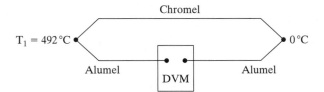

FIGURE P9.15

9.16 Prepare a table in 20°C increments between 0°C and 200°C for the output voltage of the thermocouple circuit as a function of T_1. The reference junction is boiling water at 100°C.

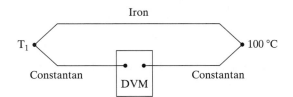

FIGURE P9.16

9.17 A type E thermocouple is placed in an oven and connected to a computer data-acquisition system. The junction box temperature is independently measured to be 30°C. The thermocouple voltage is found to be 37.0 mV. What is the temperature of the oven?

9.18 A J-type thermocouple with an ice-reference junction showed a reading of 22 mV. After the experiment, it was found that the actual temperature of the reference junction is 4°C instead of 0°C. Calculate the actual value of the measured temperature.

9.19 A set of readings with a K-type thermocouple have been performed using a reference junction that has been assumed to be at 0°C. After careful inspection, because the ice–water mixture did not really make a slush, the reference junction has been determined to be at 1°C. Calculate the systematic error in this measurement.

9.20 An RTD is connected into a Wheatstone bridge of the type shown in Figure 9.21(b). The other bridge resistors have values of 100 Ω. The supply voltage is 3 V and the bridge output voltage is −0.5 V.

    (a) Evaluate the RTD resistance neglecting the effect of lead wires.

    (b) If the lead-wire resistance of 1 Ω is considered, what is the RTD resistance?

(c) What is the error in resistance due to neglecting the lead-wire resistance?

(d) If the RTD follows the curve of Table 9.3, what is the error in temperature?

9.21 A platinum RTD with U.S. calibration has a resistance of 100 Ω at 0°C. What is the resistance at 375°C?

9.22 A nominally 100-Ω RTD (0°C) following Table 9.3 is inserted into a bridge like Figure 9.21(b). R_2 is adjusted to account for the initial lead resistances and give a zero-voltage output at 0°C. The supply voltage is 2 V. What will be the voltage output at 390°C if

(a) the lead-wire resistance is neglected?

(b) the lead-wire resistance is 2 Ω?

9.23 The RTD of Problem 9.22 is inserted into a three-wire bridge like Figure 9.21(b). The initial lead resistance is 2 Ω and the supply voltage is 2 V. What will the voltage output be at a temperature of 60°C? What will the output voltage be if the lead-wire resistance is assumed to be zero?

9.24 Argon ($M = 39.94$) is used in a pressure thermometer. The volume of the thermometer bulb is 0.1 L. The thermometer is filled with gas to 110 kPa (abs) at 20°C. What is the sensitivity (variation of pressure with respect to temperature) of this temperature sensor? Explain your assumptions clearly.

9.25 In Problem 9.24, if the temperature rises to 120°C, calculate the pressure of the gas in the thermometer.

9.26 Assuming that the bulb in Problem 9.24 is made of copper in spherical form, which readily expands with increasing temperature, calculate the error in the pressure reading if this expansion is not accounted for. Is this a systematic or a random error? The coefficient of linear thermal expansion for copper is 16.5×10^{-6} m/m-K.

9.27 Helium is used in a pressure thermometer. The volume of the thermometer bulb is 0.05 L. The thermometer is filled with gas to 110 kPa (abs) at 20°C. Calculate the pressure of the gas if the temperature is raised to 120°C. State your assumptions clearly.

9.28 Helium is used in a pressure thermometer. The volume of the thermometer bulb is 0.0.05 L. The bulb is made of stainless steel with a thermal expansion coefficient of 12×10^{-6} m/m-K. The thermometer is filled with gas to 100 kPa (abs) at 20°C. Calculate the systematic error in temperature due to neglecting the expansion of the bulb caused by the temperature rise at 150°C.

9.29 A thermocouple is used to measure the temperature of a hot gas in a duct. The thermocouple reads 800°C and the duct wall has a temperature of 700°C. If the thermocouple junction emissivity is 0.3 and the heat-transfer coefficient is 200 W/m²-C, estimate the measurement error due to radiation.

9.30 Calculate the actual temperature of exhaust gas from a diesel engine if the measuring thermocouple reads 500°C and the exhaust pipe is 350°C. The emissivity of the thermocouple is 0.7 and the convective heat-transfer coefficient of the flow over the thermocouple is 200 W/m²-C.

9.31 In Problem 9.30 a shield is used to reduce the error. If the shield temperature is about 460°C, calculate the error in the output temperature. The gas temperature remains the same.

9.32 A thermometer is used to measure the temperature of a gas flow through a pipe. The gas temperature is read at 500°C and the duct temperature is measured to be 350°C. The thermometer is cylindrical with a diameter of 1 mm and an intrusion length of 1 cm. It can be assumed that the conductivity of the thermometer material is the same as stainless steel, about 20 W/m-K. The emissivity of the thermometer tip is 0.2. The convective heat-transfer coefficient between the thermometer and the gas is estimated to be 100 W/m^2-K. Estimate the error in reading the gas temperature. Calculate the actual gas temperature?

9.33 In Problem 9.30, the measured reading of the gas temperature and the reading of the wall temperature are considered very accurate, but the emissivity of the thermocouple is uncertain to ±0.1 and the heat-transfer coefficient is uncertain to ±15%. Estimate the temperature correction for radiation and the uncertainty in that correction.

9.34 Calculate the error in reading the temperature of a gas moving at 250 m/s. The recovery factor is 0.8 and the specific heat at constant pressure of the gas is 1.2 kJ/kg-K.

9.35 Calculate the air velocity that may cause 1% recovery error in measuring static temperature of air flow at 25°C. Take the recovery factor to be 0.75.

9.36 A sling psychrometer gives dry-bulb and wet-bulb readings of 80°F. What is the relative humidity?

9.37 For Problem 9.36, what would be the mass of water in 1 kg of an air–vapor mixture?

9.38 A sling psychrometer has a dry-bulb reading of 30°C and a wet-bulb reading of 15°C. Find the relative humidity and the humidity ratio.

9.39 If a sling psychrometer gives a dry-bulb temperature of 25°C and a wet-bulb temperature of 22°C, what are the relative humidity and the humidity ratio?

C H A P T E R   1 0

Measuring Fluid Flow Rate, Fluid Velocity, Fluid Level, and Combustion Pollutants

In this chapter, we describe the most common methods for measuring fluid flow rate, fluid velocity, and liquid level. Also discussed are devices used to measure the pollutants in the exhaust flows from combustion processes.

10.1 SYSTEMS FOR MEASURING FLUID FLOW RATE

In engineering experiments and in process plants, people often need to know the quantity of a fluid flowing through a conduit. The fluid may be either a gas or a liquid, and the information is normally needed on a time-rate basis (e.g., mass flow rate in kg/s or volume flow rate in m³/s). Because fluid-flow-rate measurements are so common and because there are so many different applications, a large number of distinctly different fluid flow measurement devices (flowmeters) have been developed. Although we describe the most common flowmeters in this chapter, the user should consult appropriate references or vendors before applying the devices.

10.1.1 Pressure Differential Devices

Technical Basis   The Bernoulli equation accurately describes the flow of an incompressible fluid inside a duct or pipe if frictional and other losses are negligible. For two axial locations (sections) in a duct, labeled 1 and 2 in Figure 10.1, the *Bernoulli equation* takes the form

$$\frac{V_1^2}{2} + \frac{P_1}{\rho} + gz_1 = \frac{V_2^2}{2} + \frac{P_2}{\rho} + gz_2 \tag{10.1}$$

where V is the fluid velocity, P is the fluid pressure, z is the elevation of the location in the pipe relative to a specified reference elevation (datum), ρ is the fluid density, and g

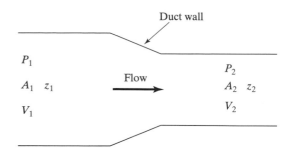

FIGURE 10.1

Flow in a duct with area change.

is the acceleration of gravity. The velocities at two axial locations in the duct with different areas are related through the conservation of mass equation,

$$\rho_1 V_1 A_1 = \rho_2 V_2 A_2 = \dot{m} \tag{10.2}$$

where A is the duct cross-sectional area and $\dot{m}$ is the fluid mass flow rate (e.g., kg/s). For an incompressible fluid, the density is a constant and Eq. (10.2) is usually written in the form

$$V_1 A_1 = V_2 A_2 = Q \tag{10.3}$$

where Q is the volume flow rate (e.g., m³/s). Equations (10.1) and (10.3) can be combined to obtain an expression for the velocity at section 2:

$$V_2 = \frac{1}{\sqrt{1 - (A_2/A_1)^2}} \sqrt{\frac{2[(P_1 + g\rho z_1) - (P_2 + g\rho z_2)]}{\rho}} \tag{10.4}$$

Substituting into Eq. (10.3) yields the equation for the flow rate:

$$Q = \frac{A_2}{\sqrt{1 - (A_2/A_1)^2}} \sqrt{\frac{2[(P_1 + g\rho z_1) - (P_2 + g\rho z_2)]}{\rho}} \tag{10.5}$$

Equation (10.5) is the theoretical basis for a class of flowmeters in which the flow rate is determined from the pressure change caused by variation in the area of a conduit. These flowmeters—the *venturi tube*, the *flow nozzle*, and the *orifice meter*—are described in the following sections. The devices are collectively called *head meters* because they depend on the change in head ($P/\rho g + z$), which has units of length. Their theory and application are presented in detail in the publication *Fluid Meters* prepared by the American Society of Mechanical Engineers (ASME, 1971). ASME Standard MFC-3M-1989 (ASME, 1989) presents the standardized characteristics of head meters. Another useful reference is Miller (1989).

Venturi Tube   The venturi tube, shown schematically in Figure 10.2, is probably the most expensive of the common flowmeters based on Eq. (10.5), but it has the lowest energy losses. As Figure 10.2 shows, the area of the pipe contracts from the initial pipe area A_1 to the minimum area A_2 (called the *throat*). The area then gradually increases to the initial pipe area, A_1. If the venturi is oriented horizontally, the elevation terms z_1

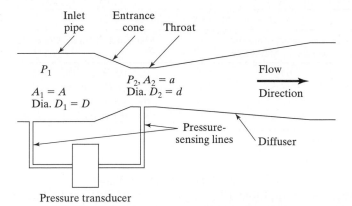

Inlet pipe
Entrance cone
Throat

P_1

$A_1 = A$
Dia. $D_1 = D$

$P_2, A_2 = a$
Dia. $D_2 = d$

Flow

Direction

Pressure-sensing lines

Diffuser

Pressure transducer

FIGURE 10.2

Flowmeter using venturi tube.

and z_2 are the same and cancel out. If the sensing lines are filled with the same fluid at the same density as is contained in the pipe, the pressure transducer measures the difference in the term $(P + g\rho z)$ at the upstream section and at the minimum area. That is, $\Delta P_{\text{trans}} = (P + g\rho z)_1 - (P + g\rho z)_2$, regardless of the flowmeter's orientation. If a manometer is used to measure the pressure difference, the analysis is slightly more complicated and the reader should consult fluid mechanics texts such as White (1999).

The volume flow rate for a venturi tube is given by

$$Q = \frac{CA_2}{\sqrt{1 - (A_2/A_1)^2}} \sqrt{\frac{2[(P_1 + g\rho z_1) - (P_2 + g\rho z_2)]}{\rho}} \qquad (10.6)$$

This equation is identical to Eq. (10.5), except that the right-hand side has been multiplied by C. The factor C, called the *discharge coefficient*, is used to account for nonideal effects.[†] C is a function of the diameter ratio, $\beta = D_2/D_1 = d/D$, and a parameter called the *Reynolds number*, which is defined as

$$\text{Re} = \frac{\rho V D}{\mu} \qquad (10.7)$$

where μ is the fluid viscosity. The Reynolds number is a dimensionless parameter, and its value is independent of the unit system.

It is possible to obtain values for the discharge coefficient by calibrating the venturi tube. However, if the venturi tube is constructed to a standardized geometric shape, the performance will be standardized and it is possible to use discharge coefficients available in the literature. Table 10.1 gives values of the discharge coefficient C for venturi tubes constructed according to the ASME geometric specification (Figure 10.3) with three different construction methods for the entrance cone. All have a smooth throat.

[†]In some cases, a coefficient K, called the *flow coefficient*, is used instead of the discharge coefficient. The flow coefficient is defined as

$$K = \frac{C}{\sqrt{1 - (A_2/A_1)^2}}$$

TABLE 10.1   Discharge Coefficients for Venturi Tubes

Rough-Cast Entrance Cone and Rough-Welded Sheet-Metal Cone	Machined Entrance Cone
$C = 0.984 \pm 1.0\%$	$C = 0.995 \pm 1.0\%$
4 in. $\leq D \leq$ 48 in.	2 in. $\leq D \leq$ 10 in.
$0.3 \leq \beta \leq 0.75$	$0.3 \leq \beta \leq 0.75$
$2 \times 10^5 \leq$ *Re $\leq 2 \times 10^6$	$2 \times 10^5 \leq$ Re $\leq 2 \times 10^6$

Source: ASME (1989).

In general, the discharge coefficient will decrease at values of Reynolds numbers lower than the ranges given in Table 10.1 but will change little at higher values. For properly constructed venturi tubes within the ranges in Table 10.1, the flow coefficient is accurate to 1.0% without calibration. This characteristic of good accuracy without calibration, shared with other ASME head meters, is a major advantage and is particularly important in large sizes, where it is often very difficult to perform accurate calibrations. Smaller meters, below the size ranges of Table 10.1, may require calibration.

Head meters are often used to determine the flow rate of compressible fluids (gases). It is common to express the compressible fluid flow rate as the mass flow rate, $\dot{m}$, rather than the volume flow rate often used with liquids. In compressible flow, the density is a function of the pressure and temperature and will change as the fluid moves through the flowmeter. Since Eq. (10.4) is based on incompressible flow analysis, it is necessary to correct it for mildly compressed flows. In this case, an equation for the mass flow rate is

$$\dot{m} = \rho_2 V_2 A_2 = \frac{CYA_2}{\sqrt{1 - (A_2/A_1)^2}} \sqrt{2\rho_1(P_1 - P_2)} \qquad (10.8)$$

The elevation (z) terms in Eq. (10.6) are usually insignificant in compressible flow, so they have been deleted in Eq. (10.8). The term Y in Eq. (10.8) is called the *expansion factor* and compensates for the fluid density change between section 1 and section 2. Y is a function of the flowing gas and the ratios A_2/A_1 and P_2/P_1. Note that the density under the radical sign is ρ_1, which is consistent with the ASME definition of Y. The venturi tube expression for Y given in ASME (1989) is

$$Y = \left(r^{2/\gamma} \frac{\gamma}{\gamma - 1} \frac{1 - r^{(\gamma-1)/\gamma}}{1 - r} \frac{1 - \beta^4}{1 - \beta^4 r^{2/\gamma}} \right)^{1/2} \qquad (10.9)$$

where γ is the gas specific heat ratio $r = P_2/P_1$ and $\beta = D_2/D_1 = d/D$. The expression is valid for r greater than 0.75. The ASME uncertainty on Y for venturi tubes is given by $\pm[4 + 100\beta^8][(P_1 - P_2)/P_1]\%$. For example, for $\beta = 0.65$ and $(P_1 - P_2)/P_1 = 0.1$, the uncertainty on Y will be $\pm 0.72\%$.

Venturi tubes are simple, reliable devices, which often do not require calibration and have relatively low insertion losses (pressure losses caused by the flowmeter). For minimum pressure loss, the diffuser angle, α_2, should be closer to 7° than 15°. On the other hand, venturi tubes are normally quite expensive to construct. Venturi tubes are

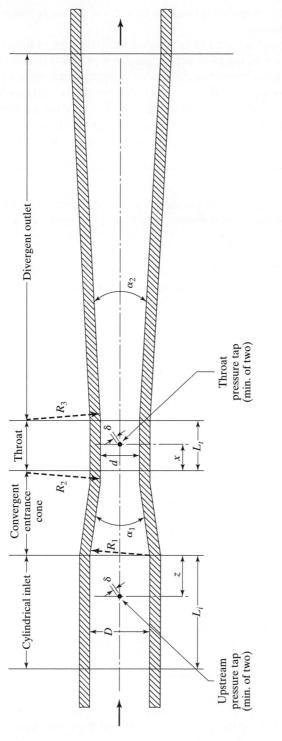

$L_i > D$ or $L_i > (D/4 + 250$ mm (10 in.))
$z = D/2$ (+0.0 D, − $D/4$), minimum of 2 taps
$L_t \geq d/3$
$x = 0.5 \, d \pm 0.02 \, d$, minimum of two taps
4 mm (5/32 in.) $\leq \delta \leq$ 10 mm (25/64 in.) and $\delta < 0.1D$ or 0.13 d

$R_1 = 1.375 \, D \pm 20\%$
$R_2 = 3.625 \, d \pm 0.125 \, d$
$5d \leq R_3 \leq 15 \, d$
$\alpha_1 = 21° \pm 1°$
$7° \leq \alpha_2 \leq 15°$

FIGURE 10.3

Dimensional specifications of venturi tube. (Based on ASME, 1989.)

317

not suitable for flows in which material tends to collect on the walls, since the calibration will be changed significantly.

A given venturi tube can be used over a rather limited range of flow rates. Examining Eq. (10.6) one can see that the pressure difference is proportional to the *square* of the flow rate. This means that a venturi designed for an appropriate pressure difference at the upper end of the flow range will have a very low pressure difference at the lower end of the flow range. For example, a venturi tube with a pressure change of 1 psi at maximum flow will have a pressure change of only 0.01 psi at 10% of maximum flow. This characteristic is shared with the flow nozzle and the orifice meter discussed later.

Example 10.1

A venturi tube, fabricated with a machined entrance cone according to ASME specifications, is inserted in a 4-in.-ID pipe and has a throat diameter of 3 in. If the flowing fluid is 60°F water, the pressure-sensing lines are filled with water, and the pressure transducer reads 4 psi, what is the water flow rate?

Solution: From Table B.2 in Appendix B, for 60°F water, $\rho = 62.34$ lbm/ft^3 and $\mu = 2.71$ lbm/hr-ft. To find the solution, we will use Eq. (10.6). Note that $A_2/A_1 = (D_2/D_1)^2$. For a machined inlet cone, Table 10.1 gives $C = 0.995$. We will check the Reynolds number later. We have

$$Q = \frac{CA_2}{\sqrt{1 - (A_2/A_1)^2}} \sqrt{\frac{2\Delta P}{\rho}}$$

$$= \frac{0.995\pi(3/12)^2/4}{\sqrt{1 - (3/4)^4}} \sqrt{\frac{2(4 \times 144)}{62.34/32.17}}$$

$$= 1.44 \text{ ft}^3/\text{sec}$$

Note that all inch units have been converted to feet and we had to divide the density by g_c to obtain consistent units. Further note that the pressure transducer in Figure 10.2 actually measures the change in $P + g\rho z$, not simply the change in pressure. The velocity of the fluid in the pipe is

$$V = \frac{Q}{A} = \frac{1.44}{\pi(4/12)^2/4} = 16.5 \text{ ft/sec}$$

and the Reynolds number is then

$$\text{Re} = \frac{\rho V D}{\mu} = \frac{(62.34/32.17) \times 16.5 \times 4/12}{2.71/(32.17 \times 3600)} = 455,000$$

The venturi is thus operating within the range of data in Table 10.1.

Example 10.2

If the diameters of the maximum and minimum areas of Example 10.1 are uncertain by ±0.002 in. and the pressure difference is uncertain by ±0.05 psi, what will the systematic uncertainty be in the flow measurement?

Solution: We will estimate the uncertainty using Eq. (7.4). Thus, we need the partial derivatives of Q with respect to C, D_1, D_2, and ΔP, the pressure change. Uncertainties in g and ρ will be considered negligible. These derivatives are obtained using Eq. (10.6):

$$\frac{\partial Q}{\partial D_1} = C\left(\frac{\pi}{4}\right)\frac{-2(D_2/D_1)^5 D_2}{[1-(D_2/D_1)^4]^{3/2}}\sqrt{\frac{2\times\Delta P}{\rho}} = -3.99 \text{ ft}^3/\text{sec-ft}$$

$$\frac{\partial Q}{\partial D_2} = C\left(\frac{\pi}{4}\right)\frac{2D_2}{[1-(D_2/D_1)^4]^{1/2}} + \frac{2(D_2/D_1)^4 D_2}{[1-(D_2/D_1)^4]^{3/2}}\sqrt{\frac{2\times\Delta P}{\rho}} = 16.84 \text{ ft}^3/\text{sec-ft}$$

$$\frac{\partial Q}{\partial C} = \frac{\pi D_2^2/4}{\sqrt{1-(D_2/D_4)^4}}\sqrt{\frac{2\times\Delta P}{\rho}} = 1.447 \text{ ft}^3/\text{sec}$$

$$\frac{\partial Q}{\partial \Delta P} = \frac{C(\pi D_2^2/4)}{\sqrt{1-(D_2/D_4)^4}}\frac{(1/2)2/\rho}{\sqrt{2\times\Delta P/\rho}} = 0.00125 \text{ (ft}^3/\text{sec)/(lb/ft}^2)$$

The uncertainties in the variables are

$$w_{D1} = \frac{0.002}{12} = 1.666\times 10^{-4} \text{ ft}$$

$$w_{D2} = \frac{0.002}{12} = 1.666\times 10^{-4} \text{ ft}$$

$$w_C = 0.00995 \text{ (1\% of 0.995, as per Table 10.1)}$$

$$w_{\Delta P} = 0.05\times 144 = 7.2 \text{ lbf/ft}^2$$

From Eq. (7.4), the systematic uncertainty in Q is

$$w_Q = \left[\left(\frac{\partial Q}{\partial D_1}w_{D_1}\right)^2 + \left(\frac{\partial Q}{\partial D_2}w_{D_2}\right)^2 + \left(\frac{\partial Q}{\partial C}w_C\right)^2 + \left(\frac{\partial Q}{\partial \Delta P}w_{\Delta P}\right)^2\right]^{1/2}$$

$$w_Q = [(3.99\times 1.666\times 10^{-4})^2 + (16.84\times 1.666\times 10^{-4})^2$$
$$+ (1.447\times 0.00995)^2 + (0.00125\times 7.2)^2]^{1/2}$$

$$w_Q = [44.187\times 10^{-8} + 7.87\times 10^{-6} + 2.07\times 10^{-4} + 8.1\times 10^{-5}]^{1/2}$$

$$w_Q = 0.019 \text{ ft}^3/\text{sec}$$

The flow rate in Example 10.1 is 1.44 ± 0.019 ft^3/sec.

Comment: The uncertainty is dominated by the discharge coefficient and the pressure-difference terms, with the dimensional terms having small contributions. Computing the derivatives is clearly algebraically complicated. It is often easier simply to insert the original formula into a spreadsheet program and determine the derivatives numerically. For example, the pipe diameter could be increased 0.1% and the change in flow rate evaluated. $\partial Q/\partial D_1$ is then $\Delta Q/\Delta D_1$.

Example 10.3

A horizontal venturi meter constructed of welded sheet metal is to be installed into a 60-cm-ID welded steel pipe. The flowing fluid is water at 40°C. The maximum flow rate is 0.563 m^3/s and the maximum allowable pressure drop is 20 kPa. Determine the throat area.

Solution: The properties of the water can be obtained from Table B.2: $\rho = 992.2$ kg/m^3 and $\mu = 0.656 \times 10^{-3}$ N-s/m^2. At this point we do not know β or Re, but we will assume that the value of C in Table 10.1, 0.984, is valid. The upstream pipe area A_1 is $\pi 0.6^2/4 = 0.2827$ m^2. We will use Eq. (10.6) to obtain the throat area:

$$Q = \frac{CA_2}{\sqrt{1 - (A_2/A_1)^2}} \sqrt{\frac{2\Delta P}{\rho}}$$

$$0.563 = \frac{0.984 A_2}{\sqrt{1 - (A_2/0.2827)^2}} \sqrt{\frac{2(20,000)}{992.2}}$$

Solving for A_2, we obtain 0.086 m^2, which corresponds to a diameter of 33 cm. This results in a β value of $33/60 = 0.55$, which is within the range of the data in Table 10.1. The pipe fluid velocity is $Q/A = 0.565/0.2827 = 2.00$ m/s. The Reynolds number is then

$$\text{Re} = \frac{\rho V D}{\mu} = \frac{992.2 \times 2.00 \times 0.6}{0.656 \times 10^{-3}} = 1.82 \times 10^6$$

which is within the range of Table 10.1. If Re had been larger, it is unlikely that the flow coefficient would change significantly.

Comment: The greater simplicity of SI units is demonstrated when comparing Example 10.3 to Example 10.1.

Flow Nozzle   The flow nozzle, shown in Figure 10.4, is similar to the venturi except that after the fluid passes through the minimum flow area, the flow area expands suddenly to the pipe area. This results in much larger fluid energy losses. As with the venturi tube, flow nozzles are constructed to dimensional standards. The ASME has

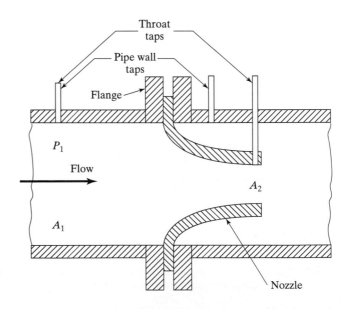

FIGURE 10.4

Flow nozzle.

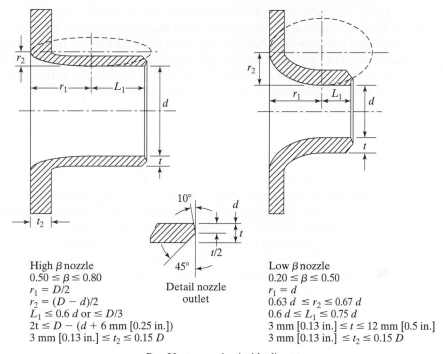

High β nozzle
$0.50 \leq \beta \leq 0.80$
$r_1 = D/2$
$r_2 = (D - d)/2$
$L_1 \leq 0.6\,d$ or $\leq D/3$
$2t \leq D - (d + 6$ mm [0.25 in.])
3 mm [0.13 in.] $\leq t_2 \leq 0.15\,D$

Detail nozzle
outlet

Low β nozzle
$0.20 \leq \beta \leq 0.50$
$r_1 = d$
$0.63\,d \leq r_2 \leq 0.67\,d$
$0.6\,d \leq L_1 \leq 0.75\,d$
3 mm [0.13 in.] $\leq t \leq 12$ mm [0.5 in.]
3 mm [0.13 in.] $\leq t_2 \leq 0.15\,D$

D = Upstream pipe inside diameter

FIGURE 10.5

Dimensional specifications for ASME long-radius flow nozzles. (Based on ASME, 1989.)

designated standard geometry for long-radius nozzles, as shown in Figure 10.5. Equation (10.6) for venturi tubes is also applicable to flow nozzles. Values and formulas for discharge coefficients, C, are given in ASME (1989) and can be used if calibration is not possible. As shown in Figure 10.4, there are two possible locations for the pressure measurements. For pipe wall taps, the upstream tap should be one pipe diameter upstream from the nozzle-inlet plane. The downstream tap should be 0.5 pipe diameter from the nozzle-inlet plane but not farther downstream than the nozzle-outlet plane. For long-radius flow nozzles with pipe wall taps, the discharge coefficient can be computed from

$$C = 0.9975 - 0.00653 \left(\frac{10^6 \beta}{\mathrm{Re}_D} \right)^{0.5} \tag{10.10}$$

where the pipe diameter D is in inches and Re_D is the Reynolds number based on the pipe inside diameter. The applicable range of pipe diameters is 2 to 30 in., the range of the Reynolds number is from 10^4 to 6×10^6, and the range of β is from 0.2 to 0.8. ASME gives a tolerance on the discharge coefficient of $\pm 2\%$. For nozzles using throat taps, the reader is referred to ASME (1989). Flow nozzles can also be used for the flow of gases, and Eqs. (10.8) and (10.9) are applicable. The tolerance on Y is $\pm[2(P_2 - P_1)/P_1]\%$.

There is a special case for flow nozzles that deserves particular mention. This is the critical flow nozzle for measuring mass flow rate in gases, including steam. If the pressure downstream of the nozzle throat is sufficiently low, then the velocity in the minimum area of the nozzle will become equal to the local speed of sound. In this case, the flow is said to be choked (critical), and further reductions in the downstream pressure will not affect the mass flow rate. For ideal gases, the absolute pressure where this occurs is given by

$$P_{\text{crit}} = \frac{P_0}{[(\gamma + 1)/2]^{\gamma/(\gamma-1)}} \tag{10.11}$$

where γ is the specific heat ratio of the gas. The 0 subscript denotes total conditions, the value of absolute pressure (and temperature) at an upstream location where the velocity is zero. As a practical matter, if the pipe area upstream of the nozzle is greater than five times the nozzle area, the upstream fluid pressure will be within 1% of the total pressure. For most gases, the critical pressure is on the order of one-half the total pressure. If the surrounding pressure on the downstream side of the nozzle is less than P_{crit}, the pressure at the nozzle throat will be P_{crit}, not the lower backpressure.

The critical flow rate for an ideal gas can be determined from the equation (Shapiro, 1953)

$$\dot{m} = \frac{A_2 P_0}{\sqrt{T_0}} \sqrt{\frac{\gamma}{R} \left(\frac{2}{\gamma + 1} \right)^{(\gamma+1)/(\gamma-1)}} \tag{10.12}$$

where R is the gas constant for the flowing gas and T_0 is the total temperature. The flow rate computed by this equation is the maximum; frictional effects will reduce the flow rate slightly and the result should be multiplied by a discharge coefficient. Although discharge coefficients are high for critical flow nozzles (on the order of 0.99 for high Reynolds numbers), ASME (1971) suggests that critical flow nozzles be individually calibrated. Many common engineering gases, such as steam and refrigerants, show significant real-gas behavior. Real-gas behavior means that they deviate significantly from ideal-gas behavior. The approach for real gases, such as steam, is given in ASME (1971). The major advantage of the critical flow nozzle is that only upstream fluid conditions need to be measured. The major disadvantage is that they cause a significant pressure loss.

Example 10.4

A flow nozzle is to be inserted in a 2-in.-diameter pipe (ID = 2.067 in.) to measure an airflow of up to 2 lbm/sec. Upstream of the nozzle, the air pressure is 150 psia and the temperature is 300°F. γ for air is 1.4. Pipe wall taps are to be used, and the maximum pressure drop across the nozzle is to be 5 psi. Determine a suitable minimum area for the nozzle.

Solution: The properties of air can be obtained from Table B.4. $\mu = 0.05748$ lbm/hr-ft. The density at 14.7 psia is 0.05221 lbm/ft^3, so at 150 psi, $\rho = 0.05221(150/14.7) = 0.533$ lbm/ft^3. We will solve this problem by solving Eq. (10.8) for A_2. To do this, we need C and Y, which require β, which in turn, requires the final answer A_2. We can, however, solve the problem iteratively.

We will first assume that $C = 1$ and $Y = 1$, estimate A_2, compute C and Y, and then reevaluate A_2. Computing A_1 to be 0.0233 ft^2, we can substitute into Eq. (10.8):

$$\dot{m} = \frac{CYA_2}{\sqrt{1 - (A_2/A_1)^2}}\sqrt{2\rho_1(P_1 - P_2)}$$

$$\frac{2}{32.17} = \frac{1 \times 1 \times A_2}{\sqrt{1 - (A_2/0.0233)^2}}\sqrt{2 \times (0.533/32.17)(5 \times 144)}$$

Solving, we obtain $A_2 = 0.0112$ ft^2, which corresponds to a nozzle minimum diameter of 1.43 in. Noting that $r = 145/150 = 0.967$ and $\beta = d/D = 1.43/2.067 = 0.692$, we obtain Y from Eq. (10.9):

$$Y = \left[0.967^{2/1.4}\left(\frac{1.4}{1.4 - 1}\right)\left(\frac{1 - 0.967^{(1.4-1/1.4)}}{1 - 0.967}\right)\left(\frac{1 - 0.692^4}{1 - 0.692^4\,0.967^{2/1.4}}\right)\right]^{1/2} = 0.974$$

We obtain C from Eq. (10.10). To find C, we need to compute the Reynolds number based on the pipe diameter. The velocity at section 1 is $V_1 = \dot{m}/\rho_1 A_1 = 2/(0.533 \times 0.0233) = 161$ ft/sec. The Reynolds number is then

$$\text{Re}_D = \frac{\rho V D}{\mu} = \frac{0.533/32.17 \times 161 \times 2.067/12}{0.05748/(32.17 \times 3600)} = 925{,}800$$

Using Eq. (10.10) for C yields

$$C = 0.9975 - 0.00653\left(10^6 \times \frac{0.692}{925{,}800}\right)^{0.5} = 0.992$$

Using Eq. (10.8) again gives us

$$\frac{2}{32.17} = \frac{0.992 \times 0.974 \times A_2}{\sqrt{1 - (A_2/0.0233)^2}}\sqrt{2 \times \frac{0.533}{32.17} \times (5 \times 144)}$$

Solving for A_2, we obtain $A_2 = 0.0116$ ft^2 or $d = 1.46$ in.

Comment: In this problem the pressure drop is small (5 psi relative to an inlet pressure of 150 psi), yet the expansion factor has a value of 0.974, an effect of more than 2%.

Example 10.5

A critical flow nozzle has been proposed for measuring airflow into a chamber designed to operate between 1 and 3 atm (absolute) pressure. Compressed air at 690 kPa (gage) and ambient temperature (20°C) are available for this application. The system needs approximately 0.1 kg/s of air. For air, $\gamma = 1.4$ and $R = 287$ J/kg-K.

(a) Is a critical flow nozzle appropriate for this application?
(b) If applicable, calculate the throat area of the nozzle.
(c) What measurements are necessary to accurately calculate the flow?
(d) How will the flow vary with changes in chamber pressure?

Solution:

(a) We have to check whether the flow across the nozzle will remain choked as the pressure varies from 1 to 3 atm. We have

$$\text{Supply pressure} = \frac{690}{101.3} = 6.8 \text{ atm (gage)}$$

$$\text{Maximum chamber-to-line pressure ratio} = \frac{P_c}{P_s} = \frac{3}{6.8 + 1} = 0.38$$

We must compare this value with the critical ratio obtained from Eq. (10.11). For air ($\gamma = 1.4$),

$$\frac{P_{\text{crit}}}{P_0} = \frac{1}{[(1.4 + 1)/2]^{1.4/(1.4-1)}} = 0.528$$

Because $P_c/P_s \leq P_{\text{crit}}/P_0$, the flow will be choked and a critical flow nozzle will be appropriate.

(b) To calculate the throat area of the nozzle, we will use Eq. (10.12):

$$\dot{m} = \frac{A_2 P_0}{T_0^{1/2}} \left[\frac{\gamma}{R} \frac{2}{(\gamma + 1)^{(\gamma+1)/(\gamma-1)}} \right]^{1/2}$$

$$0.1 = \left(A_2 \times 7.8 \times \frac{101,325}{293^{1/2}} \right) \left[\left(\frac{1.4}{287} \right) \frac{2}{(1.4 + 1)^{(1.4+1)/(1.4-1)}} \right]^{1/2}$$

$$A_2 = 5.36 \times 10^{-5} \text{ m}^2 = 0.536 \text{ cm}^2 \quad \text{diameter} = 8.3 \text{ mm}$$

(c) To measure the flow accurately, the nozzle should be calibrated. For subsequent measurements, upstream temperature and pressure should be measured. To make sure that the flow is choked, downstream pressure should also be measured for each operational condition.

(d) As long as the flow is choked in the nozzle, it will not depend on the chamber pressure. To change the flow rate, upstream pressure (and possibly temperature) should be changed. If upstream pressure is changed, one should make sure that the flow remains critical.

Example 10.6

In Example 10.5 calculate the flow rate in standard cubic meters per minute (SCMM) of airflow.

Solution: To convert the flow rate into standard conditions (usually defined as 1 atm, 20°C), we have to divide the mass flow rate of air by the density of air at standard condition (ρ_0). We will use the perfect gas relation for this purpose,

$$\rho_0 = \frac{P}{RT} = \frac{101,325 \text{ (Pa)}}{287 \text{ (J/kg-K)} \times 293 \text{ (K)}}$$

$$= 1.20 \text{ kg/m}^3$$

$$\text{SCMM} = \frac{\dot{m}}{\rho_0} = \frac{60 \times 0.1}{1.20} = 50 \text{ SCMM}$$

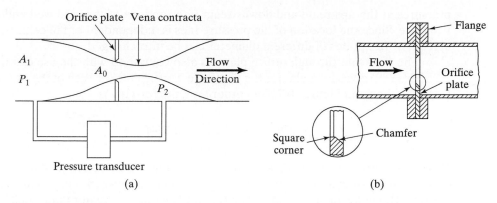

FIGURE 10.6

(a) Orifice meter system; (b) description of square-edged orifice.

Orifice Meter   The orifice meter, shown in Figure 10.6(a), is one of the most popular differential pressure devices because it is simple to install and relatively inexpensive. Fluid energy losses are significantly larger than in venturi tubes. As Figure 10.6(a) shows, the minimum fluid flow area does not occur at the orifice plate but at a location farther downstream. This minimum flow area is called the *vena contracta*. The orifice meter described in ASME (1989) is the *square-edged orifice* type, shown in Figure 10.6(b). The orifice is a circular hole cut into a thin plate. The hole is then chamfered at the downstream face to leave a thin edge. The thin edge and the inlet face should form a sharp 90°angle. The plate is then bolted between the flanges at a pipe joint. There are three standard locations for the pressure taps, as shown in Figure 10.7. For flange taps, the pressure-sensing holes should be located 1 in. upstream and 1 in. downstream of the faces of the orifice plate. For D–1/2D taps, the taps are located one pipe diameter upstream and 1/2 pipe diameter downstream of the orifice plate. Corner taps sense the

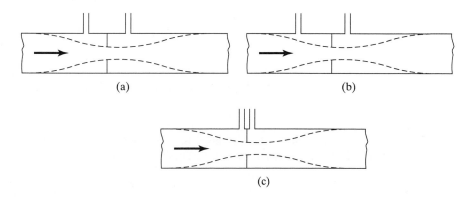

FIGURE 10.7

Location of orifice meter pressure taps: (a) flange taps; (b) $D-\frac{1}{2}D$ taps; (c) corner taps.

pressures at the upstream and downstream intersections of the pipe wall and the orifice plate. Since the location of the pressure taps is independent of the orifice-plate diameter, orifice plates of different diameter can be interchanged.

The flow rate through orifice plates is also computed with the use of Eq. (10.6). ASME (1989) gives formulas for the discharge coefficients for each of the pressure-tap locations shown in Figure 10.7. For corner taps, the formula is

$$C = 0.5959 + 0.0312\beta^{2.1} - 0.184\beta^8 + \frac{91.71\beta^{2.5}}{Re_D^{0.75}} \tag{10.13}$$

It is recommended that the diameter ratio, β, be limited to the range of 0.2 to 0.7. Tolerances on the discharge coefficient for pipe diameters between 2 and 36 in. depend on the pipe Reynolds number and β. For Reynolds numbers greater than 10,000, the uncertainty is $\pm0.6\%$ for β between 0.2 and 0.6 and $\pm\beta\%$ for β between 0.6 and 0.75. For Reynolds numbers between 2000 and 10,000, the uncertainty is $\pm(0.6 + \beta)\%$. For Reynolds numbers greater than 10,000 and $\beta = 0.65$, the uncertainty is $\pm0.65\%$. ASME (1989) gives formulas for the discharge coefficients of other pressure-tap locations. ASME (1971) also gives guidance for constructing orifice meters for pipes smaller than 2 in.

For orifice meters with compressible flows, Eq. (10.8) is applicable directly. The expression for the expansion factor Y of orifice meters is given by

$$Y = 1 - (0.410 + 0.35\beta^4)\frac{P_1 - P_2}{P_1\gamma} \tag{10.14}$$

This is valid for $P_2/P_1 \geq 0.75$. The uncertainty on Y for orifice meters is given by $[4(P_1 - P_2)/P_1]\%$. For example, if $(P_1 - P_2)/P_1 = 0.1$, the uncertainty will be 0.4%.

Orifice meters are less likely than venturi tubes or flow nozzles to be affected by fluids that contaminate the pipe or conduit walls. Deposits are less likely to form on the lip of the orifice plate than they are on the walls of venturi tubes and nozzles. However, orifice meters are more likely to be affected by abrasive fluids, which can erode the lip of the orifice plate and affect accuracy.

Example 10.7

A sharp-edge orifice with corner taps is to be used to measure water flow rate in a horizontal 2-in. pipe (2.067 in. ID). The maximum flow is expected to be about 7 cfm (cubic feet per minute). Determine the diameter of the orifice if a maximum pressure difference of about 2 psi is acceptable.

Solution: We will use Eqs. (10.6) and (10.13) for this purpose. First we have to determine the Reynolds number in the pipe. Using Table B.2, we find that $\rho = 62.34$ lbm/ft^3 and that $\mu = 2.71$ lbm/hr-ft. The pipe area is $A_1 = \pi \times 2.067^2/(4 \times 144) = 0.0233$ ft^2. The fluid velocity in the pipe is

$$V = \frac{Q}{A} = \frac{7/60}{0.0233} = 5.00 \text{ ft/sec}$$

According to Eq. (10.7),

$$\text{Re} = \frac{\rho V D}{\mu}$$

$$= \frac{(62.34/32.17) \times 5.00 \times (2/12)}{2.71/(32.17 \times 3600)}$$

$$= 69{,}000$$

The diameter of the orifice (d) must be determined by using Eq. (10.6) in a trial-and-error process because the discharge coefficient, C, is a function of $\beta = d/D$, which is initially unknown. We will assume that $C = 0.60$ and find d. We have

$$Q = \frac{C A_2}{\sqrt{1 - \left(\dfrac{A_2}{A_1}\right)^2}} \sqrt{\frac{2[(P_1 + g\rho z_1) - (P_2 + g\rho z_2)]}{\rho}}$$

Since the pipe is horizontal, $z_1 = z_2$, and we obtain

$$7/60 = \frac{0.6 \times A_2}{\sqrt{1 + \left(\dfrac{A_2}{0.0233}\right)^2}} \sqrt{\frac{2(2 \times 144)}{(62.34/32.17)}}$$

$$A_2 = 0.01 \text{ ft}^2 \qquad d = 1.35 \text{ in.}$$

This gives $\beta = 1.35/2.067 = 0.653$. Substituting into Eq. (10.13), $C = 0.615$. Using Eq. (10.6) again, the final orifice diameter is 1.35 in.

Practical Considerations in the Installation of Venturi Tube, Flow Nozzle, and Orifice Flow Meters   The preceding material provides information on the geometry of head-type flowmeters and interpretation of the test data. In installing such a meter, consideration must be given to the upstream and downstream piping geometry and the design and number of the pressure taps. There are requirements on the length of straight pipe upstream and downstream of the meter, which depend on the type of the meter and the diameter ratio (β). There are several methods to construct the pressure taps and restrictions on the tap diameters. Miller (1989) has an extensive discussion of these practical matters. ASME (1989) is also useful for practical installation considerations.

Laminar Flowmeter   One of the disadvantages of the devices we have described, which are based on the Bernoulli equation, is that the pressure drop is proportional to the flow squared. This means that a device designed for a reasonable pressure drop at the maximum flow will have a very low pressure drop at low flows. The flow in small-diameter (0.030 in., for example) tubes at moderate velocities will be in a flow regime known as *laminar flow* (with a Reynolds number based on a diameter less than about 2300). In larger tubes or at high velocities, the flow regime is known as *turbulent*. For laminar flow of a fluid in a horizontal tube, the flow rate as a function of pressure drop

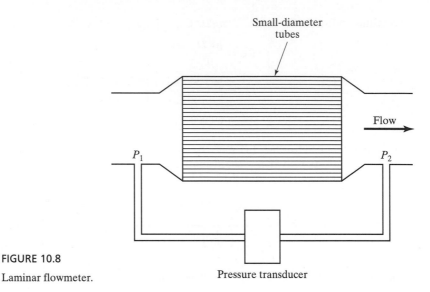

FIGURE 10.8

Laminar flowmeter.

can be computed from the following formula —White (1999):

$$Q = \frac{\pi D^4 (P_1 - P_2)}{128 \, \mu L} \tag{10.15}$$

where D is the pipe diameter, P_1 is the pressure at the upstream end, P_2 is the pressure at the downstream end, L is the pipe length, and μ is the fluid viscosity. The flow in this case is proportional to the pressure drop to the first power. A single small-diameter tube is, however, not suitable for measuring large flow rates. This problem is solved with a device known as a *laminar flowmeter* that uses a large number of small-diameter tubes in parallel, as shown in Figure 10.8. If a flowmeter of this type is designed for a pressure drop of 1 psi at maximum flow, it will have a pressure drop of 0.1 psi at 10% of maximum flow, 10 times what it would be for an orifice meter. Thus laminar flowmeters have a much wider range of measurable flows than devices based on the Bernoulli equation. Laminar flow can also be achieved in small passages of noncircular section or porous media, which are less expensive to construct than are small tubes.

Laminar flowmeters do have some disadvantages. They normally have to be calibrated after their manufacture, since small tolerance errors will have a major effect on performance. In addition, they can only be used with clean fluids since small amounts of contamination on the tube walls will significantly alter the flow–pressure-drop relationship. They are most commonly used with clean gases.

Example 10.8

A laminar flowmeter is constructed with 5000 tubes, each 1.5 mm in inside diameter and 30 cm long. For a pressure difference along the tubes of 1 kPa, what will be the flow rate of atmospheric air?

Solution: From Table B.2, at 1 atm, 20°C, $\rho = 1.20$ kg/m³ and $\mu = 1.811 \times 10^{-5}$ N-s/m². For a single tube we can use Eq. (10.15):

$$Q = \frac{\pi \times 0.0015^4 \times 1000}{128 \times 1.811 \times 10^{-5} \times 0.30} = 2.89 \times 10^{-5} \, \text{m}^3\text{/s}$$

For 5000 tubes, the flow rate is $5000 \times 2.89 \times 10^{-5} = 0.145$ m³/s. Since this device depends on the flow being laminar, we should check the Reynolds number:

$$V = \frac{Q}{A} = \frac{2.89 \times 10^{-5}}{\pi 0.0015^2/4} = 16.35 \, \text{m/s}$$

$$\text{Re} = \frac{\rho V D}{\mu} = \frac{1.2 \times 16.35 \times 0.0015}{1.811 \times 10^{-5}} = 1625$$

The Reynolds number is less than 2300; hence, the flow is laminar. This flowmeter is suitable for measuring the airflow of a small automobile engine.

10.1.2   Variable-Area Flowmeters

This class of flowmeters consists of devices in which the flow area of the meter varies with flow rate and the value of the flow rate is determined by sensing the position of the component causing the area change. The most common of these devices used in experimentation or process control is known as the *rotameter*. Rotameters can be used for both liquids and gases.

Rotameter   This device is shown schematically in Figure 10.9(a). The transparent tube, with its axis aligned vertically, contains a small weighted object called a *float*, which is often a sphere but can have other shapes. The fluid flowing over the float creates a drag

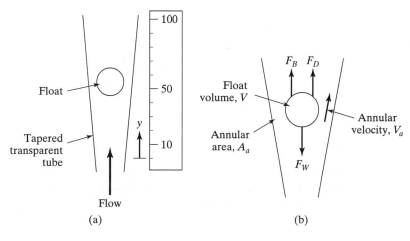

FIGURE 10.9

Sketch of rotameter.

force, which together with the buoyant force, exactly balances the weight of the float [Figure 10.9(b)]. The float moves to a flow-rate-dependent vertical position to maintain this equilibrium, and the position of the float can be measured with an attached scale. Expressed mathematically, the force balance is

$$F_D + F_B = F_W \qquad (10.16)$$

where F_D is the drag force caused by the upward-moving fluid, F_B is the buoyant force acting on the float, and F_W is the weight of the float. The buoyant force is equal to the weight of the displaced fluid:

$$F_B = \cancel{V}\rho_f g \qquad (10.17)$$

where $\cancel{V}$ is the float volume, ρ_f is the density of the flowing fluid, and g is the acceleration of gravity. The weight is given by

$$F_W = \cancel{V}\rho_b g \qquad (10.18)$$

where ρ_b is the density of the float. It is customary to describe the drag force in terms of a drag coefficient

$$F_D = \frac{C_D A_{\text{fr}} \rho_f V_{\text{ref}}^2}{2} \qquad (10.19)$$

where C_D is the drag coefficient, A_{fr} is the frontal area of the float (in the direction of the flow), and V_{ref} is a reference fluid velocity. In this case we use V_a, the velocity in the annular area between the float and the tapered tube, as V_{ref}. Substituting the force terms into Eq. (10.16), we obtain

$$\frac{C_D A_{\text{fr}} \rho_f V_a^2}{2} + \cancel{V}\rho_f g = \cancel{V}\rho_b g \qquad (10.20)$$

Solving for V_a, we obtain

$$V_a = \left[\frac{2\cancel{V}g(\rho_b - \rho_f)}{C_D A_{\text{fr}} \rho_f} \right]^{1/2} \qquad (10.21)$$

For gases, $\rho_F \ll \rho_b$, so Eq. (10.21) can be written

$$V_a = \left[\frac{2\cancel{V}g\rho_b}{C_D A_{\text{fr}} \rho_f} \right]^{1/2} \qquad (10.22)$$

The float can be designed such that C_D is approximately independent of the float position, and consequently the right-hand sides of Eqs. (10.21) and (10.22) (and hence V_a) are essentially constant. Since the volume flow rate is

$$Q = V_a A_a \qquad (10.23)$$

and V_a is constant, the flow rate is proportional to the annular area, A_a. It is possible to design the rotameter so that the annular area is a linear function of the float vertical position y. Hence, a measure of the float vertical position is a linear function of the flow rate. In some rotameters, the scale will read as a percentage of full scale; in others, the scale will read directly in units such as "standard liters per second" of the flowing fluid. The reading depends on the composition of the flowing fluid and its properties (pressure, temperature, and viscosity). Various correlations are available from the manufacturer of a given flowmeter in order to convert the reading into mass flow rate or volume flow rate.

Although it is possible to use devices that sense float position, most rotameters are not used for remote sensing of flow but are instead read visually and used to check or adjust the fluid flow rate. In fact, many rotameters have a built-in flow control valve. Rotameters have the advantages of being easy to read, having a low pressure drop, and having a scale that is linear with flow. Most rotameters must be mounted with the axis vertical, but there are rotameters that can be installed in any orientation and that use a spring instead of the weight of the float to counter the fluid drag force. While it is not common to have a sensor for remote sensing of flow rate with rotameters, it is common in the process industries to have a sensing switch that will trip an alarm if the flow exceeds some system limit.

The accuracy of the higher-quality rotameters can be better than ±2% of full scale, but many are used to give a general indication of flow and are accurate only to the order of ±10%. Rotameters are usually calibrated to produce the best accuracy for a particular application. To avoid a new calibration, a rotameter can sometimes be calibrated for one fluid and used to measure the flow of another fluid or the same fluid at another temperature or pressure. Manufacturers can supply sophisticated analysis to permit such change of application. A simpler method to permit use with another fluid, based on Eqs. (10.21), (10.22), and (10.23), is demonstrated in Examples 10.9 and 10.10. This simpler approach adversely affects accuracy. According to Omega Engineering (1993), the resulting accuracy may not be better than ±10%.

Example 10.9

A rotameter that has a stainless-steel float ($S = 7.9$) has been calibrated for water. It is desired to use it for measuring the flow rate of kerosene ($S = 0.82$). Calculate the conversion factor for the readings that are in L/min of water.

Solution: Using Eq. (10.21) and (10.23), for the same reading on the rotameter (the same value of A_a), the ratio of flow rates will be

$$\frac{Q_k}{Q_w} = \frac{V_{a,k}}{V_{a,w}} = \left(\frac{\rho_b - \rho_k}{\rho_b - \rho_w}\right)^{1/2}\left(\frac{\rho_w}{\rho_k}\right)^{1/2}$$

Considering that the specific gravity of kerosene and stainless steel are 0.82 and 7.9, respectively, we have

$$\frac{Q_k}{Q_w} = \left(\frac{7.9 - 0.82}{7.9 - 1.0}\right)^{1/2}\left(\frac{1}{0.82}\right)^{1/2}$$
$$= 1.12$$

This means that if the scale reading is multiplied by a conversion factor of 1.12, the device will give the kerosene flow rate.

Example 10.10

A rotameter is calibrated for air at standard conditions (20°C, 1 atm). It is going to be used to measure the flow of methane at 25°C, 2 atm. Calculate the conversion factor for this application. The molecular weights of air and methane are 29 and 16, respectively.

Solution: We will use Eq. (10.22) to estimate the conversion factor:

$$\frac{Q_m}{Q_a} = \frac{V_{a,m}}{V_{a,a}} = \left(\frac{\rho_a}{\rho_m}\right)^{1/2}$$

Using the perfect gas law, $\rho = PM/(\overline{R}T)$, where M and $\overline{R}$ are the molecular weight and universal gas constant, respectively. The equation for the flow ratio becomes

$$\frac{Q_m}{Q_a} = \left(\frac{P_a\, M_a\, T_m}{P_m\, M_m\, T_a}\right)^{1/2}$$

$$= \left[\left(\frac{1}{2}\right)\left(\frac{29}{16}\right)\left(\frac{298}{293}\right)\right]^{1/2}$$

$$= 0.96$$

Thus, volume flow rate readings should be multiplied by 0.96 if the fluid is the higher-pressure methane.

Comment: There is an inherent approximation in the above analysis—the equality of the drag coefficients (C_D). The Reynolds number depends on the fluid viscosity and velocity, and the drag coefficient may show some dependence on the Reynolds number. Due to the temperature dependence of viscosity, the accuracy of rotameters will deteriorate slightly with changes in temperature.

10.1.3 Turbine Flowmeters

Turbine flowmeters may be used to measure the flow rate of clean liquids. In these devices (Figure 10.10), the fluid turns a turbine wheel or propeller that is supported in the

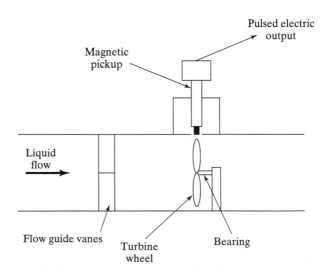

FIGURE 10.10

Turbine flowmeter.

fluid stream. The turbine wheel is either made of a magnetic material or has a magnetic material embedded in the tips. The turbine rotates, and each time a tip passes the magnetic pickup, an electrical pulse is generated. The output signal is a frequency that is proportional to the flow rate. Turbine flowmeters of this type are simple and hold their calibration very well. They can normally be used with high confidence using only the manufacturer's calibration. Accuracies as good as ±1/2% of reading are readily available. The linearity of output tends to drop off at low flows, and most become less accurate at flows less than 5 to 10% of maximum rated flow.

Since the magnetic pickup creates a slight drag on the turbine wheel, alternative sensing techniques are sometimes used. For transparent fluids, having the turbine blades intercept a light beam transmitted to a photocell is one alternative. Since the flowing liquid serves as a lubricant for the bearing, these flowmeters are not suitable for fluids that contain abrasive particles or contaminate the various surfaces. These flowmeters do generate a small pressure drop in the flowing fluid. A variation of the turbine meter, called the *paddle-wheel flowmeter*, is shown in Figure 10.11. In this device the fluid drives a small paddle wheel that is located on the side of the pipe. This flowmeter has negligible pressure loss and is relatively inexpensive.

10.1.4 Mass Flowmeters

Thermal Mass Flowmeter The thermal mass flowmeter, shown schematically in Figure 10.12, can be used to measure mass flow rates of clean gases. The flow is split into two portions, most of it going through the main tube and a small fraction going through the sensing tube. The flow through the sensing tube and through the main tube obstruction are laminar flow. Consequently, the ratio of the sensing mass flow to main mass flow is independent of flow rate. An electric heater is attached to the sensing tube and serves to heat the flowing fluid. The temperatures upstream and downstream of the heaters are measured with resistance temperature sensors. An energy balance on

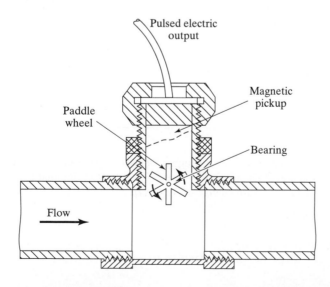

FIGURE 10.11

Paddle-wheel flowmeter.

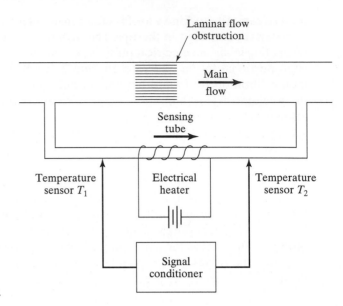

FIGURE 10.12

Thermal mass flowmeter.

the bypass flow gives the following relationship:

$$q = \dot{m}c_p(T_2 - T_1) \tag{10.24}$$

where $\dot{m}$ is the gas mass flow rate, c_p is the constant-pressure specific heat, and T_1 and T_2 are the upstream and downstream temperatures. The heater is controlled such that q is constant, regardless of the flow. Solving for $\dot{m}$ gives

$$\dot{m} = \frac{q}{c_p(T_2 - T_1)} \tag{10.25}$$

For most gases, the specific heat is known and is only a weak function of temperature. Hence, measuring the temperature change can help determine the mass flow rate. A signal conditioner converts the temperature sensor outputs to a linear voltage proportional to mass flow rate. Accuracy can be better than ±1%. In practice, the two resistance temperature sensors may also function as the heating elements. Although the power requirements are higher, some thermal mass flowmeters heat the entire conduit flow. It is also possible to use the method for liquids.

Example 10.11

It is proposed to heat the entire flow going through a 1-cm-ID pipe. The flowing fluid is air with a specific heat of 1.0057 kJ/kg-C. The mass flow rate is 0.003 kg/s. If the heat transferred to the pipe is 10 W, what will the temperature change of the gas be?

Solution: Using Eq. (10.25), we obtain

$$T_2 - T_1 = \frac{q}{\dot{m}c_p} = \frac{10}{0.003 \times 1005.7} = 3.31°C$$

This temperature difference is measurable, but there may be accuracy problems. We could use a higher power to produce a larger temperature difference. Heating only a bypass flow is a preferable approach.

Coriolis Mass Flowmeter   The *Coriolis force* is a force that occurs when dynamic problems are analyzed within a rotating reference frame. Useful flowmeters based on this effect are now widely used in the process industries. Consider a fluid flowing through the U-shaped tube shown in Figure 10.13(a). The tube is cantilevered out from a rigidly supported base. An electromechanical driver is used to vibrate the free end of the tube at its natural frequency in the y direction. The amplitude of this vibration will be largest at the end of the cantilever and zero at the base. Consider an instant in time when the tube is moving in the $-y$ direction. The fluid moving through the tube away from the base will not only have a component of velocity in the x direction but also in the $-y$ direction, and the magnitude of this y component will increase with distance from the base. As a fluid particle moves along the tube, it is thus accelerating in the $-y$ direction. This acceleration is caused by a Coriolis force in the $-y$ direction applied by the tube wall. The resultant reaction on the tube wall is a force, F in the $+y$ direction. For the fluid returning to the base, the y component of fluid velocity is decreasing in the flow direction. This results in a Coriolis force on the tube wall in the $-y$ direction. These two forces twist the U-shaped tube about the x-axis. When the tube vibration velocity is upward, the force directions are reversed and the tube is twisted in the opposite direction. Sensors measure this oscillatory twisting motion, which depends on the mass flow rate.

Practical Coriolis flowmeters use two tubes, as shown in Figure 10.13(b), but the exact shape of the tube varies widely (they are not necessarily U-shaped). The two tubes are vibrated 180° out of phase. Sensors detect the relative twisting motion of the two tubes, and signal conditioning produces an output proportional to flow. Coriolis flowmeters can accurately measure mass flow for a wide range of fluids, including gases, liquids, slurries, and two-phase (gas–liquid) mixtures. For calibrated devices, the

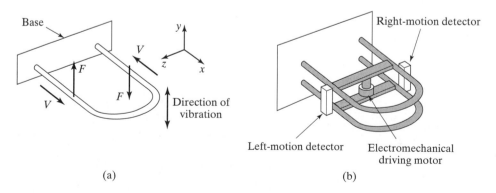

FIGURE 10.13

Coriolis mass flowmeter: (a) schematic of single tube; (b) practical implementation.

accuracy can be very good, on the order of ±0.2% of the reading. Because the tubes are vibrated at their natural frequency, Coriolis flowmeter can also be used to measure fluid density. This is because the natural frequency of vibration depends not only on the mass of the tube but also on the mass of fluid inside the tube.

10.1.5 Positive-Displacement Flowmeters

In positive-displacement flowmeters the flow causes a periodic movement of the internal components of the flowmeter, and this movement can be detected and used to determine the flow rate. Domestic natural-gas and water meters are of the positive-displacement type. In the latter applications, the mechanical motion of the meter internal components is used to drive a mechanical counter that is read by the utility. There are many types of positive-displacement meters. A whole class of positive-displacement meters are the gear-type meters. Two intermeshing gears contained inside a housing are forced to rotate by the flowing fluid. One variation of this type of flowmeter, shown in Figure 10.14, uses two two-lobed rotors. The design is geometrically similar to the common roots air blower. The shaft of one of the rotors can be used to drive a clockwork type of counter in order to obtain a reading. Alternatively, the rotation of the output shaft can be measured with a magnetic pickup or other pulse-generating device, and the pulses can be converted to flow at a remote location.

10.1.6 Other Methods of Flow Measurement

Although we have covered the most common methods for measuring flow in conduits, flow measurement is so common that many different methods have been devised. A number of other methods are listed here for completeness.

Magnetic Flowmeter   The magnetic flowmeter is based on the principle that a voltage is generated in an electrically conducting fluid if it moves through a magnetic field. It is not necessary for any components to protrude away from the pipe wall into the fluid, so this principle is useful for measuring flows of conducting liquids that are corrosive or that contain abrasive particles and are not well suited to other measurement techniques. The method will not work for pure water or most organic liquids that have low electrical conductivity.

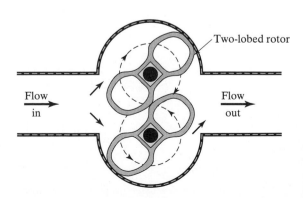

FIGURE 10.14

A positive-displacement flowmeter.

Drag Flowmeter   In this flowmeter, a vane is inserted into the flow with the blunt face normal to the flow. As a result, the fluid applies a drag force on the vane, which is sensed and used as a measure of the flow.

Elbow Flowmeter   As a fluid flows around a bend in a pipe, it creates a static pressure difference between the inside turn radius and the outside turn radius. The pressure difference can be related to the average velocity and hence the flow in the pipe. This type of flowmeter is simple to incorporate—pressure taps can be installed on an existing elbow. The pressure drop is usually smaller than in other differential-type flowmeters. In-place calibration is normally required.

Vortex Shedding Flowmeter   If a cylinder or other suitably shaped prismatic object is placed into and normal to the velocity of a flowing fluid, a series of alternating vortices will be shed from the trailing side over a wide range of fluid velocity (Figure 10.15). The frequency at which these vortices are shed is described by a dimensionless parameter called the *Strouhal number*:

$$\text{St} = \frac{fD}{V} \tag{10.26}$$

where f is the frequency of the vortex shedding, D is the diameter of the cylinder, and V is the fluid velocity. In general, the Strouhal number is a function of the Reynolds number [Eq. (10.7)]. However, for cylinders with Reynolds numbers based on the fluid velocity and the cylinder diameter in the range 500 to 100,000, the Strouhal number has an almost constant value in the range 0.2–0.21 [Roshko (1954) and Jones (1968)]. Hence, measuring the vortex frequency is a measure of the fluid velocity:

$$V = \frac{fD}{\text{St}} \tag{10.27}$$

The flowrate can then be determined from $Q = VA$. Calibrating the specific design will be required. For shapes other than cylinders, the Strouhal number will have different values. At very low or high Reynolds numbers, the vortex shedding process may not occur.

There are several methods of detecting the vortex shedding frequency. One method is to place a velocity sensor in the wake and measure the output frequency rather than the magnitude. Another method is based on the fact that the vortex shedding

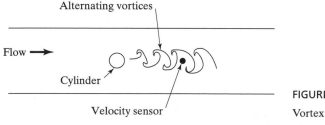

FIGURE 10.15
Vortex shedding flowmeter.

process produces oscillating lateral forces on the cylinder. These forces can be sensed with a piezoelectric force-measuring device located where the cylinder is attached to the pipe wall. Since they have no moving parts, vortex flowmeters can be used in chemically harsh environments. Accuracies can be better than ±1% of reading. Readings can be quite linear from 10% to 100% of maximum flow.

Example 10.12

Air at 1 atm flows past a 5-mm-diameter cylinder at a velocity of 30 m/s. Estimate the frequency of the vortices behind the cylinder.

Solution: From Table B.3 at 1 atm pressure and 20°C, $\rho = 1.20 \text{ kg/m}^3$ and $\mu = 1.811 \times 10^{-5}$ kg/m-s. Using Eq. (10.27) and assuming that the Strouhal number is 0.21, we have

$$f = \text{St} \cdot \frac{V}{D} = \frac{0.21 \times 30}{0.005} = 1260 \text{ Hz}$$

The Reynolds number is

$$\text{Re} = \frac{1.2 \times 30 \times 0.005}{1.811 \times 10^{-5}} = 9939$$

which is within the range that St is approximately 0.21.

Ultrasonic Flowmeter   Ultrasonic flowmeters determine the fluid velocity by passing high-frequency sound waves through the fluid. An important class of ultrasonic flowmeters are called ultrasonic transit-time flowmeters (Figure 10.16). A sound pulse is first transmitted from transducer 1 to transducer 2. The transit time is given by

$$T_{1,2} = \frac{d}{\sin \theta (C + V \cos \theta)} \qquad (10.28)$$

where C is the sonic velocity in the fluid. Next, a sound pulse is transmitted from transducer 2 to transducer 1. The transit time for this path is

$$T_{2,1} = \frac{d}{\sin \theta (C + V \cos \theta)} \qquad (10.29)$$

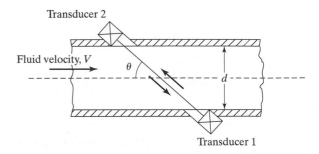

FIGURE 10.16

Transit-time ultrasonic flowmeter.

Equations (10.28) and (10.29) can be combined to obtain

$$\frac{T_{1,2} - T_{2,1}}{T_{1,2} \times T_{2,1}} = \frac{2V \sin \theta \cos \theta}{d} \tag{10.30}$$

By measuring the two transit times, the fluid velocity can be measured independent of the fluid sonic velocity. Since the fluid velocity is not uniform over the duct cross section, the measured velocity must be corrected to obtain accurate flow rates. Ultrasonic flowmeters are primarily used for liquid flow rates. Accuracies can be as good as ±1% of flow rate.

One particularly useful type of ultrasonic flowmeter is the portable clamp-on type. This type is simply clamped onto the outside of a pipe and can be used to determine the flow rate. Since the ultrasonic beam also passes through the pipe wall, correcting for the pipe wall thickness and material is required.

Ultrasonic flowmeters that make use of the Doppler shift on sound waves are also available. The principle is the same as that for Doppler radar, as discussed in Chapter 8, except that sound waves are used instead of radio waves. The ultrasonic sound is reflected from particles and bubbles in the flowing liquid. A detailed summary of ultrasonic flowmeter concepts is provided by Sanderson and Hemp (1981).

Example 10.13

An ultrasonic transit-time flowmeter is used to measure the velocity of water flowing through a 10-cm-ID pipe. The water velocity is 2 m/s, and the sonic velocity in the water is 1498 m/s. The orientation of the sensors, θ, is 30°. Compute the transit times in each direction and the difference in the transit times.

Solution: From Eqs. (10.28) and (10.29), $T_{1,2} = 133.6658898 \ \mu s$, $T_{2,1} = 133.3571553 \ \mu s$, and the difference is $0.309 \ \mu s$. These times are short, particularly the time difference, and difficult to measure. Sophisticated techniques, such as multiple reflections of the pulses, are required to make these systems function satisfactorily.

Flowmeters Using Fluid Velocity Sensors   In the next section, a number of devices are described that measure fluid velocity at a single location in a fluid flow field. Virtually all of these devices can be incorporated into a flowmeter. In a conduit, the velocity is not uniform, but varies with radial position, as shown in Figure 10.17. If the shape of

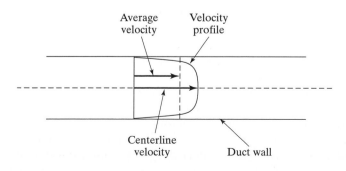

FIGURE 10.17

Velocity profile in a duct.

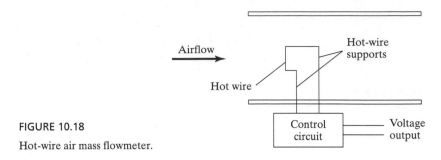

FIGURE 10.18

Hot-wire air mass flowmeter.

the distribution is known, measuring velocity at a single radial location determines the velocity everywhere. With a known velocity distribution, the average velocity can be determined by integrating. The exact relationship between the average and the measured velocity depends on the flow regime and duct cross-sectional shape, but can be estimated with reasonable accuracy in most cases. Velocity profiles for circular ducts are available in Shames (1992).

One common type of flowmeter using a velocity sensor that deserves mention is the hot-wire mass airflow meter. Many modern automobiles with computer-controlled fuel-injection systems use this device to measure engine airflow. (See Fig. 10.18.) The heated wire is actually a hot-wire anemometer velocity sensor (discussed in greater detail later in this chapter). The resistive wire is heated by passing a current through it. If the wire temperature is held constant, more heat will be transferred from the wire to the air at high air velocities than at low air velocities. Higher heat transfer requires a higher current in the wire to maintain the same temperature. Thus, the wire current is a measure of the air velocity, which can be used to determine the flowrate. Actually, the heat transferred depends on the product of density and velocity, ρV. Hence, this device measures the mass flow rate $(\dot{m} = \rho V A)$ of the gas.

10.1.7 Calibrating Flowmeters

Liquid flowmeters are generally fairly easy to calibrate. Liquid that has flowed through the flowmeter can be collected in a container over a measured period of time. The collected liquid can then be weighed or, if the container is of a simple geometry, the volume can be measured. Gas flowmeters are more difficult to calibrate. For fairly low flow rates, there is a particular type of positive-displacement meter called a *wet test meter*, described by the ASME (1971). This device consists of a set of chambers in a rotating drum and uses water as a seal. If the chambers are constructed accurately and the water level is correct, this type of flowmeter is very accurate. A flowmeter can then be calibrated by passing the same flow through both the wet test meter and the flowmeter being calibrated.

In another approach for low gas flows, the flow that has passed through the flowmeter can be collected for a measured period of time in a graduated cylinder that has been filled with water and inverted in another container of water, as shown in Figure 10.19. After the collected gas has come into equilibrium with the water, the volume can be measured. Correction will be necessary to account for water vapor that has

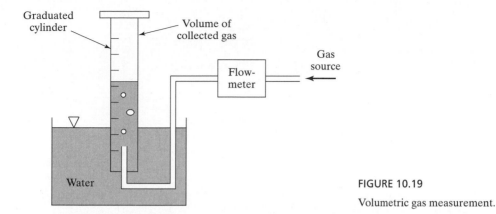

FIGURE 10.19

Volumetric gas measurement.

mixed with the gas. This process can be scaled up until the collected volume is a few liters but may be impractical for larger volumes.

For higher flow rates, one possible method is to perform a partial blowdown of a compressed tank of gas and use thermodynamics to compute the mass flow. The flow should be controlled with a valve to maintain a constant flow during the blowdown. If the gas is in equilibrium with the tank and surroundings before and after the blowdown, the tank volume and initial and final pressures can be used to compute the mass lost. The inlet conditions to the flowmeter may change during the blowdown, and it may be desirable to measure the flowmeter conditions continuously during the blowdown.

10.2 SYSTEMS FOR MEASURING FLUID VELOCITY

In Section 10.1, devices for measuring the total flow rate in a duct were presented. In this section a number of devices that can be used to measure the velocity at a point in a flow field are presented.

10.2.1 Pitot-Static Probe

The *pitot-static probe*, shown in Figure 10.20, is a very common device used to measure fluid velocity in both liquid and gas flows. A pitot-static probe normally protrudes from the leading edge of aircraft wings in order to measure airspeed. The probe is placed in the flow with the axis parallel to the flow. There are two pressure-sensing locations on the probe, the total pressure port 1 and the static pressure port(s) 2. The flow that approaches the total pressure port stops completely, so the velocity is zero. The flow that passes the static pressure port has approximately the same velocity as the velocity without the probe present. The Bernoulli equation [Eq. (10.1)] can be applied between sections 1 and 2. With $V_1 = 0$, this results in the following equation for V_2:

$$V_2 = C\sqrt{\frac{2[(P_1 + g\rho z_1) - (P_2 + g\rho z_2)]}{\rho}} \qquad (10.31)$$

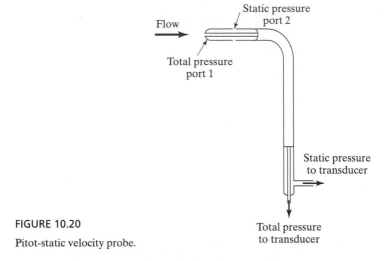

FIGURE 10.20

Pitot-static velocity probe.

The constant C, which is close to 1, accounts for the fact that the velocity at section 2 is slightly higher than it would be if the probe were not present since the probe obstructs the flow. C is determined by calibration but can generally be obtained from the probe manufacturer. Eq. (10.31) is valid for flowing liquids and gases at low velocities. For gases, z is usually negligible. Nevertheless, the pressure transducer will effectively measure the change in $(P + g\rho z)$ if the sensing lines are completely filled with the flowing fluid. For liquid flow using manometers for pressure measurement, the analysis is somewhat more complicated and the reader should consult a text on basic fluid mechanics, such as White (1999).

The *pitot probe* is similar to the pitot-static probe except that the static pressure is not measured by the probe but is sensed separately, usually at the duct wall. Pitot probes can be smaller in size and are used when better spatial resolution is needed in velocity measurement, such as the measurements needed to determine the velocity profiles in regions near boundaries.

One problem with pitot-static and pitot probes is that the axis of the probe must be oriented parallel to the flow. If the misalignment is more than 5°, an error results and this error is substantial if the misalignment is more than 15°. If possible, the probe should be rotated to give a maximum pressure differential during the velocity measurement to establish that the probe is properly aligned. There are a number of commercially available variations of the pitot-static probe, such as the *Kiel probe*, which are less sensitive to misalignment.

Example 10.14

Atmospheric air flows past a pitot-static probe, which shows a pressure difference of 2 in. of water column. Find the air velocity.

Solution: From Table B.2 the density of 70°F water is 62.26 lbm/ft^3. From Table B.4 the density of air is 0.0749 lbm/ft^3. We will use Eq. (10.31) and assume that the constant C is 1.0. First we must convert the pressure units of water column to force/unit area units:

$$\Delta P = \rho g\,\Delta h = \frac{62.26}{32.17} \times 32.17 \times \frac{2}{12} = 10.4\ \text{lbf/ft}^2$$

Substituting into Eq. (10.31) gives

$$V = 1\sqrt{\frac{2 \times 10.4}{0.0749/32.17}} = 94.5\ \text{ft/s}$$

10.2.2 Hot-Wire and Hot-Film Anemometers

The hot-wire anemometer is a rather sophisticated, but well-established, fluid-velocity-measuring device which is based on the fact that the convective heat transfer from a small-diameter heated wire is a function of the fluid velocity. This type of velocity-measurement system is discussed in detail in Lomas (1986). A typical hot-wire probe is shown in Figure 10.21. The physical size of the probe is quite small, and the wire itself is on the order of 0.04 in. long with a diameter as small as 0.0002 in. This means that the probe does a minimum to disturb the flow that it is intended to measure.

In a hot-wire anemometer, the heat-transfer situation can be viewed as a cylinder in crossflow (i.e., the flow direction is normal to the cylinder axis). The heat transfer from a hot-wire anemometer is known as *King's law:*

$$q = (T_s - T_f)(A_0 + B_0\sqrt{Re}) \tag{10.32}$$

where q is the heat transfer rate, T_s is the wire surface temperature, T_f is the fluid temperature, and A_0 and B_0 are constants that depend on the fluid and the wire diameter. The Reynolds number is based on the wire diameter and is defined by

$$Re = \frac{\rho U D}{\mu} \tag{10.33}$$

where U is the fluid velocity, D is the wire diameter, and ρ and μ are the fluid density and viscosity, respectively.

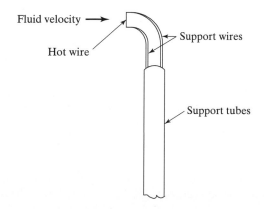

Fluid velocity

Hot wire

Support wires

Support tubes

FIGURE 10.21

Hot-wire probe.

The electrical power into the wire, which is approximately equal to the convective heat transfer from the wire (neglecting radiation and conduction), is given by

$$q = \frac{V^2}{R} \tag{10.34}$$

where V is the voltage across the sensor and R is the sensor resistance. Combining Eqs. (10.32), (10.33), and (10.34), we obtain the expression:

$$\frac{V^2}{R(T_s - T_f)} = a_0 + b_0 U^{0.5} \tag{10.35}$$

where a_0 and b_0 are constants. It is possible to operate the hot-wire sensor such that its temperature is held constant. Since the sensor resistance is a function of the sensor temperature, the resistance will also be constant. Holding temperature and resistance constant, Eq. (10.32) indicates that the voltage across the sensor wire, V, will only be a function of the fluid velocity, U.

A hot-wire sensor operated in this manner is known as a constant-temperature hot-wire anemometer. A simplified schematic for such a system is shown in Figure 10.22. In this circuit the feedback amplifier automatically keeps the Wheatstone bridge in balance by adjusting the driving voltage. The driving voltage is then the output, which is a function of velocity. A simpler hot-wire circuit is called a *constant-current hot-wire anemometer* (Figure 10.23). In this circuit the current through the wire is held constant and the resistance of the wire depends on the fluid velocity. The current is held constant

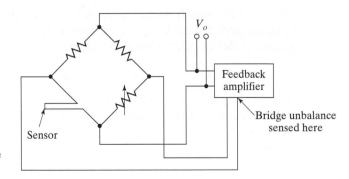

FIGURE 10.22

Constant-temperature hot-wire anemometer system.

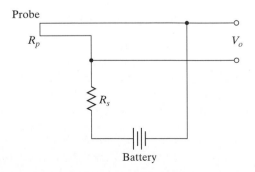

FIGURE 10.23

Constant-current hot-wire anemometer circuit.

by placing the probe in series with a much larger resistance $(R_s \gg R_p)$. The voltage drop across the sensor is the output signal.

One of the most important applications of hot-wire anemometers is the measurement of fluid turbulence, and the simple constant-current circuit is deficient in this regard. Even for probes with the smallest-diameter wires, the response to frequencies over a few hundred hertz will be significantly attenuated. Electronic compensation can extend the frequency response to about 100 kHz. Compensation circuits increase the circuit output as a function of frequency to counteract the high-frequency roll-off of the uncompensated circuit. The compensation circuits, however, require sensitive adjustments for each probe and each flow. On the other hand, constant-temperature circuits can have linear response up to frequencies of 400 kHz and are simpler to use. The constant-temperature anemometer is much more common.

Hot-wire sensors are much more sensitive to velocities that are transverse to the sensor wire axis than they are to velocities that are parallel to the wire axis. As a result, a probe can be used to determine the direction of a velocity by rotating the probe to find the peak velocity. This directional sensitivity can also be used in special probes having three orthogonal sensing wires. These three-axis probes can be used to determine the three components of velocity at a point simultaneously. Since turbulence is inherently three-dimensional, these probes have been widely used in turbulence research.

The hot-wire anemometer is suited only for measuring velocities in very clean fluids. Particles in a dirty fluid will break or contaminate the probe. An alternative based on the same principles is the hot-film probe. In this design, a thin layer of a metal (on the order of 0.0002 in.) is plated onto an insulating ceramic substrate. The size of the sensing element is larger and much more durable than the hot wire and may be used in liquids and dirty gases. Whereas hot-wire probes are used primarily in research laboratories, hot-film probes are practical for everyday use—hot-film probes are used routinely to measure air velocities in air-conditioning ducts.

10.2.3 Fluid Velocity Measurement Using the Laser-Doppler Effect

The *laser-Doppler velocimeter* (LDV), also called a *laser Doppler anemometer* (LDA), is a well-developed nonintrusive method for measuring fluid velocity. It is a sophisticated, expensive approach, but is widely used for fluid mechanics research (including turbulence measurements) and high-technology product development. This method of fluid measurement was first introduced by Yeh and Cummins (1964). In Chapter 8, we applied the Doppler effect with radio waves to solid-body velocity measurement. In this section, we discuss the measurement of fluid velocity with the use of the Doppler effect for laser-generated light waves.

Doppler Effect When a sound or electromagnetic signal of a given frequency is emitted by a moving body, the apparent frequency to a stationary observer is different from the emitted frequency. This frequency change, known as the *Doppler shift*, is named after an Austrian physicist, Christian Johann Doppler (1803–1853), who first studied the phenomenon. An example of this frequency change is the difference between the pitch (frequency) of a train horn when it approaches and when it moves away from a stationary observer. When the source of sound is moving toward an observer, the pitch of the sound is higher than when they are both stationary.

Laser-Doppler velocimeters are not used to measure the velocity of fluids directly but rather the velocity of small particles that move along with the fluid. In general principle, velocity measurement using an LDV is the same as solid-body measurement using radar. However, the frequency of laser-produced electromagnetic radiation is much higher than the frequency of radio waves. Since the wavelength, λ, of an electromagnetic wave is related to the frequency by

$$\lambda = \frac{c}{f} \qquad (10.36)$$

where c is the speed of light, the wavelength of laser light is much shorter than radio waves. This means that laser light waves can be reflected (scattered) from very small particles. Due to viscous effects, very small particles will move at essentially the same velocity as the fluid, so measuring particle velocity is equivalent to measuring fluid velocity. For successful LDV measurements, the fluid should contain small particles, usually on the order of $1\ \mu m$, with a concentration of about 10^{10} to 10^{11} scattering particles/m^3 (Durst et al., 1981). This can be accomplished by having particles that may already exist in the fluid (in water this is often the case) or by seeding the fluid with particles in the appropriate diameter and concentration range.

As with radar Doppler devices, Eq. (8.27) for the frequency shift (f_D) gives

$$f_D = \frac{2V \cos \theta}{\lambda} = \frac{2V \cos \theta}{c} f_0 \qquad (10.37)$$

where V is the particle velocity, θ is the angle between the laser beam and the particle-velocity vector, and f_0 is the frequency of the light. The velocity of fluids, even in the case of supersonic gases, is much lower than the speed of light, c, so as Eq. (10.37) shows, the Doppler frequency will be very small compared to the light frequency.

Dual-Beam Laser Velocimeter   A number of optical systems have been used to detect the Doppler frequency caused by the moving particles. The system that is currently the most popular is called the dual-beam laser velocimeter. As shown in Figure 10.24, the output of a laser is split into two beams. The lens focuses these two beams at the point at which they cross in the flow. The focal point is the location in the flow for which the velocity will be determined. Where the two beams intersect, they interfere with each

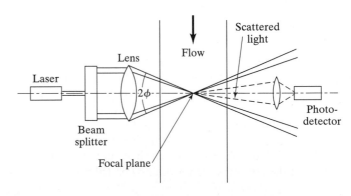

FIGURE 10.24

Dual-beam laser velocimeter.

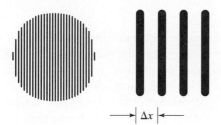

FIGURE 10.25

Fringes in sensing volume.

other, producing a fringe pattern, which consists of alternating light and dark bands as shown in Figure 10.25. This figure is a section through the flow, which is parallel to the flow and perpendicular to the bisector of the angle between the two beams. A particle that passes through this region will alternately pass through light and dark fringes. When in the light regions, the particle will scatter light, which can be collected by the lens system and detected by the photodetector.

The interference pattern can be explained by the principle of superposition of oscillations representing two electromagnetic fields. If the angle between the two wavefronts (beams) is 2ϕ, the fringe spacing will be

$$\Delta x = \frac{\lambda}{2 \sin \phi} \tag{10.38}$$

If a particle with a speed V passes through the fringe pattern, it will produce pulses of scattered light as it passes through the bright bands. The frequency of these pulses will then be

$$f = \frac{V}{\Delta x} = \frac{2V \sin \phi}{\lambda} \tag{10.39}$$

Since $\theta = 90° - \phi$, this equation indicates that this frequency is the same as the Doppler frequency given by Eq. (10.37). The pulsations are sensed by the photodetector. The output of the photodetector can then be processed to determine its frequency content, and Eq. (10.39) is used to determine the velocity.

The laser-Doppler velocimeter is a well-developed technology that is now routinely used for detailed measurement of space and time resolution of velocity fields. LDV has the advantage of being nonintrusive (no physical probe is introduced into the flow), meaning that there is no probe interference with the flow. It has a linear response, does not need calibration, and can be used for measuring flow in different directions. It is used for a very wide range of velocities, from slow moving to hypersonic flows. Details of the principles and applications of LDV systems can be found in Durst et al. (1981) and Drain (1980).

10.3   MEASURING FLUID LEVEL

There is a frequent need in the process industry to measure or control the level of liquid in tanks. Consequently, a wide variety of devices has been devised to measure liquid level, the more common of which are described here.

10.3.1 Buoyancy Devices

Figure 10.26(a) shows a float-type level-measuring device that is widely used in automobiles and trucks to measure the level of the fuel in the fuel tank. The float, which remains on the surface of the fluid, is attached to a pivoted arm that turns a rotary potentiometer. The angular position of the arm is a nonlinear function of the fluid level; however, the windings in the potentiometer are sometimes designed in a compensating nonlinear arrangement. Alternatively, the readout device may have a nonlinear scale. While the device shown in Figure 10.26(a) works well in vehicular fuel tanks, which have a large fluid surface area relative to the tank height, tanks in the process industry are often high relative to the diameter. The float-cable device, shown in Figure 10.26(b), works better in this situation. The float is attached with a cable to a spring-loaded drum such that as the fluid level changes, the drum winds up the cable. The drum is then connected to a rotary potentiometer.

Figure 10.26(c) shows another device that depends on buoyant force. The displacer is attached to the top of the tank through a force transducer. The measured vertical force is the difference between the downward weight force and the upward buoyant force. The buoyant force is $\rho_f g\ V_{\text{submerged}}$ and the submerged volume depends on the fluid level. In Figure 10.26(d), the displacer is supported by a spring. Its vertical position is measured with an LVDT transducer. The key feature of this design is that the motion of the displacer is much less than the motion of the fluid surface.

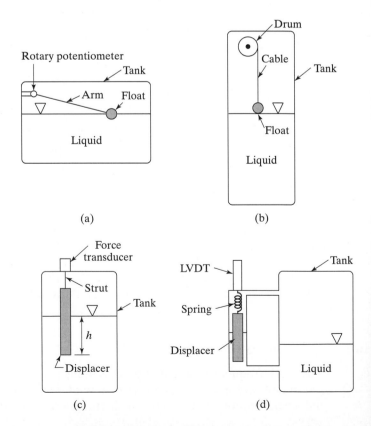

FIGURE 10.26

Float-type level systems.

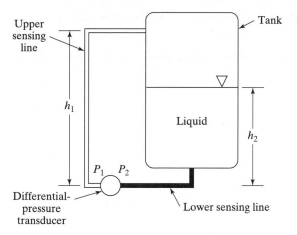

FIGURE 10.27

Differential pressure level-measuring system.

10.3.2 Differential-Pressure Devices

The difference between the pressure at the surface of a static liquid and the pressure below the surface at a depth h is given by the formula

$$\Delta P = \rho g h \qquad (10.40)$$

where ρ is the fluid density and h is the vertical distance between the fluid surface and the pressure measurement point. Thus, if the density is known, a measurement of the differential pressure determines the liquid depth, h. Figure 10.27 shows a configuration that can be used to determine the fluid level with this technique. Since the air or fluid vapor above the liquid has a very small density, the upper sensing line pressure is the same as the surface pressure. The actual h determined in this way is the depth of the pressure transducer below the fluid surface. Since the vertical position of the transducer relative to the bottom of the tank is fixed and known, the actual level can be readily determined. One problem with this approach is that the upper sensing line may become filled with condensate or other liquid, resulting in an erroneous reading. If the top of the tank is vented to the atmosphere, the upper sensing line can be eliminated so that the low-pressure side of the pressure transducer is vented directly to atmosphere. Alternatively, the upper sensing line can be deliberately filled with liquid. This will affect the transducer reading by subtracting a constant value. If the fluid contains particles or precipitates out solids, it is possible for the sensing line to become blocked over time. To overcome this problem, special transducers are available in which the diaphragm covers a hole in the tank wall (surface-mounted transducer), eliminating the lower sensing line.

Example 10.15

A differential-pressure device is used to determine the level of water in a pressure vessel (Figure 10.27). The water in the tank has a density of 60.47 lbm/ft³. The pressure transducer is attached very close to the vessel. The upper sensing line rises a vertical distance of 15 ft and is filled with cooler water with a density of 61.8 lbm/ft³. The pressure transducer reads 2.73 psi with the upper sensing line side showing the higher pressure. Determine the height of the water surface above the transducer.

Solution: The pressure difference that the transducer sees is the difference of the upper and lower sensing line pressures.

The upper sensing line pressure is

$$P_1 = P_v + \rho_1 g h_1 = P_v + \left(\frac{61.8}{32.17}\right) \times 32.17 \times 15 = P_v + 927.0 \text{ lbf/ft}^2$$

where P_v is the pressure of the vapor on top of the liquid in the tank.

The lower sensing line pressure is

$$P_2 = P_v + \rho_2 g h_2 = P_v + \frac{60.47}{32.17} \times 32.17 \times h_2$$

The pressure transducer reads the difference between these two pressures:

$$\Delta P = 2.73 \times 144 = P_v + 927.0 - P_v - 60.47 \times h_2$$

Solving for h_2, we have

$$h_2 = 8.83 \text{ ft}$$

10.3.3 Capacitance Devices

If a metal rod is inserted from the top of a tank as shown in Figure 10.28, the rod and the tank wall will form a capacitor. The rod is normally insulated to maintain a high resistance between the rod and the liquid. The capacitance value depends not only on the geometry of the rod and the tank but is also proportional to the dielectric coefficient of the fluid in the tank. Air or vapor will have a dielectric constant close to unity, but liquids will have a much higher dielectric coefficient, in the range 2 to 100 (depending on the liquid). Hence, the value of the capacitance is a strong function of the level of the fluid in the tank. With suitable signal conditioning (see Figure 8.19), this capacitance can be converted to a voltage that is a function of level.

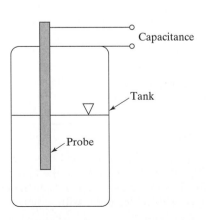

FIGURE 10.28

Capacitance level-measuring system.

10.3.4 Conductance Devices

This device consists of two vertical, electrically isolated, electrical conductors inserted into the tank. If the liquid is electrically conductive, the resistance between the two conductors will be a function of the level. This resistance can be converted to a voltage using a Wheatstone-bridge circuit. This type of device will normally not work with insulating liquids such as hydrocarbons.

10.3.5 Ultrasonic Devices

This device uses the same principle as the sonar devices used by ships to locate the bottom of a channel. A high-frequency sound pulse is directed downward toward the surface of the fluid, where it is reflected and directed back toward the receiver as shown in Figure 10.29. The distance traveled is related to the time t for the pulse to travel the path from the transmitter to the receiver; that is,

$$L = \frac{ct}{2} \tag{10.41}$$

where c is the speed of sound and L is one-half of the path traveled by the sound. Measuring the time then determines L.

Ultrasonic level sensors are very useful for corrosive liquids since the sensor does not contact the liquid directly. They can also be used to measure the level of solids such as powders. One disadvantage of this device is that the speed of sound is a function of the chemical composition and temperature of the vapor. Thus, the device requires calibration and temperature correction and may be unreliable if the vapor composition is variable.

Example 10.16

An ultrasonic level system is located 5 m above a liquid in a tank. The fluid above the liquid is air with a sonic velocity of 412 m/s. How much time elapses between the sending of an ultrasonic pulse and its reception by the receiver?

Solution: We use Eq. (10.41):

$$t = \frac{2L}{c} = 2 \times \frac{5}{412} = 0.0243 \text{ s}$$

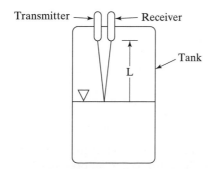

FIGURE 10.29

Ultrasonic fluid level system.

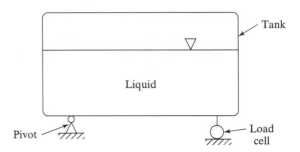

FIGURE 10.30

Tank weighing method for fluid level.

10.3.6 Weight Methods

In some cases it is possible to support the tank with load cells and hence measure the weight of the tank and the tank contents. One possible support method is shown in Figure 10.30. A significant disadvantage of this method is that the pipes carrying flows to and from the tank must be flexible where they attach to the tank, a constraint that is not always practical.

10.4   MEASURING AIR POLLUTION SPECIES

In recent years significant attention has been paid to the air pollutants generated by combustion systems such as internal-combustion engines and fossil-fuel power plants. As a result, there has been significant development of engineering instrumentation for measuring the various air pollution species produced by combustion systems. In this section, we review some of the more important instruments for measuring gas species.

The composition of a gas mixture is usually measured in terms of volumetric fractions of its components. In all air-pollution and exhaust-gas measurements, gases are treated as ideal (perfect), implying that they follow the ideal gas equation of state, or

$$P V = n \overline{R} T \tag{10.42}$$

In this equation, P, V, and T are absolute pressure, volume, and absolute temperature; n is the number of moles of the gas and $\overline{R}$ is the universal gas constant. For a mixture of gases we can define the concepts of partial volume, partial pressure, and mole fraction. The volume of a gas mixture is considered to be made up of the sums of the volumes of each component, V_i, at the mixture temperature and pressure. Thus, V_i is the partial volume of component i. Similarly, the total pressure of the mixture is considered to be the sum of the partial pressures, P_i, of each component species. P_i is the pressure of the component when it alone is at the temperature and volume of the mixture. The mole fraction of each component species is the ratio of the number of moles of that component, n_i, divided by the total number of moles of the gas mixture. For an ideal gas, it can be shown that

$$\frac{V_i}{V} = \frac{P_i}{P} = \frac{n_i}{n} \tag{10.43}$$

This equation shows that the volumetric fraction of the component species is the same as the species mole fraction and the ratio of partial pressure of the species to the total gas pressure. For details on the properties of ideal gas mixtures, you may refer to any text in thermodynamics, such as that of Van Wylen, et al. (1994).

Gas concentrations are usually quoted in volumetric percent, which is the same as mole percent. At low concentration levels (e.g., the case of atmospheric concentrations of carbon monoxide, CO and oxides of nitrogen, NO_x), concentration units of parts per million (ppm) or parts per billion (ppb) are used. These refer simply to the number of molecules per million (or billion) molecules of gas mixture. The conversion is performed as

$$1\% = 10^4 \text{ ppm} = 10^7 \text{ ppb}$$

There are numerous chemicals that are potential air pollutants, and various measurement devices are available for their measurement. Measuring air pollutants is done in either batch (grab sample) or continuous form. Long-term and continuous monitoring of gas composition is usually done by continuous analyzers. The devices to be discussed here briefly are those used for continuous measurement of the most common types of air pollutants, such as carbon monoxide, carbon dioxide, oxides of nitrogen, sulfur dioxide, volatile organic compounds, and ozone. Details of the measurement of various chemical compounds can be found in Lodge (1989).

10.4.1 Nondispersive Infrared Detectors

Nondispersive infrared (NDIR) detectors are used to measure CO, CO_2, and SO_2. The method works on the basis of infrared radiation absorption by a sample gas in a specified range of infrared frequencies. The principle behind the operation of these systems is Beer's law in radiation, which governs the absorption of radiation by a solute in a nonabsorbing solvent (for gases the compound of interest is the solute and the carrier gas is the solvent). *Beer's law* is expressed as

$$I = I_0 e^{-KCl} \tag{10.44}$$

where I is the transmitted radiation, I_0 is the incident radiation, K is the absorption coefficient, C is the molar concentration, and l is the path length. For more details on Beer's law, consult texts on radiation, such as Hottel and Sarofim (1967). As Eq. (10.44) shows, for a known incident radiation, the transmitted radiation will depend on the concentration of the compound of interest in the mixture. The difference of the incident and transmitted radiation is usually measured by passing identical light beams through the sample gas and also a reference path as shown in Figure 10.31. In this system, two identical infrared beams are generated, which pass through a chopper and then through the sample or comparison cell. The sample gas is passed through the sample cell, which absorbs the infrared radiation in a certain bandwidth. Essential to the NDIR system is a dehumidifier (usually not supplied with the system) to remove the moisture from the gas to minimize the interference effect of water vapor.

Infrared detectors are common and accurate devices for measuring CO, CO_2, and SO_2. The principles of operation and the technology for NDIR systems used for these gases are the same. The reference cell will be different depending on the type of gas to be analyzed. For most commercial exhaust and ambient gas analyzers, the minimum detectable concentration ranges from 0.1 to 0.4 ppm. NDIR measurement devices should be calibrated regularly according to the manufacturer's specifications.

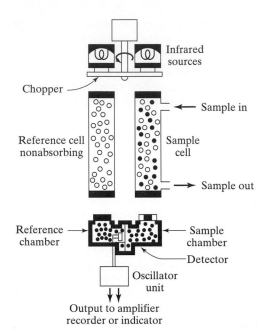

FIGURE 10.31

Schematic diagram of a typical NDIR gas analysis system.
(Based on Wark and Warner, 1981.)

10.4.2 Chemiluminescent Analyzers

Chemiluminescent analyzers are a common and accurate method for measuring oxides of nitrogen, NO and NO_2, and ozone, O_3. Chemiluminescence is defined as the production of visible or infrared radiation by the reaction of two gaseous species to form an excited species product that decays to its electronic ground state by emitting light at a specific frequency. The emitted light is then measured to determine the concentration of the desired gas species. To produce the chemiluminescent effect, a titrant[†] gas is needed to combine with the measured gas and produce the excited species, which in turn will emit light in a known wavelength. The titrant gas for measuring NO is ozone, and the titrant for measuring ozone is ethylene. The basic reactions that produce the chemiluminescent effect for measuring NO are

$$O_3 + NO = NO_2^* + O_2$$
$$NO_2^* = NO_2 + \text{light} \tag{10.45}$$

NO_2^* represents the excited state of NO_2. In these reactions the amount of light emitted is proportional to the concentration of NO, and ozone is the reacting agent for producing this reaction. Figure 10.32 shows the schematic for a chemiluminescent measurement system. To measure NO_2, a thermal catalytic converter is used to reduce NO_2 to NO, and then the concentration of NO is measured as outlined previously. For air pollution and exhaust measurements, one is usually interested in the total concentration of NO and NO_2, which is called NO_x.

[†]A titrant is a known amount of a reactant that combines with the measured gas (or solute) and fully converts it to a third chemical whose properties can be measured.

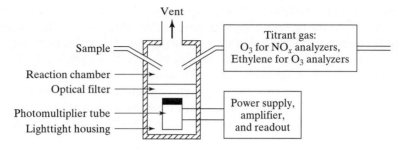

FIGURE 10.32

Schematic of a chemiluminescent measurement device. (Based on Lodge, 1989.)

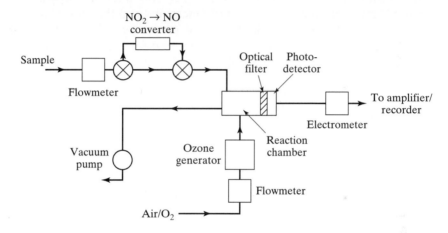

FIGURE 10.33

Schematic flow diagram of a chemiluminescent NO_x analyzer.

Chemiluminescent analyzers are accurate devices with high linearity and sensitivity. Commercially available NO_x chemiluminescent analyzers are used to measure NO_x from less than one ppm for air-monitoring applications to several hundred ppm for exhaust-gas analysis. Figure 10.33 shows the flow diagram for a commercial chemiluminescent NO_x analyzer. Like any other gas analysis equipment, chemiluminescent analyzers should be calibrated regularly with a calibration gas.

10.4.3 Flame Ionization Detectors

The flame ionization detector (FID) is used to measure total hydrocarbons and other organic compounds (e.g., aldehydes and alcohols). Carbon atoms bound to oxygen, nitrogen, or halogens produce reduced or no response by an FID. The flame ionization detector works on the principle that flames produce ions as the fuel burns with oxygen. The flame is produced by burning hydrogen (or a mixture of hydrogen and an inert gas) with high-purity air. Hydrogen flames produce very few ions compared to hydrocarbon fuels. Introducing even traces of organic matter into such a flame produces a large amount of ionization. The charged particles in the flame are collected at two terminals, resulting in a small flow of current that is measured. The response of the detector is roughly proportional to the combustible carbon content of the sample gas.

The flame ionization detector is applicable to a wide range of compounds and is highly sensitive, reasonably stable, moderately insensitive to flow and temperature, rugged, and reliable. Common gases in ambient air and exhaust stacks (e.g., Ar, CO, CO_2, H_2O, NO_x, and SO_2) do not interfere with the measurement of organic compounds by a flame ionization detector. A schematic diagram of a flame ionization detector is shown in Figure 10.34.

The flame ionization detector has the widest linear range of any gas detector in common use. Its linear dynamic range is about six to seven orders of magnitude. This wide range allows its application to ambient air for detecting organic compounds with concentrations as low as 0.001 ppm to measuring unburned hydrocarbons in the exhaust of internal combustion engines, which can be in the thousands of ppm. Generally, accuracy of an FID is ±1% of full scale on the scale 0 to 20 ppm. Flame ionization detectors should be regularly (daily) calibrated for proper accuracy. They are zeroed with hydrocarbon-free air or nitrogen (specifically prepared as a zero gas for this type of application) and calibrated with span gases (usually methane) in the range of 25 to several hundred ppm. Because the instrument is linear, zero and span gas are sufficient for calibration.

10.4.4 Other Gas-Analysis Devices

Due to air quality regulations worldwide, numerous other types of air-pollutant gas-analysis devices have been developed. These devices are mostly semiconductor or chemical-cell based. They are often portable, and they are used extensively to analyze

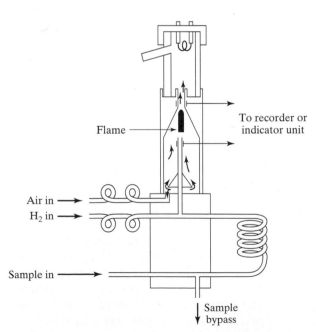

FIGURE 10.34

Schematic of flame ionization detector. (Based on Wark and Warner, 1981.)

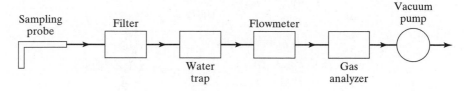

FIGURE 10.35

Typical sampling configuration for exhaust-gas measurement.

the exhaust gases of boilers, furnaces, ovens, and automobile engines. Such devices usually have a limited range, accuracy, and application, but they are generally less expensive than the devices previously described.

10.4.5 General Considerations about Sampling and Measuring Pollutant Gases

The most important consideration in any sampling of gaseous pollutants is that the sample should remain unaltered to the analysis point. Alteration can happen because of leaks, absorption, desorption, condensation, chemical reaction, addition, and other effects. These effects can alter the chemical components of the gas between the sampling point and the point(s) of analysis and can interfere with the measurement. Proper precautions should be taken to eliminate these effects. One should use chemically inert sample lines (such as glass, Teflon, or stainless steel, depending on the application), stainless-steel fittings, and stainless-steel diaphragm vacuum pumps. Care should be taken to avoid lubricants or sealant in the sampling system. Figure 10.35 shows a typical system for the continuous sampling of gaseous pollutants from an exhaust stack. It is important that a representative sample be withdrawn from the tested system. A water trap is required for measuring CO and CO_2, but it is not necessary for measuring NO_x and unburned hydrocarbons.

Example 10.17

In the exhaust of an automobile, the concentrations of CO, CO_2, and oxides of nitrogen have been measured on a dry basis and found to be as follows:

CO   250 ppm
CO_2 8.75%
NO_x 150 ppm

Calculate the partial pressure of each of these gases in the dry mixture. The pressure in the exhaust is 101 kPa (1 atm).

Solution: Using Eq. (10.43) yields

$$\frac{P_{CO}}{P_{total}} = \frac{n_{CO}}{n_{total}} = 250 \times 10^{-6}$$

$$P_{CO} = 250 \times 10^{-6} \times 101 = 2.52 \times 10^{-2}\,\text{kPa}$$

$$\frac{P_{CO_2}}{P_{total}} = \frac{n_{CO_2}}{n_{total}} = 0.0875$$

$$P_{CO_2} = 8.84 \text{ kPa}$$

$$\frac{P_{NOx}}{P_{total}} = \frac{n_{NOx}}{n_{total}} = 150 \times 10^{-6}$$

$$P_{NOx} = 1.52 \times 10^{-2} \text{ kPa}$$

REFERENCES

[1] ASME (1971). *Fluid Meters: Their Theory and Application*, 6th ed., American Society of Mechanical Engineers, New York.

[2] ASME (1989). *Measurement of Fluid Flow in Pipes Using Orifice, Nozzle and Venturi*, ASME Standard MFC-3M–1989.

[3] BENEDICT, R. (1984). *Fundamentals of Temperature, Pressure and Flow Measurements*, Wiley, New York.

[4] DRAIN, L. (1980). *The Laser Doppler Technique*, Wiley, New York.

[5] DURRANI, T., AND CLIVE, A. (1977). *Laser Systems in Flow Measurement*, Plenum Press, New York.

[6] DURST, F., MELLING, A., AND WHITELAW, J. H. (1981). *Principles and Practice of Laser-Doppler Anemometry*, 2d ed., Academic Press, New York.

[7] HOTTEL, H., AND SAROFIM, A. (1967). *Radiative Transfer*, McGraw-Hill, New York.

[8] JONES, G. (1968). Unsteady lift forces generated by vortex shedding about a large, stationary, oscillating cylinder at high Reynolds numbers, *ASME Symposium on Unsteady Flow*.

[9] LODGE, J., JR., Editor (1989). *Methods of Air Sampling and Analysis*, 3d ed., Lewis Publishers, Chelsea, MI.

[10] LOMAS, C. (1986). *Fundamentals of Hot Wire Anemometry*, Cambridge Press, Cambridge, U.K.

[11] MILLER, R. (1989). *Flow Measurement Engineering Handbook*, 2d ed., McGraw-Hill, New York.

[12] MOTT, L. (1994). *Applied Fluid Mechanics*, Merrill, New York.

[13] NORTON, H. (1982). *Sensor and Analyzer Handbook*, Prentice-Hall, Englewood Cliffs, NJ.

[14] OMEGA ENGINEERING (1993; regularly updated). *The Flow and Level Handbook*, Omega Engineering, Stamford, CT.

[15] ROSHKO, A. (1954). On the development of turbulent wakes from vortex streets, *NACA Rep.* 1191.

[16] SANDERSON, M., AND HEMP, J. (1981). Ultrasonic flowmeters: a review of the state of the art, *Advances in Flow Measurement Techniques*, BHRA Fluid Engineering, Cranfield, Bedford, England.

[17] SHAMES, I. (1992). *Mechanics of Fluids*, McGraw-Hill, New York.

[18] SHAPIRO, A. (1953). *The Dynamics and Thermodynamics of Compressible Fluid Flow*, Ronald Press, New York.

[19] STREETER, V., AND WYLIE, E. (1985). *Fluid Mechanics*, McGraw-Hill, New York.

[20] UPP, E. (1993). *Fluid Flow Measurement*, Gulf Publishing, Houston, TX.

[21] VAN WYLEN, G., SONNTAG, R., AND BORGNAKKE, C. (1994). *Fundamentals of Classical Thermodynamics*, 4th ed., Wiley, New York.

[22] WARK, K., AND WARNER, C. (1981). *Air Pollution: Its Origins and Control*, Harper & Row, New York.

[23] WHITE, F. (1999). *Fluid Mechanics*, McGraw-Hill, New York.

[24] YEH, Y., AND CUMMINS, H. (1964). Localized flow measurements with an He–Ne laser spectrometer, *Applied Physics Letters*, Vol. 4, p. 176.

PROBLEMS

10.1 A venturi tube is to be installed into a nominal-6-in. pipe, which transports a maximum flow of 0.6 cfs (ft^3/sec). What should be the throat diameter of the venturi to produce a maximum pressure difference of 5 psi? The actual diameter of the pipe is 6.065 in. The fluid is water at 80°F.

10.2 A machined entrance cone venturi tube that has been fabricated according to ASME standards is fit to a 5.25-cm-ID water pipe. The minimum diameter is 3 cm. If the pressure drop across the venturi is 14 kPa, calculate the water flow rate.

10.3 Estimate the uncertainty of flow measurement in Problem 10.1 if the uncertainty in discharge coefficient and pressure measurement are estimated to be 1% and 1.5%, respectively. The uncertainties in other parameters are considered to be negligible.

10.4 A machined entrance cone venturi is to be used to measure the airflow rate in a 4-in. pipe. The minimum area of the meter is 3 in. The pressure drop across the venturi is 1 psi. Calculate the volumetric flow [in standard cubic feet per minute (SCFM)] and mass flow rate (lbm/min) of air. The condition of air upstream of the venturi is 100 psia and 70°F.

10.5 A critical flow nozzle with a throat diameter of 5 mm is used to measure the flow of air. Air pressure and temperature upstream of the nozzle are 7 atm (gage) and 20°C. Calculate the mass flow rate and the volumetric flow rate (in SCMM) of air. At what maximum pressure downstream of the nozzle will this calculation be acceptable?

10.6 A critical flow nozzle with throat diameter of 2 mm is used to measure hydrogen flow rate to an atmospheric pressure burner. The pressure of hydrogen can be regulated. The upstream temperature is 20°C. The supply pipe diameter has a 1 cm inside diameter. Assume that the discharge coefficient is 1.0. For hydrogen the specific heat ratio is 1.40.

(a) Calculate the minimum pressure upstream of the nozzle for the flow to remain critical in the nozzle.

(b) Calculate the mass flow rate and the volumetric flow (in SCMM) of hydrogen if the pressure upstream of the nozzle is 3 atm (gage). Assume that the discharge coefficient, C, is unity.

10.7 A critical flow nozzle is to be used to measure the airflow rate into a variable pressure chamber. The air in the line is at 8 atm, 20°C.

(a) Determine the chamber pressure above which the critical flow nozzle cannot be used accurately for airflow measurement.

(b) Determine the diameter of the nozzle for a flow rate of 100 standard cubic meters per minute.

(c) Calculate the mass flow rate of air for this condition. Assume that the discharge coefficient, C, is unity.

10.8 A critical flow nozzle is used to measure the mass flow rate of air ($\gamma = 1.4$, $R = 287$ J/kg-K). The pipe diameter is 3 cm and the nozzle diameter is 1 cm. The pressure downstream of the nozzle is 100 kPa (abs) and the pressure in the pipe upstream of the nozzle is 500 kPa (abs). The upstream temperature is 400 K. Estimate the mass flow rate assuming that the discharge coefficient, C, is unity.

10.9 An orifice meter is to be installed in a 25.5-cm-ID pipe that flows at a maximum water flow rate of 0.090 m³/s. Using corner taps, determine the diameter of the orifice to give a maximum pressure drop of 65 kPa. What will be the pressure drop if the flow rate is 0.007 m³/s?

10.10 An orifice meter with corner taps in a nominal 4-in.-diameter pipe (actual diameter 4.026 in.) is used to measure the flow rate of air. The diameter of the orifice is 2 in. At the upstream section, the temperature is 100°F and the pressure is 150 psia. The pressure difference across the orifice is 6 psi. Estimate the mass flow rate of the air. γ for air is 1.4.

10.11 A square-edged orifice meter with corner taps is used to measure water flow in a 25.5-cm-ID pipe. The diameter of the orifice is 15 cm. Calculate the water flow rate if the pressure drop across the orifice is 14 kPa. The water temperature is 10°C.

10.12 What will be the water flow rate in Problem 10.11 if a flow nozzle with the same minimum area and the same pressure drop is used?

10.13 A square-edged orifice with corner taps is to be used to measure the water flow rate in a 4-in.-diameter pipe (actual diameter 4.026 in.). Find the diameter of the orifice for a water flow rate of 15 cfm and a pressure drop of 2 psi.

10.14 A square-edged orifice meter is used to measure the airflow rate in a 4-in.-diameter pipe (actual diameter 4.026 in.) with corner pressure taps. The pressure drop across the orifice is 1 psi and the diameter of the orifice is 3 in. Calculate the mass and volumetric (in SCFM) flow rate of air both with and without considering the compressibility effect. The line pressure and temperature are 7 atm (gage) and 70°F.

10.15 We would like to choose a pressure transducer to measure pressure across a machined entrance cone venturi tube that measures 20°C water flow rate in a 10-cm pipe. The β of the venturi is 0.7. The flow range is 1 to 3 m³/min. Estimate the range of the pressure transducer.

10.16 Repeat Problem 10.15 if the venturi is to be used for airflow measurement in the range 20 to 60 standard cubic meter of air per minute. The pipe air is at 7 atm, 20°C.

10.17 You are asked to choose a flowmeter to measure the flow of hot water from a boiler. The three flowmeters to consider are the nozzle, orifice plate, and venturi tube, all with the same value of β. Compare (rank) the following differential-pressure flowmeters installed in a 5-cm (2-inch) pipe in terms of size (length of the insertion into the line), pressure drop across the meter, overall pressure loss, accuracy, and cost. Assume that the same differential-pressure transducer can be used for all meters.

10.18 Based on ASME specifications, determine the size of a square-edged orifice meter (orifice size) for measuring water flow in a 5-cm (ID) pipe for a pressure drop not to exceed 40 kPa. The maximum average velocity inside the pipe is not expected to exceed 3 m/sec. Explain all your assumptions. For the chosen orifice, answer the following questions:

(a) What is the maximum flow that can be measured with the chosen orifice?

(b) What is the expected pressure drop at $\frac{1}{2}$ of the maximum flow?

(c) If the range of the pressure-differential meter for reasonable accuracy is 0.5–3 kPa, determine the turndown ratio (ratio of maximum to minimum value of the flowrate) of the selected size.

10.19 Repeat Problem 10.18 for a nozzle flowmeter.

10.20 Repeat Problem 10.18 for a machined entrance cone venturi flowmeter.

10.21 A 2.00-cm square-edge orifice is to be installed in a 5-cm (ID) pipe to measure water flow rates in the range of 100 to 300 liters per minute. Allowing for the possibility that the flow may go up to 50% above the range upper limit (considered safety factor of 50%), determine the range of a differential-pressure transducer to be used for this application.

10.22 Determine the nozzle size of a critical flow nozzle to measure a maximum airflow of 0.1 kg/sec of air at 25°C to 300°C and 1000 kPa absolute in a 5-cm pipe. Answer the following questions about the chosen nozzle:

   (a) Where will you need to install the pressure and temperature instruments?

   (b) If the upstream pressure varies between 400 to 1000 kPa, what is the maximum pressure downstream of the nozzle for the flow to remain critical?

   (c) Calculate the mass flow rate of air and its volumetric flow rate (in SCMS, standard cubic meter per second) if the air temperature and pressure are 150°C and 500 kPa.

10.23 Repeat Problem 10.22 for carbon dioxide, using $R = 188.9$ J/kg-K and $\gamma = 1.289$.

10.24 Based on ASME specifications, determine the nozzle size of a regular (not critical), low-β nozzle flowmeter needed to measure a maximum airflow of 0.1 kg/sec of air for a 25°C to 300°C temperature range and 1000 kPa in a 5-cm line. The pressure drop should not exceed 40 kPa (corresponding to the maximum flow rate). Where will you need to measure the pressure and temperature? For the chosen nozzle, calculate the mass flow rate of air if the air temperature and pressure in the pipe are 150°C and 500 kPa, respectively and the measured pressure drop is 20 kPa.

10.25 A rotameter that has been calibrated to measure airflow at 20°C, 1 atm, is going to be used to measure airflow at 20°C, 5 atm. Estimate the flow correction factor for this application.

10.26 Estimate the flow correction factor for the rotameter of Problem 10.25 if the air is at 5 atm, 200°C.

10.27 A rotameter that has been calibrated for air at standard conditions (20°C, 1 atm) is to be used to measure a propane flow at the same conditions: Estimate the flow correction factor. Propane is C_3H_8 and has a molecular weight of 44.1.

10.28 The rotameter of Problem 10.27 is to be used to measure carbon dioxide at 20°C, 5 atm (gage). Estimate the flow correction factor. Carbon dioxide is CO_2 and has a molecular weight of 44.

10.29 A rotameter with a stainless-steel float ($S = 7.8$) has been calibrated to read the water flow rate. The rotameter is going to be used to measure fuel oil with a specific gravity of 0.84. Calculate the correction factor for the flow. What might be the sources of error in using this rotameter to measure fuel-oil flow rate?

10.30 The stainless-steel float ($S = 7.8$) in a rotameter has been changed to a metal with a specific gravity of 4. The geometry of the new float is the same as before. The rotameter has been calibrated with water for the stainless-steel float. Calculate the percent change in the range and the correction factor for the flow using the new float.

10.31 A thermal mass flowmeter has been calibrated for air at 25°C and 20 kPa. It is used to measure the flow of carbon dioxide under the same conditions. The meter indicates 10 g/sec. Calculate the actual flow rate of carbon dioxide. The specific heats of air and carbon dioxide are 1.004 and 0.842 kJ/kg-K, respectively.

10.32 A thermal mass flowmeter has been calibrated for air at 25°C and 20 kPa. It is used to measure the flow of air under the same temperature, but at 5 kPa. The meter indicates 20 g/sec. What is the actual airflow rate? Explain.

10.33 A 2-cm-diameter cylinder is placed in a 15°C water flow. If the frequency of oscillations of the vortices behind the cylinder is 65 Hz, estimate the velocity of the water.

10.34 A 3-mm-diameter cylinder is placed normal to the flow in a 10-cm-diameter pipe. The frequency of vortices shed behind the cylinder is 1400 Hz. Estimate the flow rate of air in m^3/s. The air is at 1 atmosphere pressure and 400 K.

10.35 The radial velocity distribution in a circular pipe of radius r_o in laminar flow is given by

$$\frac{V}{V_{cl}} = \frac{r_o^2 - r^2}{r_o^2}$$

where V_{cl} is the centerline velocity. If V_{cl} and r_o are measured, develop expressions for the average velocity and the volume flow rate, Q.

10.36 A pitot-static probe is used to measure the velocity of air. If the air velocity is 10 m/s, what will be the ΔP measured by a connected pressure transducer? The measured density of the air is 1.1 kg/m^3. Assume that C is unity.

10.37 A pitot-static probe is used to measure the velocity of air. The air density is 1.0 kg/m^3, and the reading of a pressure transducer is 5 kPa. What is the air velocity? Assume that $C = 1$.

10.38 A pitot-static probe is used to measure the velocity of 15°C water. If the reading of the connected pressure transducer is 20 cm of water, what is the water velocity?

10.39 A differential pressure system is used to measure the level of the liquid in a tank. The sensing line to the top of the tank is filled with cold water having a density of 990 kg/m^3. When the level is at the bottom of the tank, the transducer reads -58.8 kPa (upper sensing line connection at a higher pressure than the lower sensing line connection). What is the level in the tank (relative to the tank bottom) when the transducer reads -10 kPa if the fluid in the tank has a density of 978 kg/m^3?

10.40 The level of a fluid in a tank is measured with an ultrasonic sensor. If the sound velocity of the vapor in the tank is 1500 ft/sec and the distance from the transmitter to the surface is 5 ft, what will be the transit time for the ultrasonic pulse?

CHAPTER 11

Dynamic Behavior of Measurement Systems

The general concept of the dynamic response of measuring systems was introduced in Chapter 2. In this chapter the mathematical basis for measuring dynamic response is presented and applied to important sensors and transducers.

11.1 ORDER OF A DYNAMIC MEASUREMENT SYSTEM

For static measurements, the mathematical relationship between the measurand (input) and the measurement-system output can be described with algebraic equations. For example, for a linear device, the input–output relationship can be given by

$$y = Kx \tag{11.1}$$

where y is the measurement-system output, x is the value of the measurand, and K is the instrument static sensitivity. For time-varying measurements, however, the mathematical relationship between the measurand and the output must be described with a differential equation. The equation describing the behavior of an instrument with a time-varying input is derived by applying the appropriate basic physical principles (such as conservation of energy, Newton's second law of motion, or Ohm's law) to the instrument as it interacts with its environment. In general, the input–output relationship of a linear measurement system can be expressed in the form of an ordinary differential equation (ODE):

$$a_n \frac{d^n y}{dt^n} + a_{n-1} \frac{d^{n-1} y}{dt^{n-1}} + \cdots + a_1 \frac{dy}{dt} + a_0 y = bx \tag{11.2}$$

In this equation, n is the order of the system, x is the input signal (forcing function), y is the output signal, and $a_0 \cdots a_n$ are constant coefficients that depend on the characteristics of the measurement system. x and y are, in general, both functions of time, $x(t)$ and $y(t)$. Most measuring devices can be described using Eq. (11.2) with n values of 0, 1, or 2.

In the following sections, some important dynamic characteristics of measurement systems are discussed. A more detailed presentation is given by Doebelin (1990).

11.2   ZERO-ORDER MEASUREMENT SYSTEMS

If $n = 0$, Eq. (11.2) reduces to

$$a_0 y = bx \tag{11.3}$$

Although x varies dynamically, y responds proportionally and the measurement process is essentially a static measurement. This is the ideal measurement process. No system is truly of order zero, but many systems approximate this behavior in some modes of their operation.

Consider the simple displacement potentiometer shown in Figure 11.1. In this system a fixed voltage E is applied to a resistor with length L. The instrument is used to measure displacement and generate an output voltage, e, which is proportional to the position, x, of the slider. The output can be related to the input by applying Ohm's law to give

$$e = E\frac{x}{L} \tag{11.4}$$

For a wide range of operation, the output e is unaffected by the nature of the time variation of x and shows zero-order behavior. However, if x varies extremely rapidly, inductive, capacitance, and flexural effects may invalidate Eq. (11.4), and the system can no longer be described as zero order. Strain-gage sensors can be considered as zero-order systems for most of their range of operation. However, many transducers using strain-gage sensors, such as pressure transducers, are in fact second-order systems.

11.3   FIRST-ORDER MEASUREMENT SYSTEMS

First-order systems contain only a single mode of energy storage. The simple R–C circuit is a first-order electrical system. The most common first-order measurement systems are thermal systems such as temperature-measuring devices, which include thermal capacitance and resistance to heat flow. These show first-order behavior over most ranges of operation. First-order analysis can also be used for some operational modes of mechanical systems that include friction and springs, but in which the inertial effects of mass can be neglected. (Certain applications of pressure transducers are examples.)

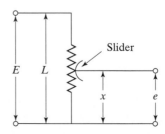

FIGURE 11.1

Linear potentiometer, zero-order system.

11.3.1   Basic Equations

For $n = 1$, Eq. (11.2) becomes

$$a_1 \frac{dy}{dt} + a_0 y = bx \tag{11.5}$$

This equation can be rewritten in the form

$$\tau \frac{dy}{dt} + y = Kx \tag{11.6}$$

K is the static sensitivity, and τ, which has units of time, is called the *time constant*.[†] The time constant, τ, is determined by the physical characteristics of the measuring system. If the time-derivative term were to be neglected, the system would show static behavior and y would have the ideal response value, Kx. However, the existence of the time-derivative term introduces a dynamic measurement error.

The solution of Eq. (11.6) is the sum of two parts. One part is called the *general solution* and is determined by the system itself and is independent of the forcing function x. The general solution is obtained by solving Eq. (11.6) assuming that the right-hand side, Kx, is zero. The other part of the solution, called the *particular solution*, is dependent on the input forcing function, x, and is a solution to the complete Eq. (11.6). In the following sections, the time responses of first-order systems to step, ramp, and sinusoidal inputs are discussed.

11.3.2   Step Input

Consider a situation in which both the input and output to the device are both zero until time $t = 0$. At $t = 0$, as shown in Figure 11.2(a), the input x suddenly rises to a value x_0. If the device were ideal (zero order), it would produce the output $y_e = Kx_0$, which is independent of time. For the actual first-order device, we seek the time variation of the device output, y. The general solution to Eq. (11.6) is

$$y_{\text{gen}} = Ce^{-t/\tau} \tag{11.7}$$

where C is a constant to be determined. The particular solution is

$$y_{\text{part}} = Kx_0 \tag{11.8}$$

The complete solution to this problem is the sum of the general and particular solutions and is given by

$$y = Kx_0(1 - e^{-t/\tau}) \tag{11.9a}$$

[†]In evaluating an instrument, the static sensitivity will have units of the physical output divided by the units of the input. For example, a thermocouple output is in millivolts, and the instrument sensitivity will be in units such as mV/°C. When the instrument is used, however, the instrument output will be converted to measurand units with a calibration function. Thus, as used, the thermocouple will have a sensitivity with units of (output degrees)/(input degrees) and the static sensitivity will effectively be unity. Dynamic measurement errors may be evaluated in either the physical output units of the instrument or in the units after the calibration function has been applied.

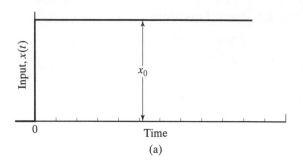

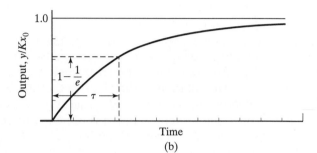

FIGURE 11.2

Response of a first-order measurement system to a step change in input: (a) forcing function; (b) system output.

where the constant C has been eliminated using the initial condition that $y = 0$ when $t = 0$. The validity of Eq. (11.9a) can be demonstrated by substituting it into Eq. (11.6) and noting that it also satisfies the initial condition.

Figure 11.2(b) shows the variation of the output y with respect to the time t. At $t = \tau$ the output has reached 63.3% of its final value, and at $t = 4\tau$ it has reached 98.2% of its final value. Since the larger the time constant of a first-order system, the longer it will take the output to approach its final value, the system should have a small time constant if rapid response is desired.

In many cases, the output does not have an initial value of zero. In this case, Eq. (11.9a) can be viewed as showing the change in output for a change in input. If the initial value of y is y_i and the equilibrium value (a long time after the step change) is $y_e (y_e = Kx_0)$, then Eq. (11.9a) will take the form

$$\frac{y - y_i}{y_e - y_i} = (1 - e^{-t/\tau}) \tag{11.9b}$$

This is the form usually used to evaluate measurement devices.

11.3.3 Ramp Input

In this case both y and x are zero at time $t = 0$. As shown in Figure 11.3(a), starting at $t = 0$, the forcing function x takes the form

$$x = At \tag{11.10}$$

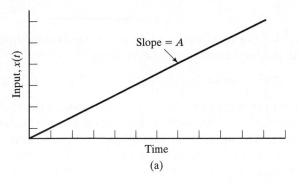

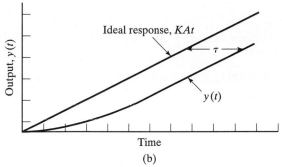

FIGURE 11.3

Response of a first-order system to a ramp input: (a) forcing function; (b) output response.

where A is the slope of the ramp. The ideal, zero-order response would then be $y_e = KAt$. As in the case of the step input, Eq. (11.6) is the applicable differential equation and Eq. (11.7) is the general solution (which is independent of the forcing function). The particular solution for the case of the ramp input is

$$y_{part} = KA(t - \tau) \tag{11.11}$$

The solution is the sum of general and particular solutions, and evaluating C using the initial condition that $y = 0$ when $t = 0$, we obtain the solution

$$y = KA(\tau e^{-t/\tau} + t - \tau) \tag{11.12a}$$

Again, this result can be verified by substituting it into the differential equation and noting that it satisfies the initial condition. Figure 11.3(b), a sketch of Eq. (11.12a), shows that there will be an error in the output with respect to the input. After a few time constants have elapsed, the error is the difference between the output value and the ideal output value one time constant earlier. Thus the error will be minimized by using devices with small values of the time constant.

In most cases, y will not be zero at time zero. If y_i is the initial value of y at $t = 0$, then Eq. (11.12a) can be expressed as

$$y = y_i + KA(\tau e^{-t/\tau} + t - \tau) \tag{11.12b}$$

This form is more general and is the form typically used.

11.3.4 Sinusoidal Input

The third input function considered for the first-order system is the continuing sinusoidal input. In this case we take the forcing function to have the form

$$x = x_0 \sin \omega t \tag{11.13}$$

For an ideal, zero-order system the output would be $y_e = Kx_0 \sin \omega t$. As with the step and ramp inputs, the solution can be viewed as the sum of the general and particular solutions. However, in the case of sinusoidal input, it is the continuing response that is of greatest interest, and the transient caused by the initial application of the forcing function can be neglected. The continuing response corresponds to the particular solution, which is given by

$$y = \frac{Kx_0}{\sqrt{1 + \omega^2 \tau^2}} \sin(\omega t + \phi) \tag{11.14}$$

where ϕ is the phase angle between the forcing function and the output response. ϕ is given by

$$\phi = -\tan^{-1} \omega \tau \tag{11.15}$$

Figure 11.4(a) presents the ratio of the actual amplitude of y to the ideal amplitude, Kx_0, versus the nondimensionalized angular frequency, $\omega \tau$. For the continuing response given, the initial input and output conditions do not matter. Figure 11.4(b) presents the phase angle. These figures show that if the product of the time constant and the forcing-function angular frequency approaches zero, the amplitude of the output will approach the ideal case and the phase angle will approach zero. Otherwise, the output magnitude and phase will have a systematic error with respect to the actual input.

This analysis considers the response to a sinusoidal input of a single frequency. As discussed in Chapter 5, periodic inputs of any general form can be broken down into a series of single-frequency sine and cosine functions using Fourier analysis. The response of the system to these individual Fourier components can then be analyzed using the previous results. The foregoing discussion applies to the dynamic behavior of any system that can be modeled by a time-dependent first-order ordinary differential equation. Further details on the solution can be found in Doebelin (1990).

11.3.5 Thermocouple as a First-Order System

Thermocouples are a very common first-order system and serve to demonstrate the application of first-order dynamic analysis. As shown in Section 2.3, the time constant for a thermometer is given by

$$\tau = \frac{mc}{hA} \tag{11.16}$$

This formula is also applicable to other temperature-measuring devices, such as thermocouples and resistance temperature detectors. To obtain a small time constant, it is necessary to have a high surface area relative to the mass. Since temperature sensors usually have relatively simple geometry (usually, cylinders or spheres), this means that

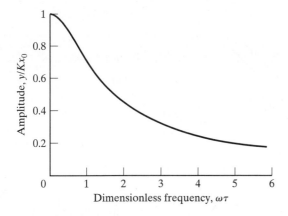

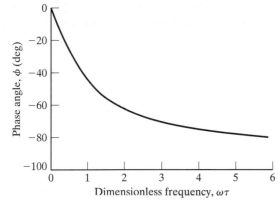

FIGURE 11.4

(a) Amplitude and (b) phase response of a first-order system to a sinusoidal input.

the sensor should be very small. It is also helpful that the heat transfer coefficient h increases as the size decreases. In Chapter 9 it was shown that for thermocouples, small size is also desirable to minimize conduction and radiation errors. Unfortunately, small temperature sensors are vulnerable to damage from impact, vibration, and corrosion.

Thermocouples, described in Chapter 9, have a wide range of applications in industry and research. For fast, dynamic response and in research projects, it is best to use thermocouples with bare junctions similar to that shown in Figure 11.5(a). Such bare-junction thermocouples are very vulnerable to damage and corrosion and also present installation problems. As a result, it is common to use thermocouples encased in a metal sheath for mechanical protection [Figure 11.5(b)]. The additional mass of the metal sheath and the insulation increases the heat storage capacity of the instrument, increases its time constant, and consequently, slows its response. In some industrial applications the shielded thermocouple is located in a fluid-filled well for further durability and mechanical and chemical protection [Figure 11.5(c)], further slowing its response. Such thermocouples might have time constants on the order of minutes. Further details on the transient response of thermocouples and methods for determining and reducing thermocouple time constants are presented in Benedict (1984). The following examples show some typical applications of first-order analysis applied to thermocouples.

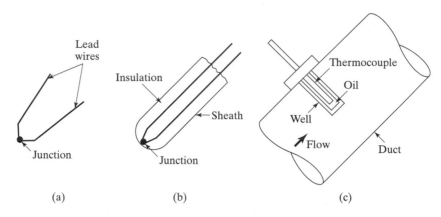

FIGURE 11.5

Common applications of thermocouples: (a) bare junction; (b) sheathed thermocouple; (c) thermowell.

Example 11.1

A Pt–Pt/13%Rd (type R) bare-junction thermocouple has an approximately spherical junction with a diameter of 0.3 mm. It is used to measure the temperature of gases in a combustion tunnel. When the flame is ignited, it produces an approximate step increase of the gas temperature of 900 K. The average heat-transfer coefficient[†] on the surface of the thermocouple is 500 W/m²-°C. The gas temperature before ignition is 300 K.

(a) Find the time constant of the thermocouple.

(b) After how much time and how many time constants will the measurement error be less than 1% of the final temperature change?

(c) If the same thermocouple is used in an aqueous environment in which the heat transfer coefficient is 6000 W/m²-°C, what will be the thermocouple time constant?

(d) Plot the thermocouple response versus time for both cases.

Solution: For a step change in input temperature, this first-order system is described by Eq. (11.9). We will take the junction properties to be those of platinum. From Table B.5, the properties of platinum are $\rho = 21{,}450$ kg/m³ and $c = 134$ J/kg-°C.

[†]To calculate the time constant, we need to calculate the heat transfer coefficient. Holman (2002) presents a comprehensive set of theoretical and empirical formulas that can be used in such applications. For flow over spheres, the nondimensionalized heat transfer coefficient can be expressed in terms of the Reynolds number ($\text{Re} = U_\infty D\rho/\mu$) as

$$\frac{hD}{k} = 2 + (0.4\text{Re}^{1/2} + 0.06\text{Re}^{2/3})\text{Pr}^{0.4}\left(\frac{\mu_\infty}{\mu_w}\right)^{1/4}$$

In this equation, h is the heat-transfer coefficient, D is the sphere diameter, and ρ, k, μ, and Pr are, respectively, the density, thermal conductivity, dynamic viscosity, and Prandtl number of the fluid. Fluid properties (except the two viscosities shown) are evaluated at the film temperature, which is the average temperature of the fluid (∞) and heat-transfer surface (w).

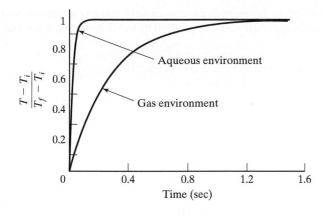

FIGURE E11.1

Thermocouple response to gas and aqueous environments.

(a) From Eq. (11.16) the time constant will be

$$\tau = \frac{mc}{hA} = \frac{\rho \forall c}{hA} = \left[21{,}450 \times \left(\frac{\pi \times 0.0003^3}{6} \right) \right] \times \frac{134}{500 \times \pi \times 0.0003^2} = 0.287 \text{ s}$$

(b) We seek the time when the left side of Eq. (11.9b) has a value of 0.99 (error of 0.01):

$$\frac{T - T_i}{T_f - T_i} = 0.99 = 1 - e^{-t/0.287}$$

Solving, we obtain $t = 1.32$ s, which corresponds to 4.6 time constants.

(c) For the aqueous environment, the time constant will be

$$\tau = \frac{mc}{hA} = \frac{\rho \forall c}{hA} = \left[21{,}450 \times \left(\frac{\pi \times 0.0003^3}{6} \right) \right] \times \frac{134}{6000 \times \pi \times 0.0003^2} = 0.024 \text{ s}$$

Due to the much higher heat-transfer coefficient in the liquid phase compared with the gas phase, this time constant is an order of magnitude smaller than the time constant for the gas phase.

(d) The actual input and the time response of the thermocouple are shown in Figure E11.1.

Comment: Certain assumptions were made in this analysis. It was assumed that the shape of the junction was spherical and that the radiation and conduction effects of connecting wires could be neglected.

Example 11.2

In a special test for determining the performance of a heat exchanger, an electric heater produces a temperature variation in an airflow that can be approximated with a 1-Hz sine wave. The air temperature is measured downstream of the heater with a thermocouple that has a time constant of 0.2 s. The maximum and minimum temperature values measured by the thermocouple are 120° and 100°C, respectively.

(a) Determine the equation for the measured variation of temperature.

(b) Determine the actual variation in the air temperature.

(c) Draw the actual and the indicated temperature as a function of time.

Solution:

(a) To determine the equation for the measured temperature variation, we need to determine the amplitude of the oscillation and the mean temperature:

$$A = \frac{T_{max} - T_{min}}{2} = \frac{120 - 100}{2} = 10°C \text{ and}$$

$$T_{av} = \frac{T_{max} + T_{min}}{2} = \frac{120 + 100}{2} = 110°C$$

The time variation of the indicated temperature is then

$$T = T_{av} + A \sin 2\pi f t \quad \text{or} \quad T = 110 + 10 \sin 2\pi t °C$$

(b) The amplitude response and the phase response of the thermocouple are determined through Eqs. (11.14) and (11.15). Both the input and output of this measuring system are measured in units of temperature, and consequently, the ideal output amplitude, y_e, would have the same value as the input amplitude, x_0. Hence, the static sensitivity, K, equals 1. In this case the amplitude of the output, y, is A, and we seek x_0, the input amplitude. Using Eq. (11.14), the output amplitude can be obtained in terms of the input amplitude:

$$A = \frac{Kx_0}{(1 + \omega^2\tau^2)^{1/2}}$$

$$10 = \frac{1x_0}{[1 + (2\pi \times 1)^2 \times 0.2^2]^{1/2}}$$

Solving gives

$$x_0 = 16.1°C$$

Using Eq. (11.15), the phase angle of the measured temperature with respect to the actual temperature will be

$$\phi = -\tan^{-1}\omega\tau = -\tan^{-1}(2\pi \times 1 \times 0.2) = -51.5° = -0.899 \text{ rad}$$

The actual variation in air temperature is then given by

$$T_{act} = 110 + 16.1 \sin(2\pi t + 0.899)$$

(c) The actual and indicated temperature variations are shown in Figure E11.2.

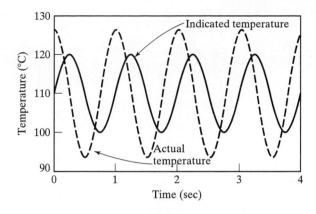

FIGURE E11.2

Variation of actual and measured (thermocouple) temperatures.

11.4   SECOND-ORDER MEASUREMENT SYSTEMS

Second-order behavior governs spring–mass mechanical systems as well as capacitance–inductance electrical systems. Load cells, pressure transducers, and accelerometers are second-order systems.

11.4.1   Basic Equations

For second-order measurement systems, the value of n in Eq. (11.2) is 2. This results in a general equation of the form

$$a_2 \frac{d^2 y}{dt^2} + a_1 \frac{dy}{dt} + a_0 y = bx \qquad (11.17)$$

It is common practice to divide Eq. (11.17) by a_0 and to make the following definitions:

$$K = \frac{b}{a_0} \qquad (11.18)$$

$$\omega_n = \left(\frac{a_0}{a_2} \right)^{1/2} \qquad (11.19)$$

$$\zeta = \frac{a_1}{2(a_0 a_2)^{1/2}} \qquad (11.20)$$

K is the static sensitivity. ω_n, the *undamped natural frequency*, is the angular frequency at which the system would oscillate if there were no damping ($a_1 = 0$). ζ is the *damping ratio*. Using these definitions, Eq. (11.17) becomes

$$\frac{1}{\omega_n^2} \frac{d^2 y}{dt^2} + \frac{2\zeta}{\omega_n} \frac{dy}{dt} + y = Kx \qquad (11.21)$$

For a system described by this equation, the ideal response (response in the absence of dynamic effects) would be $y_e = Kx$.

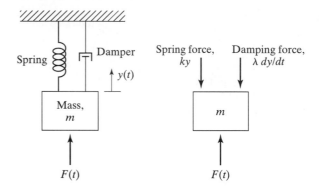

FIGURE 11.6

Spring–mass–damper system.

Consider the spring–mass–damper system shown in Figure 11.6. There are three forces acting on the mass, the spring force ky, the damping force that is proportional to the velocity, $\lambda(dy/dt)$, and the applied time-varying input force, $F(t)$. Combining these forces using Newton's second law $[\Sigma F = m(d^2y/dt^2)]$, we obtain

$$m\frac{d^2y}{dt^2} + \lambda\frac{dy}{dt} + ky = F(t) \tag{11.22}$$

which is the same form as Eq. (11.17). Comparing this equation to Eq. (11.17) and noting the definitions of Eqs. (11.18) to (11.20), we can conclude that

$$K = \frac{1}{k} \tag{11.23}$$

$$\omega_n = \left(\frac{k}{m}\right)^{1/2} \tag{11.24}$$

$$\zeta = \frac{\lambda}{2(km)^{1/2}} \tag{11.25}$$

The applied force $F(t)$ is the forcing function x, and the output y is the mass displacement.

Although many second-order systems may not have the components illustrated in Figure 11.6 in a distinct form, the basic effects of these components are present. For example, a single structural element may contain mass and also function as a spring. The solutions to second-order equations can be generalized using the nomenclature used in Eq. (11.21). The following sections summarize the response of a second-order system to step and sinusoidal inputs. More detailed solution methods and other solutions can be found in Doebelin (1990) and Thomson (1981).

11.4.2 Step Input

As with the first-order system, we will take the initial value of y to be zero and let the forcing function change from $x = 0$ to $x = x_0$ at $t = 0$. An ideal, zero-order system would show a step change in output from zero to a value $y_e = Kx_0$. The solution depends on the value of the damping ratio, ζ. The solutions are given by

$\zeta > 1$:

$$\frac{y}{y_e} = 1 - e^{-\zeta\omega_n t}\left[\cosh(\omega_n t\sqrt{\zeta^2 - 1}) + \frac{\zeta}{\sqrt{\zeta^2 - 1}}\sinh(\omega_n t\sqrt{\zeta^2 - 1})\right] \quad (11.26)$$

$\zeta = 1$:

$$\frac{y}{y_e} = 1 - e^{-\omega_n t}(1 + \omega_n t) \quad (11.27)$$

$\zeta < 1$:

$$\frac{y}{y_e} = 1 - e^{-\zeta\omega_n t}\left[\frac{1}{\sqrt{1 - \zeta^2}}\sin(\omega_n t\sqrt{1 - \zeta^2} + \phi)\right] \quad (11.28)$$

where

$$\phi = \sin^{-1}(\sqrt{1 - \zeta^2})$$

These equations are plotted in Figure 11.7. If $\zeta < 1$, the system will show oscillatory behavior and is said to be *underdamped*. If $\zeta > 1$, the system will show an asymptotic response without overshoot and is said to be *overdamped*. The condition $\zeta = 1$ is the dividing line between these two responses and is known as *critical damping*. Overall, dynamic effects are a minimum with a damping ratio of about 0.7—there is a fast response, but the overshoot is less than 5%.

Beside the damping ratio, the other important characteristic of second-order systems affecting transient response is the undamped natural frequency, ω_n. Except for the case of zero damping ratio, the output will approximate the ideal value, Kx_0, after some initial period of time. This initial period of time depends on the natural frequency ω_n

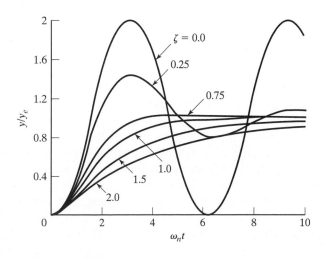

FIGURE 11.7

Response of second-order system to a step input.

and the damping ratio. For the optimum damping (0.7) an approximate steady state is reached when the nondimensional time, $\omega_n t$, has a value of about 6. This means that if we wish to measure a step change in the measurand, we should allow the instrument a settling time of about $6/\omega_n$.

Although second-order instruments are usually evaluated in terms of the undamped natural frequency, ω_n, this frequency is the frequency of free oscillations only when there is no damping ($\zeta = 0$). By examining Eq. (11.28), we see that the actual frequency of the oscillation is $\omega_n \sqrt{1 - \zeta^2}$, also known as the *ringing frequency*.

Eqs. (11.26) through (11.28) apply to a situation in which the output y has an initial value of zero. In the general case, the output will have an initial value, y_i. In this case, the left-hand side of Eqs. (11.26) through (11.28) y/y_e, can be replaced by $(y - y_i)/y_e - y_i)$, where $y_e = y_i + K x_0$ and x_0 is the change in the input.

11.4.3 Sinusoidal Input

In this case, a forcing function is applied that has a sinusoidal variation with time:

$$x = x_0 \sin \omega t \tag{11.29}$$

For an ideal system without dynamic effects, the output would be $y_e = K x_0 \sin \omega t$. The actual response of a second-order system to a sinusoidal input has a damped portion that dies out and a long-term continuing part, which, respectively, correspond to the general and the particular solutions of the second-order ODE. It is usually the continuing part of the solution that is of significance and that is used to estimate the frequency response of the system. The actual continuing response of the second-order system to a sinusoidal input is given by

$$\frac{y}{K x_0} = \frac{1}{[(1 - \omega^2/\omega_n^2)^2 + (2\zeta\omega/\omega_n)^2]^{1/2}} \sin(\omega t + \phi) \tag{11.30}$$

where the phase angle is

$$\phi = -\tan^{-1}\frac{2\zeta\omega/\omega_n}{1 - \omega^2/\omega_n^2} \tag{11.31}$$

Plots of the amplitude and phase response of second-order systems in terms of the dimensionless frequency of the input and the damping ratio are shown in Figure 11.8.

Any deviation of the amplitude ratio, $y/K x_0$, from unity indicates a dynamic measurement error. Figure 11.8(a) shows that the amplitude response will show large deviations from the ideal response unless the input frequency is small relative to the system natural frequency. The widest frequency response occurs with a damping ratio of 0.7—at frequencies less than about 30% of the natural frequency, the amplitude response is close to the ideal value (with less than 1.5% difference). However, many important second-order measurement systems have very low damping, and the dynamic error may be unacceptable at frequencies over 10 to 20% of the natural frequency. Regardless of the damping ratio, Figure 11.8(b) shows that the phase is

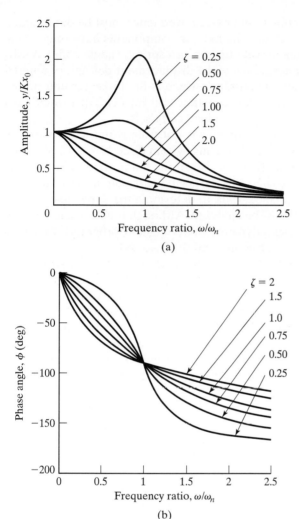

FIGURE 11.8

(a) Amplitude and (b) phase response of a second-order system to a sinusoidal input.

shifted at any frequency and the magnitude of the phase angle increases with the frequency ratio.

In this section we have discussed the response of linear instruments to a single sine wave. As with first-order systems, Fourier analysis (see Chapter 5) can be used to decompose complex waveforms into component sine and cosine waves, which can then be evaluated using the results presented here.

11.4.4   Force Transducer (Load Cell) as a Second-Order System

Systems instrumented with load cells are second-order systems, and as discussed in Section 11.4.3, the load cell will produce accurate force measurements only when the system's natural frequency is high relative to the frequency of the forcing function.

Thus, for accurate dynamic measurements, the natural frequency must be determined. A load cell itself is a second-order system—the moving components have some mass, and the flexibility of the sensing element will function as a spring. The flexibility is usually specified with the term *compliance*, which is the displacement per unit of applied force (e.g., mm/N) or stiffness (or spring constant), which is the force per unit of displacement (e.g., N/mm). Using the mass, spring constant, and Eq. (11.24), it would thus be possible to compute a natural frequency of the load cell itself.

Unfortunately, the load-cell natural frequency is not usually particularly useful. Load cells are usually connected to the force-exerting system directly or through some linking mechanism. The dynamic system then consists of the load cell itself and the attached structure. In most cases the attached structure will contribute most of the mass and at least some of the system compliance. Thus, to determine the dynamic measurement response, the complete system must be analyzed. Although time consuming, such analysis can usually be performed using dynamic finite-element programs, which include the mass and stiffness of both the structure and the load cell.

In some cases the structure has a large mass, but is also very stiff. For such situations, the load cell provides most of the system compliance, the attached structure supplies most of the mass, and the system natural frequency can be readily estimated using Eq. (11.24). The main controllable variable affecting natural frequency is the load-cell compliance, which should be low for high natural frequency. Values of compliance are either supplied directly by the manufacturer for a given load cell or can be calculated from other data supplied. For example, the displacement at maximum load may be provided. Assuming that the device is linear, the compliance is simply the maximum displacement divided by the maximum load.

Load cells using piezoelectric sensing elements usually have lower compliance than load cells using strain-gage sensing elements. Unfortunately, piezoelectric load cells are more expensive than strain-gage load cells, and consequently, the latter are more common. The compliance of strain-gage load cells usually depends on the range of the load cell (being lower for larger ranges). Since compliance depends on range, to reduce compliance it is often possible to use a load cell with a range that is much larger than that required based on simple force considerations. For example, consider a situation in which the maximum load is expected to be 1000 lb. If the compliance of a 1000-lb load cell is too high, it might be possible to use a 10,000-lb load cell, which could have a compliance only 10% as high. If this is done, the larger-capacity load cell will probably have to be calibrated individually to produce acceptable accuracy, and a lower signal-to-noise ratio in the output must be accepted.

As discussed in Section 11.4.3, measurements with acceptable dynamic error can be achieved only at frequencies that are a fraction of the natural frequency, and this fraction depends on the damping ratio, being highest for a damping ratio of about 0.7. Unfortunately, for most load-cell installations the damping ratio of the system is very small. For high-quality piezoelectric load cells, the damping ratio of the cell itself is in the range 0.01 to 0.05. For strain-gage load cells the damping ratio may be slightly larger, but it is still small. Based on Eq. (11.30) with a damping ratio of zero, Table 11.1 presents the errors in measured amplitude as a function of forcing frequency. The error will be substantial at frequencies larger than 10 to 20% of the natural frequency.

TABLE 11.1 Amplitude Errors for Second-Order Systems with Zero Damping Ratio

ω/ω_n	Amplitude Error[a] (%)
0.10	+1
0.20	+4
0.30	+10
0.50	+33

[a]The plus sign indicates that the reading is higher than the ideal value.

Example 11.3

A load cell is sought for monitoring the off-balance bearing load of a small motor. The load cell will be installed directly under the bearing at one end of the shaft. The motor is operated in the range 1500 to 6000 rpm and has a mass of 10 kg. Check to see if a low damping ratio load cell with a range of 0 to 200 N and a full-scale deflection of 0.1 mm is appropriate for this dynamic application.

Solution: We are going to use Eq. (11.30) (plotted in Figure 11.8) in this application. The stiffness (spring constant) of this transducer is

$$k = \frac{F_{max}}{\Delta x} = \frac{200}{0.1 \times 10^{-3}} = 2 \times 10^6 \text{ N/m}$$

Considering that the motor will be supported by two bearings, the effective mass for each bearing will be 5 kg. The mass of the moving part of the transducer will be neglected. Using Eq. (11.24), the natural frequency of the motor/bearing system will be

$$\omega_n = \left(\frac{k}{m}\right)^{1/2} = \left(\frac{2 \times 10^6}{5}\right)^{1/2} = 632 \text{ rad/s}$$

and the maximum forcing frequency will be

$$\omega_{max} = 2\pi N = \frac{2\pi \times 6000}{60} = 628 \text{ rad/s}$$

The ratio of the forcing frequency to the natural frequency is then

$$\frac{\omega_{max}}{\omega_n} = 1$$

According to Figure 11.8(a), at low damping ratios applying this transducer will result in measured force amplitudes far larger than the actual force amplitude. Consequently, the recommended load cell will not be appropriate for this type of measurement.

Example 11.4

For the system of Example 11.3, assuming a damping ratio of about 0.1, determine the appropriate load-cell compliance that will permit measuring the force amplitude within 1% of the actual applied force amplitude.

Solution: We want the amplitude of Eq. (11.30) to be less than 1.01. That is,

$$\frac{y}{Kx_0} = \frac{1}{[1 - \omega^2/\omega_n^2)^2 + (2\zeta\omega/\omega_n)^2]^{1/2}} = 1.01$$

$$1.01 = \frac{1}{[(1 - \omega^2/\omega_n^2)^2 + (2 \times 0.1 \times \omega/\omega_n)^2]^{1/2}}$$

Solving this equation for ω/ω_n results in

$$\frac{\omega}{\omega_n} = 0.1$$

With the maximum forcing frequency of 628 rad/s,

$$\omega_n = \frac{628}{0.1} = 6280 \text{ rad/s}$$

With this value of ω_n and the supported mass of 5 kg, the approximate spring constant of the transducer can be determined from Eq. (11.24):

$$k = m\omega_n^2 = 5 \times 6280^2 = 1.97 \times 10^8 \text{ N/m} = 1.1 \text{ lb}/\mu\text{in.}$$

The compliance of the force transducer is the inverse of the spring constant, 0.91 μin./lb.

Comment: In this case we presumed that the amplitude of the response was larger than ideal [amplitude($y/Kx_0 > 1$)]. This is true for damping ratios less than about 0.7 and ω/ω_n less than 1. For damping ratios larger than 0.7, the response amplitude will be smaller than the ideal [amplitude($y/Kx_0 < 1$)].

11.4.5 Pressure-Measurement Devices as Second-Order Systems

A pressure-measurement transducer is another common system that shows second-order behavior. In pressure transducers, there is some mass associated with the moving parts and compliance associated with the sensing element. Thus, it is possible to determine a natural frequency for the measurement system that might be used to determine suitability for a given application. Unfortunately, there are many situations in which the dynamic characteristics are significantly affected by the fluid as well as other parts of the measurement system.

Figure 11.9 shows a chamber (e.g., pipe or pressure vessel) in which we seek to measure the time-varying pressure. Ideally, we would install a flush-mounted pressure transducer, such that the diaphragm of the transducer essentially forms a portion of the chamber wall. Even if a flush-mounted transducer is used, it is not necessarily true that the pressure on the diaphragm is the same as the pressure that would exist if the transducer were not present. This is because motion of the transducer diaphragm can locally affect the chamber pressure. This effect, called a *fluid–structure interaction*, is primarily a problem with liquids. Nevertheless, the flush-mounted approach will give the best measurement that can be achieved.

Assuming that fluid–structure interactions can be neglected, the pressure transducer, being a second-order system, will show dynamic effects if the chamber pressure varies rapidly enough. As with other second-order measurement devices, dynamic errors will be

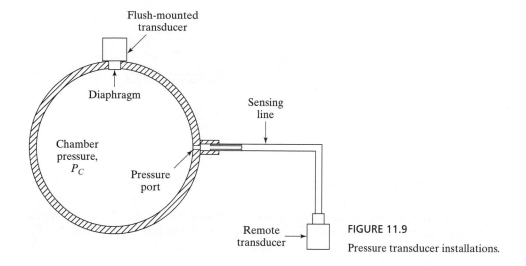

FIGURE 11.9

Pressure transducer installations.

minimized if the frequency of the pressure variations is small compared to the natural frequency of the pressure transducer. Pressure transducers with the highest natural frequency use piezoelectric sensors. This type of sensor is very stiff and the displacement of the diaphragm will be very small. Piezoelectric pressure transducers are available with natural frequencies as high as 150 kHz. These devices have very low damping ratios, and Table 11.1 can be used to determine the maximum usable frequency as a function of the acceptable error. Flush-mounted piezoelectric pressure transducers are used for high-frequency applications such as the measurement of combustion-chamber pressure in internal-combustion engines. Further details on piezoelectric pressure transducers are available in Kail and Mahr (1984).

There are a number of practical considerations that often dictate that the pressure transducer be installed in a remote location and be connected to the chamber using a length of pipe or tubing called a *sensing line*. (See Figure 11.9.) Remote location may be required so that the transducer is in a location accessible for servicing or because the region next to the chamber is environmentally hostile (e.g., very hot). If the chamber pressure is essentially static or changes very slowly, the remote transducer and the flush-mounted transducer will both accurately measure the chamber pressure. On the other hand, if the chamber pressure shows a rapid time variation, the flush-mounted transducer will usually provide a much more accurate representation of the chamber pressure. (Of course, if the pressure variation is rapid enough, neither transducer may give acceptable accuracy.) While the response of the flush-mounted transducer will be limited primarily by the dynamic characteristics of the transducer itself, the response of the remote sensor will be severely limited by the dynamic characteristics of the fluid in the sensing line and the dynamic interactions between the sensing-line fluid and the transducer.

Figure 11.10 shows a simple model of a sensing-line–pressure-transducer system. The sensing line and the sensor housing are considered essentially rigid. The diaphragm is modeled as a piston with a restoring spring. The transducer piston and

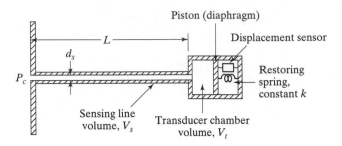

FIGURE 11.10

Model of pressure transducer–sensing line system.

spring form a spring–mass system that shows second-order behavior. The natural frequency of many designs of transducers will normally be high since the effective moving mass of the transducer is usually very small and the spring is very stiff (particularly for piezoelectric transducers). However, when the transducer is connected to the measurement location through a sensing line, the measuring-system natural frequency will be reduced significantly.

There are several effects that degrade the frequency response of remote pressure transducers:

1. The sensing line functions as an organ pipe with a resonant frequency of its own. Depending on the fluid and the length of the sensing line, this resonant frequency may be much lower than that of the transducer.

2. The fluid in the sensing line and transducer has an appreciable mass compared to the effective mass of the transducer piston, particularly for liquids. This mass increases the mass of the spring–mass system, reducing the natural frequency.

3. For the transducer to show an output, the piston must move a distance and the volume of the transducer chamber will increase. For this to happen, the fluid must flow through the sensing line, which may be a small-diameter tube with appreciable pressure drop. This is a first-order dynamic effect. This effect is not usually significant in systems designed for good dynamic response. In some cases it is a desired effect—sensing-line friction will damp out unwanted pressure fluctuation due to such effects as turbulence. This is common in process measurements.

In the following sections we examine the dynamic characteristics of sensing lines and the combined sensing-line–transducer for both gas- and liquid-filled systems. Doebelin (1990) is a useful reference for more details on this subject.

Dynamic Effects in Sensing Lines   If, for the moment, we examine only the sensing line (which is itself a second-order dynamic system) and ignore the effects of the transducer, we have the configuration shown in Figure 11.11. Acoustically, this is the same situation as an organ pipe, closed at one end. As demonstrated in most basic physics texts, this pipe has a lowest natural frequency (fundamental frequency) given by

$$f_n = \frac{\omega_n}{2\pi} = \frac{C}{4L} \tag{11.32}$$

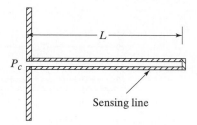

FIGURE 11.11

Sensing-line system as an organ pipe.

where C is the sonic velocity (speed of sound) of the fluid in the pipe and L is the pipe length. For the fluid, the sonic velocity is determined from

$$C = \sqrt{\frac{B}{\rho}} \qquad (11.33)$$

where ρ is the fluid density and B is the fluid bulk modulus, which has units of pressure (e.g., psi, Pa). For liquids, Eq. (11.33) is used directly to compute C. For example, water at 25°C has B = 2.22 GPa, ρ = 997 kg/m^3, and the resulting value of C is 1492 m/s. For an ideal gas, the sonic velocity is computed from

$$C = \sqrt{\gamma R T} \qquad (11.34)$$

where γ is the specific-heat ratio, R is the constant, and T is the gas absolute temperature. For air at 25°C (298 K), γ = 1.4, R = 287 J/kg-K, and the resulting value of C has a value of 346 m/s.

Using Eqs. (11.33) and (11.34) for a $\frac{1}{2}$-m-long sensing line, the line resonant frequency will be on the order of 173 Hz for air and 749 Hz for water. The sensing-line natural frequency will place limits on measurable forcing frequencies in the same manner as other second-order systems. Actually, the C used in Eq. (11.32) should be the wavespeed of the system. Wavespeed is the velocity at which a low-amplitude pressure pulse will travel down the pipe. Although not taken into consideration here, this wavespeed will not necessarily be the same as the fluid sonic velocity. In general, the pipe wall has some flexibility and, particularly for liquids, this serves to make the wavespeed less than the sonic velocity. For thick metal pipes, this effect is not large, but for thin metal pipes or plastic pipes, this effect may be significant. The effect of pipe walls on wavespeed is discussed in detail by Wylie, et al. (1993).

The fluid in the sensing line interacts with the fluid contained in the transducer and also with the spring–piston system of the sensing element. These effects serve to reduce the natural frequency further. Various investigators have considered them, and the results are summarized shortly. Although the basic principles are the same for both, the analyses for gas systems and liquid systems have been treated separately. This is because the appropriate modeling assumptions are somewhat different.

Sensing-Line Effects in Gas-Filled Systems   In a gas-filled system, the gas in the transducer chamber behaves as a spring that is acted on by the sensing-line-fluid mass. In most cases, this gas spring is much softer than the spring effect associated with the sensing piston, and as a result it is acceptable to ignore the effects of the sensing piston. The

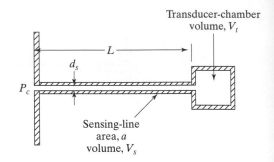

FIGURE 11.12

Model of gas-filled pressure-transducer–sensing-line system.

sensing-line fluid, together with the compressible fluid in the chamber (Figure 11.12), can be modeled as a second-order system. Hougen, et al. (1963) have determined the following expressions for the natural frequency and damping ratio for such a system:

$$f_n = \frac{\omega_n}{2\pi} = \frac{C}{2\pi L \sqrt{0.5 + V_t/V_s}} \tag{11.35}$$

$$\zeta = \frac{R_l L}{2\rho C} \sqrt{0.5 + \frac{V_t}{V_s}} \tag{11.36}$$

In these equations, V_t is the transducer-chamber volume and V_s is the volume of the sensing line. The factor R_l is the fluid resistance offered by the sensing-line wall. This resistance is not well understood for oscillatory flows such as exist in sensing lines. It is common practice to use the value for steady, laminar flow in tubes, which is given by

$$R_l = \frac{32\mu}{d_s^2} \tag{11.37}$$

where μ is the fluid viscosity.

Examining Equation (11.35), it can be seen that the natural frequency will be highest for large values of V_s and low values of L—requirements that can be met with a short, large-diameter sensing line. With pressure transducers used for dynamic measurements, it is likely that the damping ratio of the overall system will be low, close to zero. The results shown in Table 11.1 should be approximately applicable for estimating dynamic measurement error, even for strain-gage pressure transducers. If the damping ratio is larger, Eq. (11.30) can be used to estimate the dynamic measurement error.

It is also worth noting that if the term V_t in Eq. (11.35) is taken to be zero, the resulting system is the same as the organ pipe of Figure 11.12. However, the predicted natural frequency is $0.225C/L$, slightly less than the value $0.25C/L$ predicted by Eq. (11.32). This is a consequence of the modeling assumptions used for Eq. (11.35) and is not usually significant. This point is addressed further in the discussion of liquid-filled systems.

Example 11.5

A diaphragm-type pressure transducer (with a chamber diameter of 1 cm and length of 0.5 cm) is connected to a pressurized air line by rigid tubing that is 5 cm long and 0.2 cm in diameter. The initial pressure and temperature of the line are 10 atm and 25°C. Determine the natural frequency and damping ratio of the sensing-line–transducer system. Determine the approximate time lag for sensing a step change in pressure to within 99% of the true value.

Solution: We assume that the pressure change is small compared with the line pressure so that the formulation introduced in this section can be applied to the problem. This also implies that we can use the initial pressure and temperature to determine fluid properties. For more accurate calculations, average temperature and pressure should be used. From Table B.3, R (the gas constant for air) is 287 J/kg-K and γ (the specific heat ratio) is 1.4. At 298 K (25°C) and 10 atm, ρ, the density is 11.8 kg/m³, and μ, the viscosity, is 1.8×10^{-5} kg/m-s.

We will solve this problem using Eqs. (11.35) and (11.36). The volumes are as follows:

$$\text{Sensing line: } V_s = aL = L\frac{\pi d^2}{4} = 0.05\left(\frac{\pi \times 0.002^2}{4}\right) = 1.57 \times 10^{-7}\, \text{m}^3$$

$$\text{Transducer: } V_t = L_t\left(\frac{\pi d_t^2}{4}\right) = 0.005\left(\frac{\pi \times 1 \times 0.01^2}{4}\right) = 3.93 \times 10^{-7}\, \text{m}^3$$

$$\frac{V_t}{V_s} = 2.50$$

The speed of sound in air is

$$C = (\gamma RT)^{1/2} = (1.4 \times 287 \times 298)^{1/2} = 346\, \text{m/s}$$

Substituting the proper values in Eqs. (11.35), (11.36), and (11.37) yields

$$f_n = \frac{C}{2\pi L(1/2 + V_t/V_s)^{1/2}} = \frac{346}{2\pi 0.05(0.5 + 2.50)^{1/2}}$$
$$= 635\, \text{Hz}$$

$$R_l = \frac{32 \times 1.8 \times 10^{-5}}{0.002^2} = 144$$

$$\zeta = \frac{144 \times 0.05(0.5 + 2.5)^{0.5}}{(2 \times 11.8 \times 346)}$$
$$= 1.53 \times 10^{-3}$$

Because $\zeta \ll 1$, Eq. (11.28) will be used to determine the response to a step. For this low damping, the response will oscillate. The response will be within 1% of the ideal value when the amplitude of the oscillating term in Eq. (11.30) has a value of 0.01:

$$\frac{e^{-\zeta \omega_n t}}{1 - \zeta^2} = \frac{e^{-1.53 \times 10^{-3} \times 2 \times \pi \times 635t}}{1 - (1.53 \times 10^{-3})^2} = 0.01$$

Solving for t, we obtain $t = 0.75$ s.

Comment: The value of the damping ratio, ζ, is very low, although the exact value is uncertain since the value of R_l is highly uncertain. Consequently, the 1% settling time result is also uncertain.

Example 11.6

The transducer of Example 11.5 is to be used to measure pressure fluctuations in a compressed air line. The fluctuations have a frequency of 100 Hz and an amplitude of 1 psi. Estimate the error in the measurement of the unsteady pressure using this transducer.

Solution: We will assume that the fluctuations can be approximated with a harmonic function. Equation (11.30) provides the amplitude ratio of the output versus the input for a second-order system, which is the model for our application. Substituting for the parameters in Eq. (11.30) will result in

$$\omega = 2\pi f = 2 \times 3.14 \times 100 = 628 \text{ rad/s}$$

$$\omega_n = 2\pi f_n = 2 \times \pi \times 635 = 3989 \text{ rad/s}$$

$$\text{amplitude}\left(\frac{y}{Kx_0}\right) = \left[\left(1 - \frac{\omega^2}{\omega_n^2}\right)^2 + \left(\frac{2\zeta\omega}{\omega_n}\right)^2\right]^{-1/2}$$

$$= \left[\left(1 - \frac{628^2}{3989^2}\right)^2 + \frac{(2 \times 1.53 \times 10^{-3} \times 628}{3989^2}\right]^{-1/2}$$

$$= 1.025$$

This result shows that the amplitude of the output signal will be about 2.5% higher than the ideal output pressure. Since the damping ratio is so low, we could have obtained this result from Table 11.1.

Sensing-Line Effects in Liquid-Filled Systems   In liquid-filled systems the compressibility of the fluid is much smaller than for gases (although not negligible), and the springlike characteristics of the sensing element become significant. It is first necessary to define the concept of compliance for a pressure transducer. This is the change in volume of a system in response to an imposed pressure change:

$$C_v = \frac{\Delta V}{\Delta P} \tag{11.38}$$

There is a compliance associated with the transducer spring–piston shown in Figure 11.10. There is also a compliance associated with the fluid volume (V_t) in the transducer and the fluid volume in the sensing line (V_s).

It is possible to construct a very simple spring–mass second-order model of the combination sensing-line-and-transducer system. The mass (m) consists of the column of fluid contained in the sensing line. Using the nomenclature of Figure 11.10, the mass is given by

$$m = \frac{\rho L \pi d_s^2}{4} \tag{11.39}$$

where ρ is the fluid density. The restoring spring results from the compliance of the transducer diaphragm. As the column of fluid moves into the transducer cavity, there will be an increase in the cavity pressure, resulting in a force on the end of the fluid column. The effective spring constant resulting from this restoring force can be demonstrated to be

$$k = \frac{\pi^2 d_s^4}{16C_v} \tag{11.40}$$

Since the natural angular frequency, ω_n, is given by $\sqrt{k/m}$ [Eq. (11.24)], the natural frequency ($\omega_n/2\pi$) is given by

$$f_n = \frac{1}{2\pi}\sqrt{\frac{\pi d_s^2}{4\rho L C_v}} \tag{11.41}$$

This model is only approximate. Anderson and Englund (1971) note that there are actually three contributions to the system compliance—the transducer spring–piston, the transducer fluid volume, and the fluid in the sensing line—and suggest that these contributions be combined as

$$C_v = C_{vt} + \frac{V_t}{B} + \frac{V_{se}}{B} \tag{11.42}$$

where B is the fluid bulk modulus, C_{vt} is the transducer mechanical compliance, V_t is the transducer fluid volume, and V_{se} is the effective sensing-line volume, given by

$$V_{se} = \frac{4}{\pi^2}V_s \tag{11.43}$$

With some algebraic manipulation, Eq. (11.41) for the natural frequency can now be evaluated as

$$f_n = \frac{C}{4L}\sqrt{\frac{V_{se}}{BC_{vt} + V_t + V_{se}}} \tag{11.44}$$

where C is the fluid sonic velocity. [See Eq. (11.33).] In Eq. (11.43) Anderson and Englund chose the fraction of sensing-line volume, $4/\pi^2$, in order to force Eq. (11.44) to agree with Eq. (11.32) when the transducer is rigid and has zero volume (C_{vt} and V_t are each zero). Anderson and Englund also note that the damping ratio of the system is given by

$$\zeta = \frac{32\mu}{\pi d_s^3}\sqrt{\frac{\pi L C_v}{\rho}} \tag{11.45}$$

but observe that ζ is very small for practical liquid-filled sensing systems.

Examining Eq. (11.44), it can be concluded that the natural frequency will be highest when L is short and the sensing line volume is large relative to the transducer volume. The latter requirement is met with a large-diameter sensing line, the same requirement as for the gas-filled systems. Unlike gas systems, where the mechanical compliance, C_{vt}, has a negligible effect, C_{vt} is significant for liquid systems. Unfortunately, manufacturers rarely supply compliance values. Anderson and Englund (1971) give some typical values for strain-gage pressure transducers that may be useful. It may, however, be necessary to determine the compliance experimentally.

Anderson and Englund devote considerable attention to a major problem with the dynamic response of liquid-filled systems, the problem of gas bubbles. Small gas bubbles are highly compressible and can dramatically increase the compliance of a liquid-filled system, thus reducing the natural frequency. It is necessary to remove all gas from the sensing-line–transducer system when installing it initially. Since many liquids have dissolved gases, it may be necessary periodically to remove bubbles that are formed from gases coming out of solution.

Example 11.7

A pressure transducer is to be used to measure pressure transients in a water line. The sensing line is 0.5 cm in diameter and 10 cm long. The transducer volume is 2.8 cm^3, and the manufacturer specifies the change in volume of the transducer to be 0.003 cm^3 for the 150 psi range. The water temperature is 20°C. Determine the natural frequency and the damping ratio of this pressure-measurement system.

Solution: The properties of water can be evaluated at 20°C from Table B.1. $\rho = 998.2$ kg/m^3, $\mu = 1.005 \times 10^{-3}$ kg/m-s, and $B = 2.20$ GPa. The following variables must be computed:

$$C = \left(\frac{B}{\rho}\right)^{0.5} = \left(\frac{2.20 \times 10^9}{998.2}\right)^{0.5} = 1484 \text{ m/s}$$

$$C_{vt} = \frac{\Delta\text{volume}}{\Delta P} = \frac{0.003 \times 10^{-6}}{(150/14.7) \times 101,325}$$
$$= 2.90 \times 10^{-15} \text{ m}^5/\text{N}$$

$$V_s = 0.1 \times \pi \times \frac{0.005^2}{4} = 1.96 \times 10^{-6} \text{ m}^3$$

$$V_{se} = \left(\frac{4}{\pi^2}\right)V_s = \left(\frac{4}{\pi^2}\right)1.96 \times 10^{-6} = 7.96 \times 10^{-7} \text{ m}^3$$

$$C_v = C_{vt} + \frac{V_t}{B} + \frac{V_{se}}{B} = 2.90 \times 10^{-15} + \frac{2.8 \times 10^{-6}}{2.20 \times 10^9} + \frac{7.96 \times 10^{-7}}{2.20 \times 10^9}$$
$$= 4.53 \times 10^{-15} \text{ m}^5/\text{N}$$

The natural frequency is obtained from Eq. (11.44):

$$f_n = \frac{C}{4L}\left[\frac{V_{se}}{(BC_{vt} + V_t + V_{se})}\right]^{0.5}$$
$$= \frac{1484}{4 \times 0.1}\left(\frac{7.96 \times 10^{-7}}{2.20 \times 10^9 \times 2.91 \times 10^{-15} + 2.8 \times 10^{-6} + 7.96 \times 10^{-7}}\right)^{0.5}$$
$$= 1048 \text{ Hz}$$

The damping ratio is determined from Eq. (11.45):

$$\zeta = \frac{32 \times 1.005 \times 10^{-3}}{\pi \times 0.005^3}\left(\frac{\pi \times 0.1 \times 4.52 \times 10^{-15}}{998.2}\right)^{1/2}$$
$$= 0.97 \times 10^{-4}$$

The damping ratio is very low; hence, Table 11.1 can be used to determine an acceptable upper limit on the frequency of pressures to be measured.

11.4.6 Second-Order Systems for Acceleration and Vibration

While the material on second-order systems in Section 11.4 can be applied directly to force-measuring devices such as load cells and pressure transducers, it requires interpretation for application to systems for measuring vibration and acceleration. This is because the input to systems described by Eq. (11.22) is a time-varying force, whereas the input (measurand) for acceleration is a time-varying acceleration or displacement. As a result, the mathematical model described in Section 11.4.1 must be reexamined.

A model for a linear accelerometer is shown in Figure 11.13(a). It consists of a mass (m) contained within a rigid housing. The mass is connected to the housing with a spring that has constant k. The mass can move relative to the housing, and the relative displacement, y, is measured with a displacement-measuring device. When there is relative motion, there is a damping force given by $\lambda \, dy/dt$. The input to the system is $s(t)$, the base displacement, and the output is y, the relative motion between the accelerometer's mass and the base.

To find the relation between s (or its derivatives) and y, we apply Newton's second law of motion to the free-body diagram of the mass [Figure 11.13(b)]. If we take z to be the absolute displacement of the mass m, then

$$z = s + y \tag{11.46}$$

Newton's second law gives

$$m\frac{d^2z}{dt^2} + \lambda\frac{dy}{dt} + ky = 0 \tag{11.47}$$

Upon substituting for z in terms of s and y, we obtain

$$m\frac{d^2y}{dt^2} + \lambda\frac{dy}{dt} + ky = -m\frac{d^2s}{dt^2} \tag{11.48}$$

This equation then has the same form as Eq. (11.17), the general second-order equation, with the forcing function being the base acceleration (the second derivative of the base displacement).

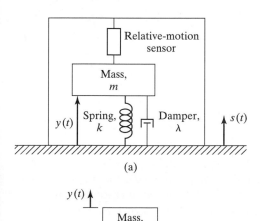

(a)

(b)

FIGURE 11.13

Basic mechanical model for an accelerometer: (a) schematic diagram; (b) free-body diagram.

In Chapter 8, two basic types of acceleration-measuring devices were introduced. The first type, consisting of piezoelectric, strain-gage, and servo accelerometers, is used to measure accelerations having frequencies that are much lower than the device natural frequency. When operated in the usable frequency range, the sensing mass has about the same acceleration as the input acceleration. The second type, the vibrometer, has a very low natural frequency and is used to measure displacements and accelerations with frequencies higher than the device natural frequency. In the vibrometer, the mass remains approximately stationary while the base moves. Equation (11.48) can be used to analyze both types. The subject is discussed here briefly. For more details, see the books by Thomson (1983) and Doebelin (1990).

In the case of accelerometers, it is best to view the acceleration, $a(t) = d^2s(t)/dt^2$, as the forcing function. Rewriting Eq. (11.48) in this form, we obtain

$$m\frac{d^2y}{dt^2} + \lambda\frac{dy}{dt} + ky = -ma(t) \tag{11.49}$$

This equation can also be compared with Eq. (11.17), the general equation for a second-order system:

$$a_2\frac{d^2y}{dt^2} + a_1\frac{dy}{dt} + a_0y = bx \tag{11.17}$$

These two equations have the same form with $a(t) = x$, $a_2 = m$, $a_1 = \lambda$, $a_0 = k$, and $b = -m$. The definitions of natural frequency, ω_n, and the damping ratio, ζ, are the same as given by Eqs. (11.24) and (11.25). The static sensitivity, K, will have the value $-m/k$. The response of an ideal accelerometer would be

$$y_e = Ka(t) = -\frac{m}{k}a(t) \tag{11.50}$$

That is, the ideal output response is proportional to the input acceleration. The results of Section 11.4.2 for step inputs and the results of Section 11.4.3 for sinusoidal inputs are directly applicable to accelerometers.

For the step input (the sudden input of a constant acceleration a_0), the response is described by Eqs. (11.26) through (11.28) and the response is shown graphically in Figure 11.7. The damping ratio, ζ, for piezoelectric accelerometers is very small, being in the range of 0.01 to 0.05. Figure 11.7 shows that for such low values of the damping ratio, there will be significant oscillations in the output. It is possible to damp the output electronically. Strain-gage accelerometers often have damping close to the ideal value of 0.7 and will not show significant oscillations. On the other hand, they have a relatively low natural frequency, so the response times will be fairly large.

For a continuing sinusoidal input (of the form $a_0 \sin \omega t$), the response is described by Eqs. (11.30) and (11.31), and the amplitude and phase responses are shown graphically in Figures 11.8(a) and 11.8(b). For piezoelectric accelerometers with the low values of damping, the amplitude will show significant dynamic error if the forcing frequency is greater than about 0.1–0.2 times the natural frequency. However, piezoelectric accelerometers can have very high values of natural frequency (up to 150 kHz) and consequently can be used to measure forcing frequencies up to 30 kHz. Table 11.1 can be used to estimate the dynamic amplitude error as a function of frequency ratio for piezoelectric accelerometers. Strain-gage accelerometers have higher values of damping ratio and consequently can have acceptable amplitude errors at higher fractions of the natural frequencies—up to 0.4. However, the natural frequency of strain-gage accelerometers is much lower (usually less than 1 kHz), so they are usually not useful for forcing frequencies in excess of a few hundred hertz.

The natural frequency of a strain-gage accelerometer depends on the range of accelerations that it can measure. The higher the acceleration range, the higher the natural frequency. For example, one type of strain-gage accelerometer has a natural frequency of 500 Hz for a range of ±10g and a natural frequency of 4 kHz for a range ±1000g (g represents the acceleration of gravity, 9.8 m/s²). This results in a trade-off between optimum selection of range and acceptable natural frequency.

For the low-natural-frequency vibrometer, the input can be viewed as a displacement, $s(t)$. We apply a sinusoidal input to the base having the form

$$s(t) = s_0 \sin \omega t \tag{11.51}$$

Substituting this into Eq. (11.48), we obtain

$$m\frac{d^2y}{dt^2} + \lambda\frac{dy}{dt} + ky = -m\frac{d^2(s_0 \sin \omega t)}{dt^2} = m\omega^2 s(t) \tag{11.52}$$

This equation has the same form as Eq. (11.17),

$$a_2\frac{d^2y}{dt^2} + a_1\frac{dy}{dt} + a_0 y = bx(t)$$

with $x = s(t)$. The static sensitivity [Eq. (11.18)] can be written as

$$K = \frac{b}{a_0} = \frac{m\omega^2}{k} = \left(\frac{\omega}{\omega_n}\right)^2 \tag{11.53}$$

where

$$\omega_n = \left(\frac{a_0}{a_2}\right)^{1/2} = \left(\frac{k}{m}\right)^{1/2}$$

The solution for a sinusoidal input, Eq. (11.30), is applicable. Using the K given by Eq. (11.53) and noting that $x_0 = s_0$, Eq. (11.30) for this application becomes

$$\frac{y}{s_0} = \frac{(\omega/\omega_n)^2}{[(1 - \omega^2/\omega_n^2)^2 + (2\zeta\omega/\omega_n)^2]^{1/2}} \sin(\omega t + \phi) \tag{11.54}$$

where the phase angle is the same as that given by Eq. (11.31).

Figure 11.14, which is a plot of Eq. (11.54), shows that at frequencies a few times higher than the natural frequency of the transducer, the output displacement y has an

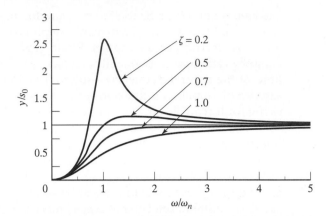

FIGURE 11.14

Amplitude response of a vibrometer to a harmonic input.

amplitude close to that of the base displacement. In this situation the mass remains essentially stationary while the base moves. These devices are used to measure time-varying displacements at frequencies that are high relative to the natural frequency, the opposite of most second-order measurement devices. This device is the type used to measure ground motion in earthquakes.

Example 11.8

A high-quality strain-gage accelerometer is used to measure the acceleration of a vehicle. The acceleration of the vehicle suddenly changes from zero to a constant value. The natural frequency of the accelerometer is 1 kHz, and its damping ratio is 0.5. If the accelerometer shows an acceleration of 10 m/s^2 after 0.5 s from the start, estimate the error in measuring the acceleration of the vehicle.

Solution: We will assume that the error in the output is only due to the dynamic behavior of the system. For a step change in input, Eq. (11.28) describes the response for this problem. The device is underdamped and will show oscillatory behavior. To estimate the error, we evaluate the amplitude of the oscillatory term in Eq. (11.28):

$$\omega_n = 2\pi f_n = 2\pi 1000 = 2000\pi$$

$$\frac{e^{-\zeta \omega_n t}}{(1 - \zeta^2)^{1/2}} = \frac{e^{-0.5 \times 2000\pi \times 0.5}}{(1 - 0.5^2)^{1/2}}$$

$$= 0$$

This amplitude of the oscillation shows that, for all practical purposes, the output will be the same as the equilibrium value at $t = 0.5$ s.

Example 11.9

Consider an accelerometer with natural frequency of 500 Hz and damping ratio of 0.7. Determine the input vibration frequency above which there will be an amplitude distortion greater than 1%.

Solution: Following Eq. (11.30) and Figure 11.8, we have

$$\frac{1}{[(1 - \omega^2/\omega_n^2)^2 + (2\zeta\omega/\omega_n)^2]^{1/2}} = 0.99$$

Solving this equation, the limiting frequency is given by

$$\frac{\omega}{\omega_n} = 0.40$$

and the limiting input frequency is then $0.4 \times 500 = 200$ Hz. Below this frequency the amplitude distortion will be less than 1%.

Comment: Depending on the value of the damping ratio, the amplitude error may be positive or negative relative to the equilibrium value. For example, for $\zeta = 0.25$, the error is always positive below the natural frequency. In this case, the equation

$$\frac{1}{[(1 - \omega^2/\omega_n^2)^2 + (2\zeta\omega/\omega_n)^2]^{1/2}} = 0.99$$

would have no solution for a frequency less than the natural frequency. The right-hand side should be 1.01 for determining the maximum frequency for a 1% error. The dividing line between curves that always decrease with increasing ω/ω_n and those that show a positive peak is a damping ratio of about 0.7.

Table 11.2 lists some common accelerometer types, a range for their natural frequency, and their damping ratio. Typical applications for these instruments are also presented.

11.5 CLOSURE

We have briefly discussed the response of some common instruments to time-varying inputs. All instruments except those of zero order will produce some dynamic error. This error should be taken into consideration at the time of selection or design of the measurement system and in the final analysis of the results. In the examples and discussions in this section we have dealt only with the dynamic behavior of the mechanical components of the transducers. We have not discussed the dynamic behavior of the signal-transmission and signal-conditioning components of the measurement systems. In certain high- and low-frequency ranges, these external effects may be significant. The subject is beyond the scope of this book. The interested reader should consult advanced texts, such as Doebelin (1990), and consult the manufacturers of the equipment.

TABLE 11.2 Common Accelerometers and Applications

Sensor Type	ω_n (kHz)	ζ	Range (g)	Application
Resistive potentiometer	<0.1	0.5–0.8	±50	Low-frequency vibration
Strain gage	<1	0.6–0.8	±1000	General purpose
Differential transformer	<1	0.6–0.7	±700	General purpose
Piezoelectric	<150	≈0.01	±100,000	Wide range
Servo feedback	<0.2	0.6–0.9	±50	High accuracy

REFERENCES

[1] ANDERSON, R., AND ENGLUND, D. (1971). *Liquid-Filled Transient Pressure Measuring Systems: A Method for Determining Frequency Response*, NASA TN D-6603, Dec.

[2] BENEDICT, R. (1984). *Fundamentals of Temperature, Pressure and Flow Measurement*, Wiley, New York.

[3] DOEBELIN, E. O. (1990). *Measurement Systems*, McGraw-Hill, New York.

[4] DORF, R. C. (1989). *Introduction to Electric Circuits*, Wiley, New York.

[5] HOLMAN, J. P. (2002). *Heat Transfer*, McGraw-Hill, New York.

[6] HOUGEN, J., MARTIN, O., AND WALSH, R. (1963). Dynamics of pneumatic transmission lines, *Control Engineering*, Sept., pp. 114–117.

[7] KAIL, R., AND MAHR, W. (1984). *Piezoelectric Measuring Instruments and Their Applications*, Kistler Instrument Corp., Amherst, NY.

[8] THOMSON, W. T. (1981). *Theory of Vibration with Applications*, 2d ed., Prentice Hall, Englewood Cliffs, NJ.

[9] WYLIE, E., STREETER, V., AND SUO, L. (1993). *Fluid Transients in Systems*, Prentice Hall, Englewood Cliffs, NJ.

PROBLEMS

11.1 If the thermocouple of Example 11.1 is used to measure the temperature of air that increases at a rate of 10°C/s,

(a) determine the time after which the thermocouple will follow the gas temperature steadily (slope is within 1% of the ideal value).

(b) calculate the continuing error in the temperature measurement after this time.

11.2 Determine the time that temperature measuring devices (such as a thermocouple) will reach 98% of their final value when exposed to a step change in temperature if they have time constants of 0.1, 1, and 10 s.

11.3 If the temperature-measuring devices of Problem 11.2 are exposed to a fluid with sinusoidal temperature variation of 0.1 Hz, determine the ratio of the response amplitude to the input amplitude and the phase lag for each case.

11.4 Which of the devices of Problem 11.2 will you recommend for measuring the temperature in an air-conditioning control system of a building that has a temperature variation of 0.5°C/min?

11.5 A commercial thermocouple that has an effective spherical junction diameter of 2 mm is intended to be used for some transient measurements. You may assume that the thermocouple material has properties similar to those of copper. Estimate the time constants of this thermocouple for gaseous and liquid environments that have convective heat-transfer coefficients of 100 W/m²-°C and 3000 W/m²-°C, respectively.

11.6 A chromel–alumel thermocouple junction, which can be approximated with a sphere, has an effective diameter of 1 mm. It is used to measure the temperature of a gas flow with an effective heat-transfer coefficient of 500 W/m²-°C.

(a) Determine the time constant of this thermocouple.

(b) If the gas temperature suddenly increases by 100°C, how long will it take the thermocouple to attain a temperature rise within 1% of the gas temperature rise?

(c) Find the answer to the previous questions if the diameter of the bead is doubled. (Assume that the heat-transfer coefficient remains the same.)

11.7 The following set of data are produced by a temperature-measuring device (first order) that is suddenly immersed into a mixture of ice and liquid water. Initially, the device is in ambient air (20°C). Determine the time constant of this temperature-measuring device.

Time (s)	0.1	0.5	1	2	3
Temperature (°C)	16.7	8.1	3.3	0.6	0.1

11.8 Two copper–constantan thermocouples with 1- and 2-mm diameters are dipped into boiling water from ambient (20°C). The heat-transfer coefficient to the thermocouples is 3000 W/m^2°C. Determine the output from each thermocouple in 1, 5, and 10 s. Estimate the time beyond which they will read the same temperature (within 0.1°C). Assume the properties of copper–constantan to be the same as those of copper, and assume that the junction is spherical.

11.9 A thermometer with a time constant of 10 s is used to measure the temperature of a fluid that is heated by steam. The temperature rise can be approximated with a ramp of 10°C/min. Determine the time delay in the measurement indicated and the instantaneous error in the temperature reading.

11.10 Air is heated by an electric heater. The air temperature output (in °C), as measured by a thermocouple, can be approximated with

$$T = 200 + 10 \cos 0.5t$$

The time constant of the thermocouple is approximately 5 s.

(a) Determine the average, maximum, and minimum measured temperatures.

(b) Estimate the actual air temperature as a function of time.

11.11 The following equation describes the behavior of a second-order system:

$$\frac{5d^2y}{dt^2} + \frac{dy}{dt} + 1000y = x(t) \quad x(t) = 25 \text{ for } t \geq 0$$

Determine the natural frequency and the damping ratio of the system. Convert it to standard form as shown in Eq. (11.21), and find the equilibrium response (response of the system in the absence of the dynamic effect) of the system.

11.12 Repeat Problem 11.11 for the equation

$$\frac{d^2y}{dt^2} + \frac{2dy}{dt} + 100y = 50 \sin 50t$$

11.13 In Problem 11.12, calculate the amplitude response (y/y_e) and the phase response (ϕ) of the system.

11.14 A mass–spring–damping system has the following characteristics: $m = 20$g, $\lambda = 5$ N-s/m, and $k = 10$ N/cm. Calculate the natural frequency and the damping ratio of this system. Is the system overdamped or underdamped?

11.15 A force transducer with stiffness of 10^6 N/m is used to measure a dynamic force. The effective mass of the connected parts is 2 kg.

(a) Calculate the natural frequency of the force-measuring system.

(b) Assume that the effective damping ratio is 0.1. If a harmonic force with a frequency of 25 Hz is measured with this system, calculate the approximate error in the amplitude and the phase lag of the measurement.

11.16 A piezoelectric force transducer has a stiffness of 5 lbf/μin. The mass under the dynamic load is 50 lb. Calculate the frequency under which the system produces a reading within

0.5% of the low-frequency response of the transducer. (This is what is used instead of a static response for piezoelectric transducers.) Take the damping ratio to be 0.05.

11.17 A strain-gage load cell has the range 0 to 500 lb and a maximum deflection of 0.010 in. Calculate the compliance and the stiffness of the transducer. If the effective mass of the loading structure is 20 lb, calculate the undamped natural frequency of the measuring system. If the system is suddenly brought under a load of 100 lbf, calculate the time after which the response of the system will be within 0.5% of the static response. Assume the damping ratio to be 0.1.

11.18 A diaphragm pressure transducer, which has a chamber volume of 2 cm³, is connected to a pressurized, natural-gas (CH_4) line by rigid tubing. The length of the tube is 20 cm, and its diameter is 0.5 cm. The transducer is used to detect sudden changes in pressure. The conditions in the line are 10 atm and 25°C. Calculate the settling time until the indicated pressure change is 90% of the actual pressure change. Methane is a perfect gas with molecular weight 16 and viscosity of 1.15×10^{-5} kg/m-s, and its ratio of specific heat is 1.3.

11.19 In Problem 11.18, calculate the settling time if the line carries compressed air instead of natural gas (under similar conditions).

11.20 Consider a flush-mounted pressure transducer with a natural frequency of 10 kHz and a damping ratio of 0.1. What is the maximum frequency (less than transducer natural frequency) of a detected sinusoidal pressure fluctuation with less than 1% error in the amplitude measurement?

11.21 In Problem 11.18, what is the maximum frequency of a detected sinusoidal pressure fluctuation with less than 1% error in the amplitude measurement?

11.22 Calculate the natural frequency and the damping ratio of a pressure transducer used to measure the pressure fluctuations in a water line. The water temperature is 10°C. The cavity of the transducer has a volume of 2 cm³, and its compliance is 3×10^{-5} cm³/psi. The connecting tube is 15 cm long and 0.5 cm in diameter.

11.23 The arrangement of Problem 11.22 is used to measure the pressure fluctuations with 10 Hz frequency. Calculate the error introduced in the measured pressure amplitude.

11.24 If the pressure-measurement arrangement of Problem 11.22 is used for detecting sudden changes in pressure of a water line, calculate the settling time until the response of the transducer is 95% of the imposed pressure change.

11.25 In Problem 11.22, if the uncertainty in measurement of the transducer compliance is 25%, calculate the uncertainty in the calculation of the natural frequency. Other parameters in the equation for calculating natural frequency of the measurement system can be determined with much higher accuracy.

11.26 A pressure transducer is used to measure pressure transients in a water line. The connecting line is 0.5 cm in diameter and 25 cm long. The transducer volume is 2.8 cm³, and the manufacturer specifies the change in volume of the transducer to be 0.003 cm³ for the 150 psi range. Determine the natural frequency and the damping ratio of this pressure-measurement system. Compare your results with those of Example 11.7.

11.27 Redo Example 11.7 for the case that the water temperature is 100°C. Compare your results with those of Example 11.7.

11.28 Consider an accelerometer with natural frequency of 800 Hz and damping ratio of 0.6. Determine the vibration frequency above which it will cause an amplitude distortion greater than 0.5%.

11.29 Consider a piezoelectric accelerometer that has a natural frequency of 10 kHz and a damping ratio of 0.02. Determine the amplitude distortion in measurement of vibration with a frequency of 1500 Hz.

CHAPTER 12

Guidelines for Planning and Documenting Experiments

In this chapter we provide guidelines for designing and documenting experimental programs.

12.1 OVERVIEW OF AN EXPERIMENTAL PROGRAM

The best outcome of an experimental program will be achieved when a systematic approach is taken. The experimental process is a special case of the design process, a subject that has been studied and documented extensively. Although different studies have different names for the various activities, terms the steps or phases for an experimental program are generally as follows:

1. Problem definition.
2. Experiment design.
3. Experiment construction and development.
4. Data gathering.
5. Analysis of data.
6. Interpreting results and reporting.

12.1.1 Problem Definition

When an engineering experimental program is initiated, it can be assumed that a technical need was established and that nonexperimental approaches were determined to be inadequate or inappropriate. However, the required output of the experimental program may be poorly defined. Consider a test to evaluate the performance of a complex engineered system. A low-cost, success-oriented testing program could be undertaken to demonstrate adequate performance using a limited number of tests and minimal instrumentation. On the other hand, a more conservative testing program could be selected, using many instruments on subsystems, to determine the source if poor performance occurs. There is thus a trade-off between a high-risk, low-cost approach and a safe, high-cost approach. Decisions of this type should be taken with great care—but they frequently are made with little thought. Engineers frequently spend insufficient time defining the problem, and they initiate the design process without even being aware that they have eliminated many better options.

12.1.2   Experiment Design

This step is a major portion of any experimental program and may include the following major components:

1. Searching for information (usually, a literature survey).
2. Determining the experimental approach.
3. Determining time schedule and costs.
4. Determining the analytical model(s) used to analyze the data.
5. Specifying the measured variables.
6. Selecting instruments.
7. Estimating experimental uncertainties.
8. Determining the test matrix (values of the independent variables to be tested).
9. Performing a mechanical design of the test rig.
10. Specifying the test procedure.

These components of the design process are usually interactive. For example, an experimental approach might have to be abandoned or modified significantly if the uncertainty analysis results in unacceptable accuracy.

If possible, several experimental approaches should be conceived at the outset, analyzed, and at some point, the best approach selected. Initial brainstorming sessions are productive in this regard and should include qualified persons who will not take part in later aspects of the program. In most test programs, the design phase is performed in two parts: a preliminary design and then a final design. The preliminary design phase is a scoping type of study, including cost estimates, and results in a document called a proposal. This proposal is reviewed by the project's sponsoring organization, which normally controls the funding for the project. It might be another component of the same organization, an outside corporation, or a government agency. The sponsoring organization may have proposals prepared by several testing groups and select only one group to complete the experimental program.

12.1.3   Experiment Construction and Development

This phase is probably the most expensive portion of the program. The required equipment is procured, and the rig is constructed. A series of shakedown tests are run to establish the adequacy of the instrumentation and the approach. The results of these shakedown tests will usually result in modifications to the equipment or the experimental approach. In some cases pilot tests are performed. These are tests on subsets or scale models of the final test rig. Pilot tests are performed to determine the validity of the approach prior to major expenditures on the final test rig.

12.1.4   Data Gathering

Once the test rig has been constructed and debugged, the test data can be gathered following the specified test matrix.

12.1.5   Data Analysis

In most test programs, data analysis is performed using computer programs developed during the design phase (the analytical model). At least some of the data analysis is performed while the data gathering is in progress to check the validity of the data. With computerized data-acquisition systems, preliminary data analysis is often almost immediate. In many experiments, the data analysis will continue well after the testing is complete. New theories may be developed, and alternative data computations may be performed.

12.1.6   Interpreting Data and Reporting

After the data have been analyzed, the experimenter must interpret the data. Logical reasons should be developed to explain the trends of the data, and reasons must be found to explain anomalous data. Comparison and validation may be performed with results from previous or similar experiments. The data should be used to respond to all the project objectives. When interpretation has been completed, the results must be documented in a final report that transmits the results to the sponsoring organization.

12.2   COMMON ACTIVITIES IN EXPERIMENTAL PROJECTS

Most aspects of an experimental program depend on the particular program; however, there are a number of techniques that are general and may be used in many programs:

1. Dimensional analysis and determining the test rig scale.
2. Uncertainty analysis.
3. Shakedown tests.
4. Determining the test matrix and test sequence.
5. Scheduling and cost estimation.
6. Design review.
7. Reporting.

These techniques are discussed in greater detail in the sections that follow.

12.2.1   Dimensional Analysis and Determining the Test Rig Scale

A very powerful technique available in experimental work, called *dimensional analysis*, permits a reduction in the number of variables to be considered in an experiment, simplifies the presentation of results, and provides a reliable method to design scale-model tests. Dimensional analysis is the process by which the dimensional variables describing a physical problem can be combined to form a set of *dimensionless parameters*.

In Chapter 10 we introduced the Reynolds number, a dimensionless parameter used to design pressure-differential flowmeters. The *Reynolds number* is defined as

$$\text{Re} = \frac{\rho V L}{\mu}$$

where ρ is a fluid density, V is a fluid velocity, L is a length scale, and μ is a fluid viscosity. While all the component variables in the Reynolds number have dimensions and have values that depend on the unit system, the Reynolds number itself is dimensionless and is independent of the unit system. Another well-known dimensionless parameter is the *Mach number*, which is used in aerodynamics and other compressible-fluid flows. The Mach number is the ratio of the local fluid velocity to the local speed of sound.

For any particular problem, a complete set of dimensionless parameters can be determined by standard techniques, even if the actual equations governing the problem are not known. The only requirement for forming a complete set of dimensionless parameters is that the set of relevant dimensional parameters be complete. E. Buckingham, in the early part of the twentieth century, promoted dimensional analysis and presented a theorem that demonstrates key features of this technique. This is the *Buckingham* Π *theorem*, where the symbol Π (the uppercase Greek letter pi) is the symbol used to represent a dimensionless parameter. The theorem is presented in most introductory fluid-mechanics texts. The following presentation is based on the presentation in Streeter and Wylie (1985).

If there are m dimensional variables describing a physical phenomenon—A_1, $A_2, \ldots, A_m$ (e.g., density, velocity, and length), then there exists a functional relationship between these variables:

$$f(A_1, A_2, \ldots, A_m) = 0$$

This simply means that if our list of variables is complete and relevant, there exists a solution in nature to the problem. If there are n dimensions contained in this set of dimensional variables (e.g., length, mass, and time), then there exists a set of $(m - n)$ dimensionless variables $\Pi_1, \Pi_2, \ldots, \Pi_{m-n}$ describing the same physical problem. These dimensionless parameters are related by another functional relationship:

$$F(\Pi_1, \Pi_2, \ldots, \Pi_{m-n}) = 0$$

Neither of the functions f and F are necessarily mathematical functions and may simply represent the relationships that can only be determined from experiments. It should be noted that in some cases, n is less than the number of dimensions. (See Shames, 1992.)[†]

[†]This occurs when some of the dimensions in a problem always occur in a fixed relationship to each other. For example, consider the deflection of a spring, $F = k\delta$, where F is the force, δ is the deflection, and k is the spring constant. In a mass–length–time set of dimensions (M-L-T), F has dimensions of M-L/T^2, k has dimensions of M/T^2 and δ has dimensions of L. There are three dimensions. However, M occurs only in the form M/T^2; hence, there are really only two dimensions in the problem, L and M/T^2. As a result, m is 3 and n is 2, leading to a single dimensionless parameter, $\Pi = F/k\delta$.

There are two important consequences of this theorem. First, a physical problem can be functionally described using a suitable set of dimensionless parameters. Second, the set of dimensionless variables has fewer members than the set of dimensional variables.

There are several techniques that can be used to determine a set of dimensionless parameters from the dimensional variables. One approach, called the *method of repeating variables*, is presented in fluid mechanics texts such as that of Streeter and Wylie (1985). The resulting set of dimensionless parameters describing a physical problem is not unique; there are usually alternative sets, but some are preferable for practical and historical reasons. As an example, consider the problem of the pressure drop for the laminar flow of a fluid through a horizontal tube. Using dimensional variables, this problem can be expressed as

$$f(\Delta P, V, D, L, \rho, \mu) = 0$$

where ΔP is the pressure drop, V is the fluid velocity, D is the pipe diameter, L is the pipe length, and ρ and μ are the fluid density and viscosity, respectively. There are six dimensional variables, and since in this case there are three dimensions, length, mass, and time, the Π theorem tells us that we can alternatively describe the problem with three dimensionless parameters. This would take the form

$$F(\Pi_1, \Pi_2, \Pi_3) = 0$$

The most common set of dimensionless parameters used in this problem are the Reynolds number (Re), the friction factor (f) and the length-to-diameter ratio:

$$\Pi_1 = \text{Re} = \frac{\rho V D}{\mu}$$

$$\Pi_2 = f = \frac{\Delta P \left(\dfrac{D}{L}\right)}{\rho \left(\dfrac{V^2}{2}\right)}$$

$$\Pi_3 = \frac{L}{D}$$

To see how these parameters can be used, consider a test in which the length-to-diameter ratio is held constant and we seek the variation of the friction factor versus the Reynolds number. We can vary the Reynolds number by changing either the velocity, the diameter, the density, or the viscosity—and it does not matter which variable is varied. We can effectively evaluate the effect of varying viscosity simply by varying the fluid velocity during the test. This versatility greatly simplifies the design of the experiment. Data from the experiment can be applied to most common fluids—for example, water, air, or mercury—and pipes of any diameter.

One key consequence of nondimensionalizing a problem is that the number of variables is reduced. In the laminar flow example, the number of variables is reduced from six to three. This greatly simplifies not only the testing (fewer tests need be performed), but also the presentation of the results. In the example, a complete set of data can be presented on a single graph. We can use the Reynolds number as the horizontal axis and the friction factor as the vertical axis. The length-to-diameter ratio can be a parameter labeling separate curves.[†] In dimensional form, with six variables, a large number of graphs would be required. Since the use of dimensionless parameters compacts the presentation of results, they are also often used for graphical presentations of the results of nonexperimental analysis. Figure 11.8 shows the results for a second-order dynamic system in dimensionless form.

Dimensionless parameters can also be used to determine an appropriate scale for a test by a technique known as *similitude*. We usually talk about similitude in terms of the prototype (the real device) and the model (the device to be tested). The concept of similitude is usually stated in two parts: geometric similitude and dynamic similitude. The prototype and the model are said to be geometrically similar if all the corresponding dimensions have the same ratio. For example, if the length of the model is 1/10 that of the prototype, the width of the model must be 1/10 the width of the prototype.

The prototype and the model will be dynamically similar if the corresponding dimensionless parameters have the same values for the prototype and the model. In our laminar-pipe-flow example above, we will have dynamic similitude if

$$\mathrm{Re}_p = \frac{\rho_p V_p D_p}{\mu_p} = \frac{\rho_m V_m D_m}{\mu_m} = \mathrm{Re}_m$$

$$\frac{L_p}{D_p} = \frac{L_m}{D_m}$$

In this pipe friction case, if we make two of the three applicable dimensionless parameters the same, the third, the friction factor, will automatically be the same since it is functionally related to the Reynolds number and the length-to-diameter ratio. Consider a 1/10-scale model laminar-pipe-flow test. Both the model diameter and length will be 1/10 the corresponding sizes of the prototype. Using the same fluid (same density and viscosity), similarity of the Reynolds number requires that the model velocity be 10 times the prototype velocity.

In the laminar-pipe-flow example we were able to specify an exact scale-model test. Exact scale-model tests are possible in many problems involving fluid mechanics, aerodynamics, and dynamics. Unfortunately, it is not always possible to design scale-model tests using available fluids and materials. A common example of this is the scale-model testing of the fluid drag on ships. In this case the dimensionless parameters are the drag coefficient, the Reynolds number, and another parameter called the *Froude number*. It turns out to be impossible to scale both the Reynolds number and the Froude number at the same time using common fluids. In this case, testing is performed by scaling the test with the Froude number and using analysis to account for the Reynolds-number effect. There are many situations in which scale-model tests involve significant compromises, and scale-model tests are sometimes impractical. Nevertheless, scale-model testing has significant economic advantages, and the experimenter should give it consideration.

[†]The standard solution for laminar pipe flow does not include the L/D effect because it is for infinite L/D.

12.2.2 Uncertainty Analysis

Uncertainty analysis will be a significant component of virtually any engineering experiment. It is usually performed in two separated stages: in the design phase and later in the data analysis phase. The actual techniques are discussed in detail in Chapter 7. In the design phase, uncertainty analysis is an integral part of specifying the instrumentation. The instrumentation must have sufficient accuracy so that when the data are finally analyzed, the results will be useful for the intended purpose. The analysis should include instrument systematic and random errors and estimates of various loading and installation errors. Unfortunately, some sources of random error will be unknown in the design phase and can be obtained only from the tests themselves (or shakedown or auxiliary tests). At the time that this preliminary uncertainty analysis is performed, it will be necessary to have at least a preliminary version of the analytical model (procedure) that will be used to analyze the test data.

When the testing has been completed, it will be necessary to perform a final uncertainty analysis. At this time, there should be additional information on random errors based on the test data. Some parts of the test matrix may be replicated (i.e., performed several times, but not at the same time), giving additional information on random errors. Instrument calibrations may reduce systematic error limits, and there may be information to reduce previous estimates of various loading and installation errors. In some cases, test experience may cause an increase in uncertainty intervals if a previously unknown phenomenon is identified.

12.2.3 Shakedown Tests

One major characteristic of experimental programs is that the unexpected occurs. A satisfactory uncertainty analysis and high-quality instruments will not guarantee that the results will be acceptable. Unforeseen fluid and thermal leaks, errors in wiring and plumbing, and other unanticipated problems will usually appear. Successful experimental programs require attention to details and persistence by the experimenters. To assure the quality of an experiment, it is common to perform special tests and measurements with the purpose of ensuring that the experiment satisfies the basic requirements. These measurements, usually called *shakedown tests*, depend on the experiment; however, there are several procedures that are widely used. It is often necessary to install supplementary instrumentation for these tests.

Testing Instrumentation in a Static Mode   When the test rig is not operating, it is often possible to know from independent means what the readings of certain instruments should be. For example, temperature-measuring devices should measure a known temperature (such as ambient), and it is often possible to determine the expected readings of other instruments, such as pressure gages and force transducers.

Testing Special Cases   In many cases, tests can be performed and the results compared to analysis or preexisting data sources. For example, steady-state readings may be made on a rig intended for transient measurements. In other cases, selected subsets of the independent variables may result in situations with known outcomes. For example, in an experiment designed to determine certain properties of a new material, another material with established properties might be tested.

Overall Balance Checks It is often possible to use a general physical principle to perform an overall check on the validity of an experiment. In thermal experiments, energy will be conserved. With adequate instrumentation, it is often possible to measure all energy flows. Deviations from true conservation might indicate a significant error in the measurements. Other overall balance check possibilities include force balances, conservation of mass and conservation of momentum, conservation of atomic species, and Kirchhoff's law. Overall balance checks can be performed not only in the shakedown phase but also in the data-taking phase as continuing monitors of correct performance.

12.2.4 Test Matrix and Test Sequence

Test Matrix An experiment seeks to determine the relationship between one or more dependent variables, called *responses*, and a set of independent variables, often called *factors*. The experimenter normally specifies and sets values for the factors, and the experiment then determines the response. If there is only a single independent variable, the experiment is a *one-factor experiment*; two independent variables make a *two-factor experiment*; and so on. The set of combinations of factors for a given experiment is known as the *test matrix*, and the test matrix is usually determined at least partially before the start of testing. For single-factor experiments, determining the test matrix is usually fairly straightforward. It is necessary to specify the values of the factor to be tested (both range and spacing), the order in which the tests will be performed, and the number of replications of each test. The expected form of the test results and the method used to analyze the data can affect these decisions.

Replicating a test condition differs from repeating the test condition. In simple repetition, the test condition may be set up, the data taken, and then the process repeated immediately. In replication, the test condition will be repeated at a different, later time possibly using different instruments. Replication is preferable to repetition in providing a basis to estimate the random error of the measurements. It should be noted that if there is large random error in a single-factor experiment, it is not really a single-factor experiment; there certainly exist one or more uncontrolled extraneous factors that affect the response significantly.

Each factor will be varied over some range from a minimum value to a maximum value. The spacing of values of the factor within this range deserves some attention. For a narrow range of the factor (say, values from 2 to 3), it might be reasonable to space the values of the factor uniformly within the range. For wide ranges of the factor (say, 1 to 1000), the response data will probably be presented versus the logarithm of the factor. In this case it is probably best to space the values of the factors in uniform increments of the logarithm of the factor.

For two or more factors, determining the test matrix is more complicated and there are more alternatives. Here, we discuss two- and three-factor experiments; however, the methods are essentially the same for larger numbers of factors. Consider first a two-factor experiment where the response R is a function of the factors X and Y. First we need to determine suitable ranges for X and Y; then we can set up a test matrix as shown in Table 12.1 (for five values of each factor). With this matrix, we will perform 25 test runs. It will be necessary to replicate one or more test conditions to demonstrate

TABLE 12.1   Conventional Test Matrix

X Values	Y Values				
	Y_1	Y_2	Y_3	Y_4	Y_5
X_1	x	x	x	x	x
X_2	x	x^*	x	x^*	x
X_3	x	x	x	x	x
X_4	x	x^*	x	x^*	x
X_5	x	x	x	x	x

repeatability of the test. If the random error is high, it may be necessary to replicate each test condition one or more times to obtain usable results.

The results of this experiment can then be plotted as shown in Figure 12.1, and curves can be fitted to the data or a correlating function can be determined. When one examines Figure 12.1, it is apparent that the shape of the constant Y curves are all similar. Consequently, some of the data may be unnecessary, and it may not be necessary to make measurements for all positions in the test matrix shown in Table 12.1. We might confidently eliminate some of the test conditions, shown by the circled points in Figure 12.1 and values with asterisks in Table 12.1.

The type of test matrix shown in Table 12.1 is likely, with high confidence, to uncover the complete response over the selected range of the factors in most engineering experiments. For example, consider the measurement of the quasi-steady pressure rise of a hydraulic centrifugal pump as a function of flow rate and shaft speed. The measurements in this experiment have low random error (which minimizes replication), and the measurements at each test condition can be completed rapidly. However, with the test matrix shown in Table 12.1, there are rather a large number of test conditions (even if a few conditions are eliminated, as we have noted), and this number will increase if many replications are required and will increase rapidly if the number of factors is increased. In our example, if a third factor with five values is included, the test matrix will increase from 25 test conditions to 125. This approach may not be practical for many tests. For example, consider the test of a solid rocket booster to determine the effects of fuel type, chamber pressure, and nozzle geometry. There will be a high cost

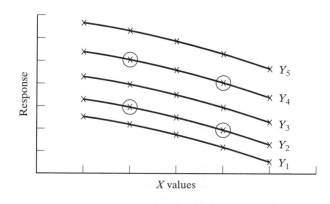

FIGURE 12.1

Results of conventional test matrix.

per test, and since there may be a high random error, replications may be required. In such cases, other approaches should be considered.

One possibility is the one-factor-at-a-time test matrix. The response of some processes can be characterized by one of the following functional forms:

$$R = F(X)G(Y)$$
$$R = F(X) + G(Y)$$

In each case, the function governing the response to factor X is completely independent of the function determining the response to factor Y. That is, the factors do not interact. If the experimenter has reason to believe that the process follows one of these functional forms, a simpler test matrix can be specified as shown in Table 12.2. The data from the X_3 row can be used to determine $G(Y)$, and the data from the Y_3 column can be used to determine $F(X)$. With this test matrix, there are only 9 test conditions rather than 25, and the number of test conditions would increase to only 13 if a third factor were included. The main disadvantage of the one-factor-at-a-time approach is that it will not permit the evaluation of interactions between factors or even determine the existence of interactions.

Modern statistical techniques offer other options for the analysis of test data, and the determination of the test matrix. There are two situations where statistical techniques are clearly the best choice: situations with a large number of factors and situations in which the precision of the measurements is poor (high random error). Manufacturing-process and product-lifetime tests often have high random error and are best analyzed with statistical tests. Even if a conventional test matrix such as Table 12.1 is used, statistical methods can reduce the uncertainty. However, statistical methods often make it possible to obtain useful results with a smaller number of test conditions. For example, consider the matrix shown in Table 12.3 for three factors and five values of each factor. This type of matrix, sometimes called a *Latin square*, has the same number of values of each factor but has only 25 test conditions. Even with replication, the total number of test conditions will be less than if we extended the matrix of Table 12.1 to three factors, which would result in 125 conditions. It would be difficult to reduce the data resulting from the Table 12.3 to create a graph such as Figure 12.1 using nonstatistical techniques; however, it is possible to use available statistical methods. One statistical method that can be used is multiple regression, introduced in Chapter 6. This method can produce a functional presentation of the results if the factors are continuous variables. Another method is *analysis of variance* (ANOVA),

TABLE 12.2 One-Factor-at-a-Time Test Matrix

X Values	Y Values				
	Y_1	Y_2	Y_3	Y_4	Y_5
X_1			x		
X_2			x		
X_3	x	x	x	x	x
X_4			x		
X_5			x		

TABLE 12.3   Latin Square Test Matrix

X Values	Y Values				
	Y_1	Y_2	Y_3	Y_4	Y_5
X_1	Z_1	Z_2	Z_3	Z_4	Z_5
X_2	Z_2	Z_3	Z_4	Z_5	Z_1
X_3	Z_3	Z_4	Z_5	Z_1	Z_2
X_4	Z_4	Z_5	Z_1	Z_2	Z_3
X_5	Z_5	Z_1	Z_2	Z_3	Z_4

which can be used to analyze not only situations in which the factors are continuous variables, but also cases where the factors are not continuous. Evaluating the performance of a set of multistep manufacturing processes is an example of an experiment in which the factors are not continuous. ANOVA and other statistical methods are available as commercial software packages. Advanced statistical methods are beyond the scope of this book but are discussed widely elsewhere (e.g., Hicks, 1982; Montgomery, 1991; Ghosh, 1990; and Mason, 1989). In general, statistical techniques can either reduce the size of the test matrix, improve the precision of the results, or both.

Test Sequence   Considerable attention must also be placed on the sequence in which the various test conditions are applied. It is best if the various runs are completed in a random order. There are aspects of the measurement process that can change over the course of completing a test matrix, and if the data are taken sequentially, these effects may appear as artificial trends in the response. This can occur even if we control the experiment carefully. Changes that can occur in the measurement process include thermal drift of instruments, deterioration in instruments and the tested device, and inadvertent changes in the technique of the experimenter. If, on the other hand, test conditions are applied in a random order, there will be scatter in the data but no artificial trends.

There are some experiments where a random ordering of the runs is not possible or is not desirable. In transient (time-varying) experiments, the data must be taken in a time-sequenced manner. In tests to destruction such as materials-strength tests, the data must be taken sequentially since the test sample changes during the test and the sequence of these changes is the desired response. In Chapter 2, a procedure for calibrating an instrument was presented in which the runs were taken sequentially. Sequential testing was necessary in this case to separate the effects of hysteresis. In that particular case, the disadvantages of sequential testing were partially mitigated by replicating the experiment several times. There are also situations in which random ordering of the runs would be prohibitively expensive. Consider a test of a nuclear reactor in which the pressure-vessel pressure and the power level are factors. Since it takes several hours to change vessel pressure (a process that also shortens the pressure-vessel life), it would be necessary to complete all power-level variations before changing the vessel pressure.

If completely random testing is practical, the sequence of runs can be determined by assigning a random number to each test condition and then sequencing the runs according to the values of the random numbers. Random-number generators are available in common spreadsheet programs. A question occurs as to sequencing runs if the complete test matrix is to be replicated. It is recommended that all test conditions from all replications be randomized. This means that two or more runs at one test condition might be completed before the first run of another test condition.

12.2.5   Scheduling and Cost Estimation

Scheduling and cost estimation are vital parts of any experimental program. For major projects in large corporations, these activities are performed primarily by planning specialists using sophisticated computer programs. Two of these methods, known by the acronyms CPM (critical-path method) and PERT (program evaluation and review technique), are discussed by Wiest (1979). For smaller projects, engineers will perform the scheduling and cost-estimation activities themselves. A readily available scheduling program is Microsoft® Project. While the general principles of scheduling and cost estimation are the same for large and small projects, the material presented here is oriented toward small projects. Useful references on scheduling include those of Ruskin (1982) and Gibbings (1986).

Scheduling   The first step in determining a project schedule is to define a set of tasks. A task is a well-defined activity of small enough effort that the engineer can visualize the work involved and make a reasonable estimate of the amount of effort required. Past experience is extremely important in determining tasks and the effort involved. Tasks should be quite specific and should not include activities of different types. For example, a task titled "Design of Experiment" is far too general, whereas "Perform Uncertainty Analysis" might be quite manageable. A task should have well-defined objectives, inputs, and outputs. For each task, several attributes must be determined—the qualifications of the person(s) performing the work, the effort necessary to perform the task, the duration (calendar time) required to complete the task, and the destination tasks [subsequent task(s) which use the outputs].

People with a variety of professional skills work in engineering organizations. There are engineers, engineering assistants, draftspersons, technicians, computer programmers, coordinators, technical writers, and persons with other skills. For each task it is necessary to determine which skills are required. In some cases, particularly in small organizations, it may be necessary to select particular individuals. In large organizations, the actual person will frequently be selected only when the required work is initiated.

Effort is the amount of work that must be done to complete the task. It is usually measured in units of labor-hours. A labor-hour is the work performed by one person working for 1 hour. Normally, a person performs 40 hours of work per week or 1900 to 2000 hours per year. For example, five people working full time on a task for 4 weeks will provide 800 labor-hours of effort. The required number of labor-hours must be determined for each skill type required for the task.

The required duration (calendar time) between the start of the task and the end of the task must be determined. If one engineer can complete the task in 40 hours, the minimum duration is 1 week. However, if that same engineer can only work half-time on the task, the duration for the task is 2 weeks. The duration for a task can often be reduced by assigning more individuals to the task. However, the more people that are involved in a task, the less efficient their performance will be. Twice as many people cannot usually complete a task in one-half the duration for a single person.

Finally, the destination task(s) must be determined for each task. The output of one task will supply input to one or more other tasks. Specifying of the destination tasks is necessary to determine the sequence in which tasks are performed. As an example, a set of tasks has been prepared for a project to test a small steam-to-air heat exchanger for a client. A proposal has been completed, and the tasks represent the work required to finish the project. The list of tasks is presented in Table 12.4.

The next step is to prepare a chart called an *activities network*, as shown in Figure 12.2. In this chart each of the tasks is represented by an arrow, and the arrows are connected so that each task points to a junction connecting to the next task in sequence. The network will frequently branch since a task may supply inputs to more than one subsequent task. In the example, task B provides input directly to tasks C and D and E. This branching can become quite extensive in large projects. In Figure 12.2, junctions c and e (and also junctions m and l) are connected by an arrow with a dotted line. The arrows with dotted lines represent dummy activities. A dummy activity has zero duration and zero effort and is used as a convenience to clarify the activities network. Without dummy activities, arrows for different activities might overlap. Task P, project management, is a continuing task and is not shown on the activities network.

Underneath each task arrow in Figure 12.2 is a number that indicates the duration (in weeks) for each task. It is now possible to determine the duration of the entire project. It is simply necessary to add up the duration values for each of the sequential tasks. For parallel branches, the duration is the duration of the longest path. For example, the duration from junctions b to f is that of the b–d–e–f path, 3.0 weeks. The path that controls the total project time is called the *critical path*. The critical path has a duration of 19 weeks in this example and is indicated with heavy lines in Figure 12.2. Changes in duration of any tasks on the critical path will change the duration of the complete project. These tasks are those that must be monitored most closely by the project manager.

It is common to present the tasks on another chart—a bar chart called a *Gantt chart*. For our example, this is presented in Figure 12.3. The tasks are listed in rows, and calendar time is listed in the columns. A bar is drawn between the starting time and the

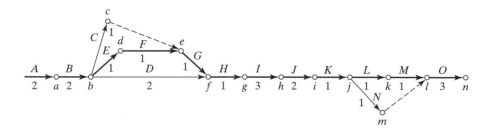

FIGURE 12.2

Activities network for heat-exchanger project.

TABLE 12.4   Tasks for Heat Exchanger Project

Task ID	Title	Objective	Effort (labor-hours)	Duration (weeks)	Destination Tasks
A	Determination of experimental approach	Evaluate possible configurations of rig and select best	Sr. Engr. 40 Jr. Engr. 60 Engr. Assist. 30	2	B
B	Specification of instrumentation	Identify in-house or commercially available instrumentation and determine actual characteristics	Sr. Engr. 10 Jr. Engr. 30	2	C, D, E
C	Instrument uncertainty analysis	Estimate uncertainty intervals for each of the measurements for the given application	Sr. Engr. 10 Jr. Engr. 20	1	G
D	Design drafting of test rig	Prepare drawings of test rig used for fabrication	Sr. Engr. 10	2	H
E	Determination of test matrix	Specify values of independent variables to be used in test program	Sr. Engr. 20	1	F, G
F	Preparation of data-reduction program	Prepare computer program to reduce test data	Sr. Engr. 10 Jr. Engr. 40	1	G, H
G	Preliminary results uncertainty analysis	Extend uncertainty analysis to uncertainty intervals for final results	Sr. Engr. 20	1	H
H	Design review	Present experiment design to senior engineering staff and management; closure of open items	Sr. Engr. 40 Jr. Engr. 30	1	I
I	Procurement of materials and instrumentation	Allow time for materials to be ordered and delivered to test site	None	3	J
J	Construction of test rig	Assemble test rig and install instrumentation	Jr. Engr. 40 Sr. Tech. 80 Jr. Tech. 80	2	K
K	Shakedown tests	Demonstrate validity of rig, fix problems	Sr. Engr. 10 Jr. Engr. 40 Sr. Tech. 40 Jr. Tech. 40	1	L, N
L	Final tests	Collect all data corresponding to test matrix	Jr. Engr. 40 Sr. Tech. 40 Jr. Tech. 40	1	M
M	Data reduction and final uncertainty analysis	Compute results from data; prepare plots and tables	Sr. Engr. 10 Jr. Engr. 40	1	N
N	Preliminary preparation of report	Start writing final report	Sr. Engr. 10 Jr. Engr. 30 Engr. Asst. 10	1	O
O	Final preparation of report	Write final report; interpret results	Sr. Engr. 30 Jr. Engr. 50 Engr. Asst. 30	3	End
P	Program management	Coordinate project	Sr. Engr. 60		

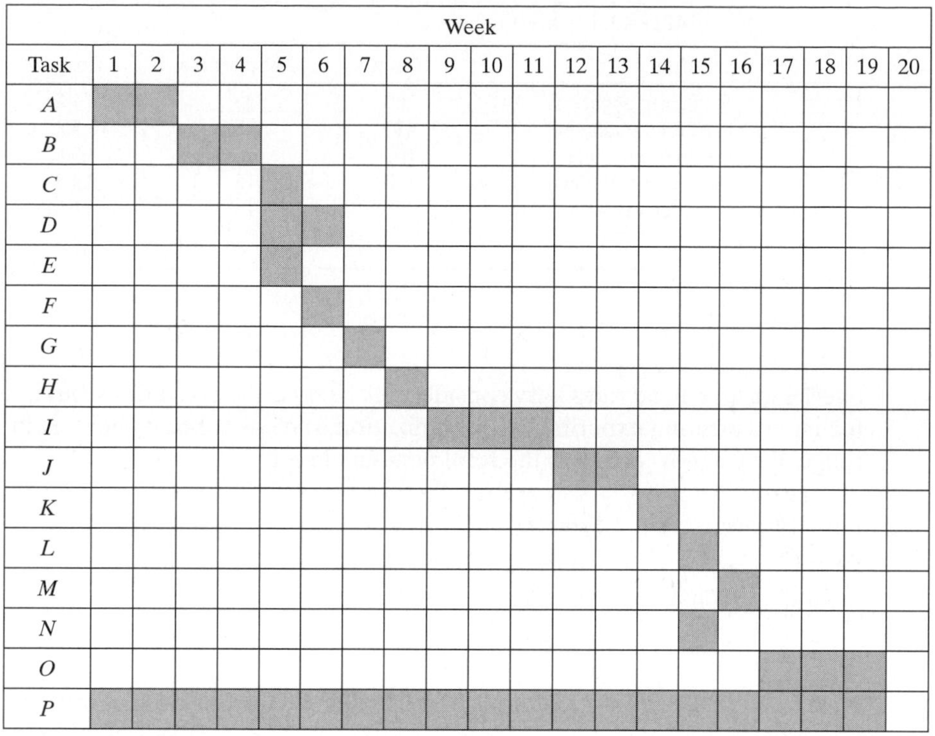
Week
Task
1 2 3 4 5 6 7 8 9 10 11 12 13 14 15 16 17 18 19 20
A
B
C
D
E
F
G
H
I
J
K
L
M
N
O
P

FIGURE 12.3

Gantt chart for heat-exchanger project.

finishing time for each task. This chart provides the project manager with a concise picture of the progression of the project.

Cost Estimation The task descriptions provide the basis to determine salary costs. For estimation purposes, it is often assumed that all people with a given title have the same salary, which is the average of all such persons with that title in the organization. Thus we can add up the hours of each job title, multiply by the applicable hourly rate, and determine the direct salary costs. In our example the total hours and costs for each category are shown in Table 12.5.

The direct salary cost is only part of the total costs. To this must be added the cost of employee benefits and organization overhead. Benefits include vacations and holidays, health benefits, and social security and other payroll taxes. These normally are treated as a percent of direct salary. A typical number might be 30% of direct salaries. Overhead costs are other costs of business that are not charged directly to the contract. These costs include costs of buildings, furnishings, and equipment and the salaries of higher managers and support personnel, such as payroll, purchasing, and maintenance staff. Proposals are often prepared at no direct cost to the client and are charged to

TABLE 12.5   Direct Labor Costs

Title	Hours	Cost per hour	Cost
Senior engineer	260	$40	$10,400
Junior engineer	420	25	10,500
Senior technician	160	25	4,000
Junior technician	160	15	2,400
Engineering assistant	70	15	1,050
			$28,350

overhead. Overhead rates vary considerably between organizations, but a rate of 100% for an engineering experimental organization, such as we have here, is in the normal range. We can now arrive at the total personnel cost:

Direct costs	$28,350
Benefits (30%)	8,505
Overhead (100%)	28,350
	$65,205

Next, the cost of equipment purchased specifically for this project must be added. For this project, the equipment charges are $100,205. The total cost of labor and materials is then $100,205. To this total cost is added a fee (profit) for the organization: 10% is a common fee. The final total cost that will be proposed to the client is thus $110,225.

12.2.6 Design Review

A simple yet effective tool that can be used in any design project as a quality-assurance technique is the design review. In a design review, the design team makes a presentation to a group of engineers (called the committee) who are competent in the subject area but are not working directly on the project. The committee is usually made up of senior engineers and technically competent engineering managers. One of these committee members is appointed chairman. A committee is formed to review specific material, and the review may not cover all aspects of a project.

In the first step, the design team prepares written material and submits it to each member of the review team. At a subsequent time, a meeting between the committee and the design team is scheduled. In this meeting, which may last from a few hours to several days, the design team makes a formal oral presentation of the material to be reviewed. During this meeting the committee members can ask questions and make informal comments on the subject material. After the meeting, the committee meets as a group and discusses the presentation. They will establish two lists, recommendations and open items. The list of *recommendations* is simply a list of comments about the reviewed material that the design team should consider but which the design team is free to ignore. The list of open items is more important. Open items are issues that if not resolved might jeopardize the success of the test program. The committee chairman documents the recommendations and open items in a letter to the design team.

The *open items* are issues to which the design team is compelled to respond and satisfy the concerns of the committee. The design team must then show why the open item is not a concern or alter the design to mitigate the concern. These actions will be documented in a letter to the committee chairman. If the chairman considers the closure of the open items to be satisfactory, after consulting with the other committee members, the chairman will write a letter to the design team leader stating that the design review is closed. Occasionally, an open item is of such major significance that the review process cannot be closed in a reasonable period of time. In this case the review will be terminated without closure. A new review will be scheduled at a later date to give the design team sufficient time to correct the problem.

The expectation of a review motivates the design team to perform quality work and identify problems that can be rectified before the review. During the review, competent engineers on the committee will frequently identify significant problems in the project which design team members have overlooked. The result is a test program with substantially reduced risk of major delays or failure.

12.2.7   Documenting Experimental Activities

Documentation is an important aspect of any engineering experimental program. Before the activity starts, there will be some kind of proposal or planning document in which the objectives are stated, the scope of the project is identified, an experimental approach is suggested, and preliminary estimates are made of the costs and schedule. During the course of the experimental program, there will be a series of progress reports. When the experiment and the analysis of the data have been completed, there will be a final report.

The actual level of documentation will, of course, depend on the complexity and duration of the experimental program. In some routine testing programs, the documentation may involve little more than filling out some forms. In most research and development programs, the three documentation components—proposal, progress reporting, and final reporting—will each be significant efforts with the greatest error in the proposal and final reporting.

There are really no universally correct ways to perform any of the documentation tasks. Many organizations have standardized formats for reports, and members of the organization are required to use those formats. In this chapter, we describe some common formats for each stage of reporting that can be used if there are no other specific requirements. Other reporting approaches usually differ in details rather than general concepts presented here.

Formal Report of an Experimental Program   The document completed at the end of a major engineering activity (or at the completion of major milestones) is generally a formal report. The formal report is usually very complete and has a rather well-defined (formal) structure. Formal reports are used for many different engineering activities, many of which have no experimental activities, and hence there are many variations of the format. In this section we discuss a common format for writing formal reports for experimental programs.

In writing a technical report, it is wise to keep in mind the purposes that the report is to serve. The most common purposes are

1. to communicate to others the results and conclusions drawn from the laboratory or field work.
2. to create a permanent record of laboratory or field work that may be used at a later time in connection with other work.

The formal report consists of the following sections:

1. Title page
2. Abstract
3. Table of contents
4. Summary
5. Introduction
6. Apparatus and test procedure
7. Results
8. Discussion
9. Conclusions and recommendations
10. References
11. Appendices

Title Page.   The title page should contain the test title, the date of the report, and the names of the author and coauthors. Other information that may appear on the title page includes the name of the organization that performed the project, the sponsoring organization, and the report and contract number if such numbers exist. If there is room on the title page for the complete abstract, it may also be placed there.

Abstract.   The abstract is a very brief, stand-alone summary of the material presented in the report. The abstract contains only enough information about the work being reported to enable someone who is searching for information to determine quickly the applicability of the report to his or her problem. Try to think what information you would want if you were performing a library search. The abstract should normally not be longer than 100 to 200 words. While the abstract should refer only to material described in the report, it should never directly reference material in the report. It should never say something such as "is shown in Figure 15." There should be no graphs, tables, or pictures in the abstract, since these cannot readily be stored in most computer data-retrieval systems.

The abstract should start with a statement of what was done or what the objective of the experiment was. In one sentence or so, it should then describe how the results were obtained. That is, there should be some description of the experimental approach and apparatus and any unusual instrumentation. Then, some key results should be stated. For example, "A performance test was performed on a Fulmer model C-15A rotary air compressor. Instrumentation was provided to measure shaft speed, airflow rate, and air inlet and outlet pressures and temperatures. The required shaft power was measured with an electric-motoring dynamometer. The highest efficiency was observed to be 65% at a shaft speed of 600 rpm. At this condition, the airflow was 10 scfm and the shaft power was 2.7 hp."

Table of Contents. This is simply a list of the sections of the report and the page numbers where they can be located. A minor point can be noted here: In books, the title that appears at the top of the table of contents is most commonly "Contents," not "Table of Contents." Reports sometimes also have separate pages listing all the figures and all the tables that appear in the report.

Summary (or Executive Summary). Like an abstract, the summary (often called the executive summary) presents an overview of the complete report. The summary is normally longer and more complete than an abstract and may, if necessary, include graphs and tables. While an abstract is used by the reader to determine whether or not to read the report, the summary will normally be read by persons who are interested in the work described. The summary states the objectives, describes briefly the techniques used, and states the most important results and conclusions. It is not a good practice to refer to figures and tables that appear later in the main portions of the report since it is awkward for the reader to search through the report for information. If figures and tables are used, it is best to duplicate them in the summary section of the report. Another reason for duplicating figures and tables is that the summary is sometimes copied as a separate document and distributed without the remainder of the report.

The summary lays down the framework of the remainder of the report and makes it easier to follow. If, however, the reader is not using the results directly, the summary may be the only portion of the report read. Such people might include higher management and persons in an organization not directly connected with the subject. The length of the summary depends on the size of the experimental program. For a short program such as an undergraduate student report, the summary might be as short as one or two pages. For a large program involving thousands of labor-hours, the summary could be tens of pages long (and the report itself could be presented in several volumes).

Introduction. The purpose of the introduction is to prepare the reader for the body of the report by giving clear statements of the background and objectives of the project. The following items are often components of a good introduction:

1. The objectives should be clearly stated, and information should be provided to convince the reader that the work is valuable.
2. A review of applicable literature may be included. If the review is extensive, it may be included as a separate section of the report following the introduction.
3. The limits (scope) of the work should be outlined here (what will be tested and what will not be tested).
4. If the test was performed to verify an existing theory, or if an unusual method was used to analyze the data, the theory is often discussed in the introduction. If

a theory was developed on the basis of the test results, it is better presented in the discussion section of the report. It is acceptable to present existing theory as a separate section of the report, following the introduction.

5. The outline of the presentation in the remainder of the report might be discussed, particularly if it is different from normal practice.

Apparatus and procedure.   The apparatus should be described both in words and in sketches or pictures. Block diagrams should be provided that show all (or primary) sensor locations. Photographs of the apparatus are normally included. Detailed engineering drawings may be referenced here but should be included in an appendix. A list of the instrumentation should be included directly or referenced here and placed in an appendix. The list should include the manufacturer, model number, and serial number of each instrument used. If the test procedure is simple, it can be included here. If the test procedure is lengthy, it should be referenced here and included in an appendix.

Results.   The primary results from the test should be included here in graphical or tabular form. A short text should be included to state what results are included and on which graph or table they are included. Each table and each figure should be assigned a unique number. The number normally appears adjacent to the title, as in the following example:

Figure 10   Beam stress vs. time
Table 3   Stresses at strain gages 1–20

If the numerical results are not too extensive, they should be presented in tabular form. Somewhat more extensive tabular results might be referenced and placed into an appendix. Only include results that are of direct interest to the reader, and do not include intermediate calculations. With computerized data-acquisition systems, huge quantities of data can be obtained, and it may not be practical to present them in tabular form at all. In this case, only summary types of numerical results (e.g., averages, minimums, and maximums) can be tabulated.

The format of the graphical results is important. Figure 12.4 represents an acceptable graph prepared using a spreadsheet computer program. The following notes apply to graphs:

1. With the exception of a page number, nothing should be written within 1/2 inch of the top, bottom, and one side of the page or within 1 inch of the side of the page that will be bound (the left-hand side when the pages are printed on one side).

2. The graph should be placed in the report so that it can be read in the same direction as a written page or can be read by rotating the report 90° clockwise from that orientation.

3. Discrete data points should be plotted as points surrounded by a plot symbol such as a circle or a square. If there are a large number of points so that the plot symbols

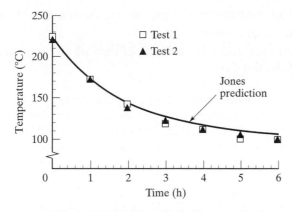

FIGURE 12.4

Temperature at center of casting.

overlap (such as might exist for time-varying data taken with a computerized data-acquisition system), it is better to eliminate the plot symbols and connect the data points with straight lines.

4. Continuous theoretical predictions or correlations should be presented as lines or curves, and it is recommended that they not be plotted as discrete points. If no theory is presented or if the theory agrees poorly with the data, it is common to draw a best-fit curve through the data points. If the data show a trend (such as rainfall versus month of the year) and the y-axis data are not functionally related to the x-axis, theoretical and best-fits curves are not meaningful. In this case it is common to connect the individual data points with a series of straight-line segments. Bar graphs are also appropriate in presenting trend data.

5. There should be a legend identifying different types of data points, as on the example.

6. The graph should have a title (caption). This can be at either the top or bottom of the graph or on the grid portion of the graph. The title location should be the same for all figures in a given report.

7. All curves should be drawn smoothly, either with computerized functions or with french curves.

Discussion.   This is a major part of the report and will vary in scope, length, and complexity according to the nature of the investigation. Basically, the discussion evaluates the results, interprets them, and investigates their significance. It is the bridge that leads the reader from the results to the conclusions. It is possible to integrate the discussion and results sections into a single section. The discussion commonly includes the following items:

1. A statement about each result presented, including its significance and how it relates to the project objectives.

2. A discussion of any results that are unexpected.

3. A discussion of the causes of experimental uncertainties.

4. A comparison of results with theories or preexisting experimental results.
5. Personal opinions to explain results.
6. If applicable, a description and a comparison of data of any new theory developed on the basis of the test data.

Conclusions and Recommendations. There should always be specific conclusions and recommendations that answer the objectives of the experiment or explain why the objectives were not met. No new results should be presented in this section of the report. This section can include a listing of the main results. Recommendations for future work or methods to improve the experiment can be included in this section of the report.

References. A listing of documents mentioned in other portions of the report should be presented in a consistent form that includes the document title, author(s)' names, name of publisher, publication date, and other relevant information to allow the reader to locate the original document.

Appendices. Material that is placed in appendices is normally of interest to some but not necessarily all readers. If it is of interest to most readers, it should go in the main body of the report. Some material is placed in appendices to provide a permanent record. The following may be included in the appendices:

1. Original laboratory data sheets.
2. Sample calculations.
3. Test procedure, if not included in the body of the report.
4. Detailed drawings of the test facility.
5. Information of interest to some readers but not central to the report.
6. Manufacturers' instrument specifications.

In student reports, the laboratory data sheets should be included in the report. In professional practice, these items are usually kept as records separate from the report.

Technical Memorandum The technical memorandum, sometimes called a letter report, is a much less formal method of transmitting information than the formal report. It is also a very common method for transmitting information within an organization. It is frequently not a permanent record, and the original may be destroyed within a few months or years of issuance. Some of the more common uses are

1. Reporting intermediate results of a project (progress reports).
2. Reporting final results in a preliminary form prior to the issuance of a formal report.
3. Reporting the results of studies or evaluations for which a permanent record is not required.

As with other types of documentation, a suitable format may be specified by your organization. The format suggested below is one that is commonly used.

The technical memorandum will usually have a header section like the following:

DATE:

TO:

FROM:

SUBJECT:

REFERENCES:

Normally, the author signs his or her name next to the typed name. The references are related documents referred to in the memo.

The text of the letter will contain the information you are transmitting. The exact contents depend on the situation and may depend on letters or reports transmitted previously. Generally, the letter should contain at least the following:

1. Purpose of the memo.
2. Results to be conveyed.
3. Significance of the results.

In most cases the text will be continuous, broken up only into paragraphs, without subheadings (although subheadings are useful in long letters). Materials that are not included in the text are called *attachments*. Attachments are normally sequenced as Attachment A, Attachment B, and so on, and are so referenced in the body of the letter. The formats for graphs, figures, and tables are similar to those for a formal report, but these items may be of lesser quality in a memorandum. For example, neat freehand drawings or graphs may be acceptable, and tables may be direct printouts from data-acquisition systems. In longer memoranda (more than two pages), it is usually a good idea to start with a short summary, one or two paragraphs long. This summary should clearly state the purpose of the memo and the most important conclusions.

Proposals   In many cases the sponsoring organization requesting a proposal will specify the desired format. If the sponsoring organization is within the same company, the proposal process may be quite informal: sketches, costs, and schedules presented in an oral presentation. Proposals to outside sponsoring organizations are normally more formal and have at least some common characteristics. The following format is only one of many possible ones:

1. Title page
2. Abstract
3. Table of contents
4. Introduction and background
5. Description of the proposed project, method of approach, and work scope
6. Task descriptions, schedule, and costs
7. Personnel and capabilities
8. References
9. Appendices

The objective of the proposal is to convince the sponsoring organization that the proposer is uniquely capable of performing the project with acceptable costs.

The comments given above regarding the title page, abstract, table of contents, references, and appendices for a final report also apply to a proposal. The remaining sections are different and are discussed in the following pages.

Introduction and Background.   In this section the author discusses the background leading to the proposed experimental program. In a proposal this section will frequently be much more extensive than in a final report since the experimenter is seeking to convince the funding agency of his or her competence in the field of the proposal. The general outline of the remainder of the report is presented here.

Description of the Proposed Project, Method of the Approach, and Work Scope. This section is the heart of the proposal. In at least preliminary form, the author will describe the general approach to the experiment, the apparatus and instrumentation, and the test procedure. In addition, it is important to discuss what will be done with the data. How will they be manipulated? How will they be presented? Will some attempt be made to correlate the data or establish a new theory? When the reader has completed this section, he or she should have a good understanding of what is going to be done. In this section of the proposal, the proposer shows that he or she knows the methodology to solve the problem and understands the scope and limitations of the problem and the proposed work.

Task Descriptions, Schedule, and Costs.   This section should include a complete set of tasks, required effort, schedule, and costs.

Personnel and Capabilities.   It is important to the customer to have confidence that the proposing organization and its personnel are competent to perform the proposed work. The customer will normally want to know something about the key personnel who will perform the work in the proposed experimental program. The key personnel will be listed in this section, together with relevant experience. Résumés of the key personnel will usually be included in an appendix.

This section may also include discussions of the capabilities of the organization itself. Relevant facilities and relevant prior projects may be described.

12.3   CLOSURE

The material in this chapter is presented only as a framework for planning and documenting experiments. Organizations and individuals often develop their own methods and styles, which are effective and should be respected.

REFERENCES

[1] BARTEE, E. (1968). *Engineering Experimental Design Fundamentals*, Prentice Hall, Englewood Cliffs, NJ.

[2] DAY, R. (1988). *How to Write and Publish a Scientific Paper*, Oryx Press, Phoenix, AZ.

[3] DOEBELIN, E. (1995). *Engineering Experimentation: Planning, Execution, Reporting*, McGraw-Hill, New York.

[4] GHOSH, S. (1990). *Statistical Design and Analysis of Industrial Experiments*, Marcel Dekker, New York.

[5] GIBBINGS, J. (1986). *The Systematic Experiment*, Cambridge University Press, Cambridge, MA.

[6] HELGESON, D. (1985). *Handbook for Writing Technical Proposals That Win Contracts*, Prentice Hall, Englewood Cliffs, NJ.

[7] HICKS, C. (1982). *Fundamental Concepts in the Design of Experiments*, Holt, Rinehart, and Winston, New York.

[8] KLINE, S. (1965). *Similitude and Approximation Theory*, McGraw-Hill, New York.

[9] LIPSON, C. AND SHETH, N. (1973). *Statistical Design and Analysis of Engineering Experiments*, McGraw-Hill, New York.

[10] MASON, L., GUNST, F. AND HESS, J. (1989). *Statistical Design and Analysis of Experiments*, John Wiley, New York.

[11] MONTGOMERY, C. (1991). *Design and Analysis of Experiments*, John Wiley, New York.

[12] ROBERSON, J. AND CROWE, C. (1965). *Engineering Fluid Mechanics*, Houghton Mifflin, Boston.

[13] RUSKIN, A. AND ESTES, W. (1982). *What Every Engineer Should Know About Project Management*, Marcel Dekker, New York.

[14] SHAMES, I. (1992). *Mechanics of Fluids*, McGraw-Hill, New York.

[15] SCHENCK, H. (1979). *Theories of Engineering Experimentation*, McGraw-Hill, New York.

[16] STREETER, V., AND WYLIE, E. (1985). *Fluid Mechanics*, McGraw-Hill, New York.

[17] WHITE, F.M. (1999). *Fluid Mechanics*, McGraw-Hill, New York.

[18] WIEST, J. AND LEVY, F. (1979). *A Management Guide to PERT/CPM*, Prentice Hall, Englewood Cliffs, NJ.

Answers to Selected Problems

CHAPTER 2

2.2 Systematic error $= -0.0179$ in., max. random error $= 0.0003$ in.
2.10 50 m/s
2.15 with offset, max. error 6 V (high), w/o offset, max. error ± 4 V
2.22 20 cm, 21 cm, 5%
2.25 mVout $= -0.27 + 0.408$ force, accuracy $+0.35 - 0.20$ mV (depends on line fit), repeat ± 0.08 mV
2.32 75.1 C

CHAPTER 3

3.1 120 dB
3.6 $f_c = 10$ kHz, $\psi = -45°$
3.13 $R_1 = 8879\ \Omega$, $R_2 = 1124\ \Omega$
3.15 1.86 octaves
3.20 **(a)** -24 dB, **(b)** -34 dB, **(c)** -14 dB

CHAPTER 4

4.1 10010011
4.3 01111001, 10000111
4.6 15 bits, 16 for 2's complement
4.8 **(a)** 3123 **(b)** 589 **(c)** 4095 **(d)** 0
4.13 $\pm 0.047\%$
4.17 010101000011 (1347 dec.)

CHAPTER 5

5.1 $b_1 = 0.6366$, $b_2 = -0.3183$, $a_0 = a_1 = a_2 = 0$
5.11 $f_{\text{alias}} = 10$ Hz

5.13 40 Hz
5.19 90.3 dB

CHAPTER 6

6.3 $\bar{x}$ = 49.6 cm, med = 49.55 cm, S = 0.5 cm, mode = 49.2 and 49.3 cm
6.5 0.056 (5.6%)
6.7 2.7×10^{-5}
6.12 **(a)** 0.2286 **(b)** 0.0286 **(c)** 0
6.14 **(a)** 0.8145 **(b)** 0.9995
6.23 P(>5) = 0.384
6.25 P(0) = 0.189
6.28 **(a)** 0.0668 **(b)** 0.958 **(c)** 0.042
6.33 **(a)** 38 readings **(b)** 95 readings **(c)** 100 readings **(d)** 8 readings
6.41 replace at 3395 hours
6.47 μ = 50,000 ± 980 miles
6.50 μ = 91.0 ± 0.57
6.51 40 persons (height controls)
6.52 2.20 < σ < 4.73
6.59 no data rejected
6.61 0.9804
6.63 **(a)** r_{xy} = −0.848 **(b)** r_{xy} = −0.99984

CHAPTER 7

7.1 series, w_R = ±0.22 Ω, w_{Rmax} = ±0.3 Ω, parallel, w_R = ±0.05 Ω, w_{Rmax} = ±0.07 Ω
7.2 K = 0.0601 ± 0.003 (w_{Kmax} = 0.004)
7.2E K = 0.601 ± 0.003 (w_{Kmax} = 0.004)
7.5 w_ω = ±1.3 rad/s
7.12 P_i = 1.2°C $P_{\bar{x}}$ = 0.4°C
7.15 w = 0.027 kg
7.22 **(a)** $P_{\bar{x}}$ = 0.11°C, $w_{\bar{x}}$ = 0.2°C **(b)** P_i = 0.43°C, w_i = 0.5°C
7.24 P_x = 2.3 mV
7.39 C_D = 10.67 ± 1.03

CHAPTER 8

8.1 E = 14×10^6 psi, v = 0.25
8.3 3.75 mV
8.12 12.5 mV
8.16 **(a)** 8667 μstrain **(b)** 14.4 mV
8.24 **(a)** 4.43 pF **(b)** −22.1 pF/mm **(c)** −0.443 pF/mm
8.31 100 rpm

8.37 29.4 mV
8.41 100 N-m, 36.5 kW

CHAPTER 9

9.1 7.65 m
9.3 10.38 in Hg, 11.76 ft H_2O
9.6 17.3 kPa
9.8 0.0255 cm/Pa, 1.96 Pa
9.14 476.7°C
9.18 406.5°C
9.20 **(a)** 200 Ω **(b)** 201 Ω **(c)** 0.5% **(d)** 200 Ω → 261.7°C, 201 Ω → 264.5°C
9.24 34.1 Pa/K
9.30 541 K
9.34 20.8°C
9.38 $\phi = 17\%$, $\omega = 4.6$ g/kg dry air

CHAPTER 10

10.1 2.01 in.
10.3 1.25%
10.5 $\dot{m} = 0.038$ kg/s, $Q = 1.9$ SCMM
10.9 $A_2 = 0.01273$ m^2
10.15 7.2 − 65 kPA
10.25 $Q_{5atm}/Q_{1atm} = 0.45$
10.29 $Q_{oil}/Q_{water} = 1.10$
10.36 55 Pa
10.39 5.09 m

CHAPTER 11

11.1 **(a)** 1.32 sec **(b)** $\Delta T = 2.86$°C
11.3 $\tau = 0.1$, y/ye = 0.998, $\phi = 0.063$ rad, $\tau = 1$, y/ye = 0.847, $\phi = 0.56$ rad, $\tau = 10$, y/ye = 0.157, $\phi = 1.41$ rad
11.5 $\tau_{gas} = 11.45$ sec, $\tau_{liq.} = 0.38$ sec
11.7 $\tau = 0.55$ sec
11.9 10 sec, 1.67°C
11.11 $\omega_n = 14.1$ rad/sec, $\zeta = 0.007$
11.15 $y/Kx_0 = 1.05$, $\phi = -2.7°$
11.17 982.4 rad/sec, 54 msec
11.20 1 kHz
11.26 $f_n = 626$ Hz, $\zeta = 1.6 \times 10^{-6}$
11.28 154 Hz

APPENDIX A

Computational Methods for Chapter 5

A.1 EVALUATING FOURIER-SERIES COEFFICIENTS USING A SPREADSHEET PROGRAM

Example 5.1 evaluated Fourier coefficients using direct integration. However, general experimental time-varying functions are not readily described with mathematical functions, and it is necessary to evaluate the coefficients numerically. Special programs are available to do this, but standard spreadsheet programs can readily be used. A solution to Example 5.1 has been set up in Table A.1 in the form of a spreadsheet. In column A, the time of one cycle has been discretized into 20 equal intervals. Column B is then a computation of $f(t)$. Columns C, D, and E contain corresponding values of $f(t) \sin \omega t$, $f(t) \sin 2\omega t$, and $f(t) \sin 3\omega t$, respectively. The coefficients b_1, b_2, and b_3 are computed by applying the trapezoidal rule to integrate the data in columns C, D, and E. The trapezoidal rule, in general terms for a function $g(t)$ with equally spaced values of time from t_0 to t_n, is given by

$$\int_a^b g(t) \, dt = \frac{b - a}{2n} [g(t_0) + 2g(t_1) + \cdots + 2g(t_{n-1}) + g(t_n)]$$

where n is the number of intervals. This equation can be used to compute a value of the integral in Eq. (5.2). Noting that the period, T, is 0.1 s, coefficient b_1 can then be computed from

$$b_1 = \left(\frac{2}{0.1}\right)\left(\frac{0.1 - 0}{2 \times 20}\right)(\text{C5} + 2 \times \text{SUM}(\text{C6} .. \text{C24}) + \text{C25})$$

TABLE A.1   Evaluating Fourier Series Coefficients Using a Spreadsheet Program

	A	B	C	D	E
1			Computing Fourier-Series Coefficients		
2					
3	Time	$f(t)$	$f(t) \sin wt$	$f(t) \sin 2wt$	$f(t) \sin 3wt$
4	(sec)				
5	0	0	0	0	0
6	0.005	0.3	0.092702	0.176331	0.2427
7	0.01	0.6	0.352662	0.570627	0.570644
8	0.015	0.9	0.728101	0.855966	0.278187
9	0.02	1.2	1.141254	0.705414	−0.70523
10	0.025	1.5	1.5	0.000139	−1.5
11	0.03	1.8	1.711933	−1.05785	−1.05826
12	0.035	2.1	1.699016	−1.99713	0.648547
13	0.04	2.4	1.410829	−2.28265	2.282371
14	0.045	2.7	0.83456	−1.58738	2.184743
15	0.05	3	0.000278	−0.00056	0.000834
16	0.055	−2.7	0.834084	−1.58657	2.183861
17	0.06	−2.4	1.410469	−2.28237	2.282783
18	0.065	−2.1	1.698787	−1.99737	0.649657
19	0.07	−1.8	1.71183	−1.05839	−1.05745
20	0.075	−1.5	1.5	−0.00042	−1.5
21	0.08	−1.2	1.141323	0.705054	−0.70577
22	0.085	−0.9	0.728199	0.855863	0.277711
23	0.09	−0.6	0.352752	0.570696	0.570541
24	0.095	−0.3	0.092755	0.176421	0.242798
25	0.1	1.17E-15	−2.2E-19	−4.3E-19	−6.5E-19
26			B1 =	B2 =	B3 =
27			1.894099	−0.92339	0.58885

The resulting value for b_1 is 1.8941. Coefficients b_2 and b_3 are −0.9234 and 0.58885, respectively. These differ slightly from the exact values computed in Example 5.1 (b_1 = 1.9098, b_2 = −0.9549, and b_3 = 0.6366). Part of the cause of the discrepancy is that the discretization resulted in inexact values of $f(t)$ between t = 0.05 and t = 0.055. $f(t)$ actually has two values at t = 0.05, +3 and −3, but only the +3 value was included here. The approximation would be better if the function were discretized into finer increments. If the time interval were halved, the coefficients b_1, b_2, and b_3 would improve to 1.9059, −0.9471, and 0.6248. Alternatively, two values could be included at t = 0.05, but then the integration would have to account for unequal time intervals.

Another factor that affects the values of the coefficients is aliasing. There are 20 samples in 0.1 s, giving a sampling rate of 200 Hz. This means that the maximum frequency in the signal that will not cause an alias will be 100 Hz. The ramp function of Example 5.1 actually has component frequencies from 100 Hz up to infinity, which will result in alias frequencies. The frequency components higher than 100 Hz have lower amplitudes than those less than 100 Hz. a_{11}, the coefficient for 110 Hz, the lowest frequency that will cause an alias, has a value of 0.173. A higher sampling rate will reduce the effects of the alias frequencies, but for this function, there will always be errors due to the alias frequencies. In an actual experiment, however, the signal should be filtered so that alias-causing frequencies have been eliminated.

A.2   EVALUATING THE FAST FOURIER TRANSFORM USING A SPREADSHEET PROGRAM

Major spreadsheet programs include the capability to evaluate the fast Fourier transform (FFT) of sets of time-varying data. A spreadsheet program was used to generate Figure 5.12, which is the result of a FFT performed on the function

$$f(t) = 2 \sin 2\pi 10t + \sin 2\pi 15t$$

using 128 samples over a period, T, of 1 s (Table A.2). Columns A and B of the table represent the time and the values of the function. The FFT function was then called for the function values in column B, rows 6 through 132, and the results are printed in column E, rows 6 through 132. The first 65 of these values of F correspond to frequencies from 0 to $(N/2)\Delta f$. Since $T = 1$ s, $\Delta f = 1/T = 1/1 = 1$ Hz and the frequencies range from 0 to $(128/2)1 = 64$ Hz. These frequencies have been calculated and are presented in column D. The values of F are complex numbers, meaning that each frequency component has both a magnitude and a phase. The magnitudes of the F's have been calculated, and these values are presented as $|F|$ in column G. It is these magnitude values, $|F|$, that have been plotted versus f in Figure 5.12.

REFERENCE

[1]   TAYLOR, J. (1990). *Computer-Based Data Acquisition Systems*, Instrument Society of America, Research Triangle Park, NC.

TABLE A.2   Evaluating Fourier transform using a spreadsheet program

	A	B	C	D	E	F	G
1	Computation of the Fast Fourier Transform of Equation 5.11						
2							
3	time	f(t)		frequency	F(f)		|F(f)|
4	(sec)			(Hz)	FFT(B6..B132)		IMABS(E?)
5	0	0		0	-0.000093013		0.0001
6	0.007874016	1.625446203		1	0.006278664-0.259589043i		0.2597
7	0.015748031	2.667732223		2	0.026070125-0.532621808i		0.5333
8	0.023622047	2.785235096		3	0.06151133-0.835241839i		0.8375
9	0.031496063	2.008004094		4	0.11712642-1.190282419i		1.1960
10	0.039370079	0.69993179		5	0.201542303-1.634998709i		1.6474
11	0.047244094	-0.621862166		6	0.33205384-2.239400297i		2.2639
12	0.05511811	-1.618026841		7	0.548045863-3.169359097i		3.2066
13	0.062992126	-1.796695932		8	0.96279636-4.84131515i		4.9361
14	0.070866142	-1.547379543		9	2.101458222-9.359683871i		9.5927
15	0.078740157	-1.037278361		10	30.93163356-123.5001836i		127.3148
16	0.086614173	-0.538125825		11	-2.621645108+9.474122167i		9.8302
17	0.094488189	-0.18244974		12	-1.042423442+3.43647556i		3.5911
18	0.102362205	0.074963979		13	-0.210209183+0.636084884i		0.6699
19	0.11023622	0.377649474		14	1.12698183-3.147555389i		3.3429
20	0.118110236	0.824743277		15	21.52550289-55.80959618i		59.8169
21	0.125984252	1.357602951		16	-4.109068125+9.921101093i		10.7384
22	0.133858268	1.747682748		17	-2.442209989+5.509966777i		6.0269
23	0.141732283	1.704321868		18	-1.909765804+4.038110677i		4.4669
24	0.149606299	1.04879596		19	-1.641318195+3.260930777i		3.6507
25	0.157480315	-0.144143641		20	-1.477585442+2.76451273i		3.1346
26	0.165354331	-1.520410179		21	-1.366667883+2.412600994i		2.7728
27	0.173228346	-2.567359762		22	-1.288345541+2.146226883i		2.5022
28	0.181102362	-2.832769292		23	-1.225478515+1.935318112i		2.2907
29	0.188976378	-2.139030043		24	-1.177784011+1.762741327i		2.1200
30	0.196850394	-0.685751478		25	-1.139446996+1.617943168i		1.9789
31	0.204724409	1.015667908		26	-1.108007431+1.494020194i		1.8600
32	0.212598425	2.350299746		27	-1.081801653+1.386240165i		1.7684
33	0.220472441	2.855976874		28	-1.059666358+1.291241959i		1.6704
34	0.228346457	2.408861042		29	-1.040757697+1.206558362i		1.5934
35	0.236220472	1.2564925		30	-1.024450609+1.130334767i		1.5255
36	0.244094488	-0.123835396		31	-1.010272031+1.061146032i		1.4662
37	0.261968504	-1.229655587		32	-0.997867243+0.997878671i		1.4112
38	0.25984252	-1.759224846		33	-0.986918147+0.939643133i		1.3627
39	0.267716535	-1.695423349		34	-0.97722826+0.88572488i		1.3189
40	0.275690551	-1.253291311		35	-0.968600327+0.835638253i		1.2792
41	0.283464567	-0.723837838		36	-0.960894592+0.788599893i		1.2431
42	0.291338583	-0.306870425		37	-0.953980366+0.744505972i		1.2101
43	0.299212598	-0.024772664		38	-0.947756171+0.702916414i		1.1800
44	0.307086614	0.242047323		39	-0.942142405+0.663541913i		1.1624
45	0.31496063	0.62800425		40	-0.937063351+0.626134396i		1.1270
46	0.322834646	1.146759846		41	-0.932458475+0.590482116i		1.1037
47	0.330708661	1.62785038		42	-0.928279638+0.556397438i		1.0823
48	0.338582677	1.789617949		43	-0.924479485+0.523718596i		1.0625
49	0.346456693	1.386338134		44	-0.921023019+0.492303656i		1.0443
50	0.354330709	0.380378304		45	-0.917875402+0.462026881i		1.0276
51	0.362204724	-0.979437537		46	-0.915010072+0.43277327i		1.0122
52	0.37007874	-2.221173783		47	-0.912399083+0.404446291i		0.9980
53	0.377952756	-2.839824009		48	-0.910025567+0.376949787i		0.9850
54	0.385826772	-2.52735772		49	-0.9078677+0.350207448i		0.9731
55	0.393700787	-1.330561971		50	-0.905911662+0.324144162i		0.9622
56	0.401574803	0.346376173		51	-0.904140763+0.298691362i		0.9522
57	0.409448819	1.896190321		52	-0.902542919+0.273786739i		0.9432
58	0.417322835	2.769865418		53	-0.901107565+0.249373317i		0.9360
59	0.42519685	2.693739049		54	-0.899825484+0.225396679i		0.9276
60	0.433070866	1.773097055		55	-0.898687462+0.201809764i		0.9211
61	0.440944882	0.420154828		56	-0.897687763+0.178562999i		0.9153
62	0.448818898	-0.845473755		57	-0.896819085+0.155614972i		0.9102
63	0.456692913	-1.624022998		58	-0.896075189+0.132923276i		0.9059
64	0.464566929	-1.782063572		59	-0.89645472+0.110444676i		0.9022
65	0.472440945	-1.45648146		60	-0.894951031+0.088147238i		0.8993
66	0.480314961	-0.929729056		61	-0.894562609+0.065987512i		0.8970
67	0.488188976	-0.454607411		62	-0.894287042+0.043934353i		0.8954
68	0.496062992	-0.127124876		63	-0.894122213+0.021949422i		0.8944
69	0.503937008	0.127096153		64	-0.894066364		0.8941
70	0.511811024	0.464564351			-0.894122213-0.021949422i		
71	0.519685039	0.929671923			-0.894287042-0.043934353i		
72	0.527559055	1.456430346			-0.894562609-0.065987512i		
73	0.535433071	1.782060221			-0.894951031-0.088147238i		
74	0.543307087	1.624072942			-0.89645472-0.110444676i		
75	0.551181102	0.84558845			-0.896075189-0.132923275i		
	ROWS 76 TO 128 DELETED FOR COMPACTNESS						
129	0.976377953	-2.785195742			0.11712642+1.190282419i		
130	0.984251968	-2.667797388			0.06151133+0.835241839i		
131	0.992125984	-1.625598291			0.026070125+0.532621808i		
132	1	-0.000185752			0.006278664+0.259589043i		

428

APPENDIX B

Selected Properties of Substances

TABLE B.1 Properties of Saturated Water, SI Units

Temperature (°C)	Density, ρ (kg/m^3)	Viscosity, $\mu \times 10^3$ (N-s/m^2)	Specific heat, c (kJ/kg-K)	Thermal Conductivity, k (W/m-K)	Prandtl Number, Pr	Bulk Modulus, B $\times 10^{-7}$ (N/m^2)
0	999.9	1.792	4.194	0.56	13.3	204
10	999.7	1.308	4.202	0.58	9.4	211
20	998.2	1.005	4.190	0.60	7.0	220
30	995.7	0.801	4.179	0.62	5.4	223
40	992.2	0.656	4.177	0.63	4.3	227
50	988.1	0.549	4.178	0.65	3.5	230
60	983.2	0.469	4.183	0.66	3.0	228
70	977.8	0.406	4.187	0.66	2.6	225
80	971.8	0.357	4.197	0.67	2.24	221
90	965.3	0.317	4.206	0.68	1.98	216
100	958.4	0.284	4.233	0.68	1.75	207

Source: Based on V. Streeter and E. Wylie, *Fluid Mechanics*, McGraw-Hill, New York, 1985; and J. Sucec, *Heat Transfer*, Wm. C. Brown, Dubuque, IA, 1985.

TABLE B.2   Properties of Saturated Water, British Units

Temperature (°F)	Density, ρ (lbm/ft^3)	Viscosity, μ (lbm/hr-ft)	Specific Heat, c (Btu/lbm-°F)	Thermal Conductivity, k (Btu/hr-ft-°F)	Prandtl Number, Pr	Bulk Modulus, B $\times 10^{-3}$ (lbf/in^2)
32	62.42	4.33	1.009	0.327	13.37	293
40	62.42	3.75	1.005	0.332	11.36	294
60	62.34	2.71	1.000	0.344	7.88	311
80	62.17	2.08	0.998	0.355	5.85	322
100	61.99	1.65	0.997	0.364	4.52	327
120	61.73	1.36	0.997	0.372	3.65	333
140	61.39	1.14	0.998	0.378	3.01	330
160	61.01	0.97	1.000	0.384	2.53	326
180	60.57	0.84	1.002	0.389	2.16	313
200	60.13	0.74	1.004	0.392	1.90	308
210	59.88	0.69	1.005	0.393	1.76	301

Source: Based on V. Streeter and E. Wylie, *Fluid Mechanics*, McGraw-Hill, New York, 1985; and A. Chapman, *Heat Transfer*, Macmillan, New York, 1967.

TABLE B.3   Properties of Dry Air at One Atmosphere Pressure, SI Units[a]

Temperature (°K)	Density, ρ (kg/m^3)	Viscosity, $\mu \times 10^5$ (N-s/m^2)	Specific Heat, c_p (kJ/kg-K)	Thermal Conductivity, k (W/m-k)	Prandtl Number, Pr
100	3.6010	0.6924	1.0266	0.009246	0.770
150	2.3675	1.0283	1.0099	0.013735	0.753
200	1.7684	1.3289	1.0061	0.01809	0.739
250	1.4128	1.5990	1.0053	0.02227	0.722
300	1.1774	1.846	1.0057	0.02624	0.708
350	0.9980	2.075	1.0090	0.03003	0.697
400	0.8826	2.286	1.0140	0.03365	0.689
450	0.7833	2.484	1.0207	0.03707	0.683
500	0.7048	2.671	1.0295	0.04038	0.680
550	0.6423	2.848	1.0392	0.04360	0.680
600	0.5879	3.018	1.0551	0.04659	0.680

Source: Based on J.P. Holman, *Heat Transfer*, McGraw-Hill, New York, 1990.

[a]Gas constant $R = 287.0$ J/kg-K; specific heat ratio = 1.400; molecular weight = 28.97. μ, k, c_p, and Pr are not strongly pressure dependent and may be used over a wide range of pressures. The density is proportional to pressure and may be corrected by multiplying by the factor $P/101,325.0$, where P is the absolute pressure in pascal.

TABLE B.4 Properties of Dry Air at One Atmosphere Pressure, British Units[a]

Temperature (°F)	Density, ρ (lbm/ft^3)	Viscosity, μ (lbm/hr-ft)	Specific Heat, c_p (Btu/lbm-°F)	Thermal Conductivity, k (Btu/hr-ft-°F)	Prandtl Number, Pr
−100	0.11028	0.03214	0.2405	0.01045	0.739
−60	0.09924	0.03513	0.2404	0.01153	0.733
−20	0.09021	0.3800	0.2403	0.01260	0.725
20	0.08269	0.04075	0.2403	0.01364	0.718
60	0.07633	0.04339	0.2404	0.01466	0.712
100	0.07087	0.04594	0.2406	0.01566	0.706
140	0.06614	0.04839	0.2409	0.01664	0.700
180	0.06201	0.05077	0.2403	0.01759	0.696
220	0.05836	0.05308	0.2418	0.01853	0.693
260	0.05512	0.05531	0.2424	0.01945	0.689
300	0.05221	0.05748	0.2431	0.02034	0.687
340	0.04960	0.05959	0.2439	0.02122	0.685
380	0.04724	0.06165	0.2447	0.02208	0.683
420	0.04509	0.06366	0.2457	0.02293	0.682

Source: Based on A. Chapman, *Heat Transfer*, Macmillan, New York, 1967.

[a]Gas constant R = 53.34 ft-lbf/lbm-°R; specific heat ratio = 1.400; molecular weight = 28.97. μ, k, c_p, and Pr are not strongly pressure dependent and may be used over a wide range of pressures. The density is proportional to pressure and may be corrected by multiplying by the factor $P/14.7$, where P is the absolute pressure in psia.

TABLE B.5 Selected Properties of Thermocouple Metals

Material	Density (kg/m^3)	Thermal Conductivity,[a] (W/m-K) (at 100°C)	Specific Heat (J/kg-K)
Alumel	8,600	30	526
Constantan	8,920	21	395
Chromel	8,730	19	450
Copper	8,930	377	385
Iron	7,870	68	447
Platinum	21,450	72	134
Platinum/13% rhodium	19,610	37	—
Platinum/10% rhodium	19,970	38	—

Source: Based on data from D. McGee, *Principles and Methods of Temperature Measurement*, Wiley, New York, 1988.

[a]For estimation purposes only. Thermal conductivity is a fairly strong function of temperature.

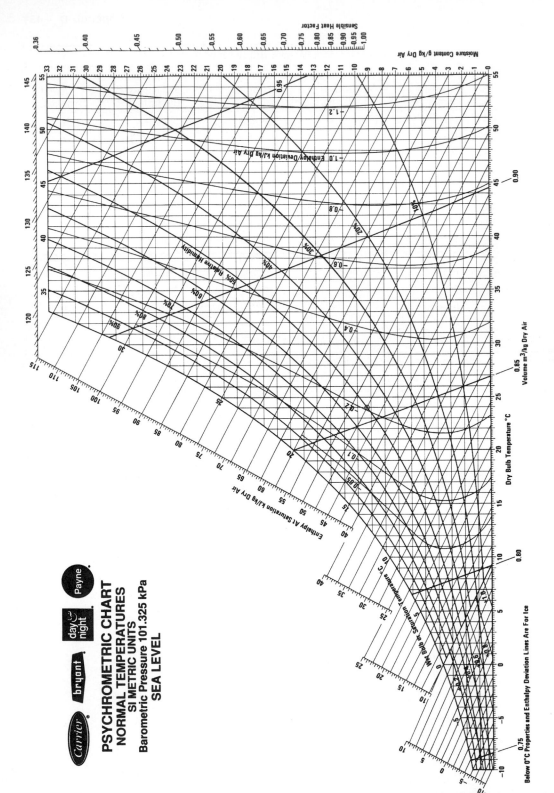

FIGURE B.1 Psychrometric chart, SI units. Reprinted by permission from Carrier Corp., Syracuse, NY.

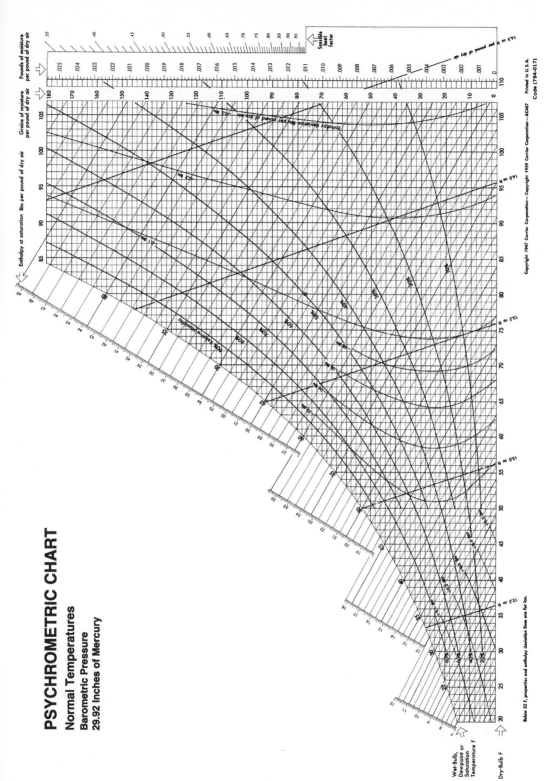

PSYCHROMETRIC CHART

Normal Temperatures
Barometric Pressure
29.92 Inches of Mercury

Copyright 1947 Carrier Corporation — Copyright 1959 Carrier Corporation — AC467   Printed in U.S.A.
Code (794-017)

FIGURE B.2   Psychrometric chart, British units. Reprinted by permission from Carrier Corp., Syracuse, NY.

Glossary

Accelerometer   Device used to measure the acceleration of a solid body.

Accuracy   Closeness of the reading of an instrument to the actual quantity being measured. Usually expressed as ±percentage of full-scale output or reading. In instrumentation, usually includes only those errors caused by hysteresis, nonlinearity, and nonrepeatability.

Active Filter   Electronic filter that combines passive elements, such as resistors and capacitors, with active elements, such as operational amplifiers.

Alias Frequency   False frequency component that appears in the recorded signal when a signal is sampled at a frequency less than twice the maximum frequency in the signal.

Analog Multiplexer (MUX)   Electronic switching device that can be used to connect any one of a set of inputs to a single output.

Analog-to-Digital (A/D) Conversion   Process whereby an analog input signal is changed into a digital code.

Anemometer   Instrument for measuring or indicating the velocity of an airflow. Sometimes used in place of a velocimeter in the general sense.

ANOVA (Analysis of Variance)   Statistical technique for correlating test data.

Aperture Time   Time required for an A/D converter to convert an input voltage or for a sample-and-hold device to capture a voltage. Since the input voltage may change during the aperture time, a measurement error may result.

ASCII (American Standard Code for Information Interchange)   Standard method of encoding alphanumeric characters into 7 or 8 binary bits.

Assembly Language   Machine-oriented computer language in which mnemonics represent each instruction.

Attenuation   Reduction in signal amplitude that may be frequency dependent.

Bandpass Filter   Filter that allows a selected range of frequencies to pass but attenuates frequencies outside the range selected.

Bandwidth   Range of frequencies over which a system will operate with close-to-constant gain.

Bias Error   *See* Systematic Error.

Binary   Base-2 numbering system in which the only allowable digits are "0" and "1."

Bipolar   Signal that includes both positive and negative values.

Bit   Smallest unit of binary information. A bit will have a value of "1" or "0." *Bit* stands for "*bi*nary dig*it*." Eight bits make up a byte.

Bourdon Gage   Very common pressure measurement device, which works on the basis of deformation of a curved and flattened tube when the pressure inside the tube changes.

Bus   Set of conductors inside a computer used to communicate between computer components. The bus may include connectors, which can be used to install accessory cards such as data-acquisition systems.

Byte   Binary number consisting of 8 binary bits.

Calibration   Test during which known measurand values are applied to the measurement device under specified conditions and the corresponding output readings are recorded.

Calibration Curve   Graph representing calibration data.

Calibration Cycle   A set of data consisting of the instrument output versus the known input values.

Central Processing Unit (CPU)   Portion of the computer that controls its operation and performs arithmetic and other data-manipulation operations.

Cladding   Material surrounding an optical fiber that causes the light rays to be reflected inside the fiber due to its smaller refractive index.

Common-Mode Rejection Ratio (CMRR)   Ratio of a device's differential voltage gain to common-mode voltage gain, usually expressed in decibels. Devices with high CMRR values are usually better able to reject noise signals.

Comparator   Circuit that compares two input voltage values and produces one of two possible output voltages, which identifies greater input value.

Compensation   Application of specific materials or devices to counteract and alleviate a known error.

Compliance   In structures, the deformation of an elastic medium per unit of exerted force. It is the inverse of stiffness. Also used in pressure transducers to define the change in chamber volume per unit of applied pressure.

Confidence Interval   Estimated interval containing a population parameter.

Confidence Level   Probability that a random variable x lies in a specified interval. The same as *degree of confidence*.

Confidence Limits   Two values that define the confidence interval.

Continuous Random Variable   Random variable that can attain any value continuously in some specified interval.

Conversion Time   In a data-acquisition system, the time required for an analog-to-digital converter to generate a digital output from an analog input or the time for a digital-to-analog converter to determine an analog output from a digital input.

Coriolis Force   Apparent force that occurs when the motion of a body is studied in a rotating reference frame.

Correlation Coefficient, r   Measure of how well a curve fits a set of data. A value of 1 indicates perfect relationship, and a value of 0 indicates no relationship.

Counter   Hardware circuit for counting pulses.

Crosstalk   Undesirable phenomenon in which the signal on one path interferes with other signals on other paths.

Current Loop   Method of transmitting signals over wires in the form of a modulated current rather than voltage. Current-loop systems are less sensitive to noise pickup than voltage systems, and they are used to transmit data over long distances.

Damping Ratio   Parameter used to define the damping characteristics of second-order linear dynamic systems.

Data Acquisition   Actual capture of information from real-world sources such as sensors and transducers.

Data-Acquisition Board   Data-acquisition system incorporated on printed circuit board that is compatible, electrically and mechanically, with a particular computer system.

Data-Acquisition System (DAS)   Automated system that takes data from measurement devices, processes them, then stores or records the data.

Dead-Weight Tester   Device, usually used for calibration purposes, that measures pressure by applying weights to a piston of known area.

Decibel (dB)   Logarithmic measure of the ratio of two signal levels.

Degree of Confidence   *See* Confidence Level.

Degrees of Freedom   Number of independent measurements available for estimating a sample statistic. The degrees of freedom are reduced by one for each previously calculated statistic used to calculate a new statistic.

Density Function, $f(x)$ Function which yields the probability that the continuous random variable takes on any one of its permissible values.

Digital Data   Information in the form of a signal with discrete possible values. Digital data are usually represented using binary code.

Digital Encoder   Device that converts linear or angular displacement directly into a digital signal.

Digital-to-Analog (D/A) Conversion   Process whereby a digital value or code is changed into an analog signal. The electronic system performing the conversion is called a converter.

Dimensional Analysis   Process of combining the variables into nondimensional groupings for the purpose of data presentation or scale-model testing.

Discrete Random Variable   Random variable that can have values only from a definite number of discrete values.

Disk Drive   Computer peripheral device that stores information by altering the magnetic properties of the surface of a rotating disk.

Display   Device with a computer or instrument that visually presents data to the user.

Distortion   Usually undesirable artifact of processing a signal by such devices as amplifiers and filters.

Doppler Effect   Change in the frequency of a wave when generated by or reflected from a moving object. May apply to sound or electromagnetic waves.

Drift   Change in output of a measurement device over a set period of time due to several factors such as temperature or deterioration of components.

Dry-Bulb Temperature   Temperature of a gas as measured with a temperature sensor. *See also* Wet-Bulb Temperature.

Dynamic Calibration   Calibration process in which a known time variation of input is applied to a device and the output is recorded versus time.

Dynamic Measurement   Measurement in which the measurement system input varies with time or the measurement system output depends on time.

Dynamic Range   Ratio of the largest possible output of a data converter to the smallest output it can resolve.

Dynamometer   In the most common usage, a device that is used to measure the torque of rotating shafts.

Electromagnetic Interference   Unwanted background noise generated in a measurement system by interference sources such as radio waves.

Elemental Error   Individual source of measurement error.

Error   Difference between the value indicated by the measurement system and the true value of the measurand being sensed.

Excitation   External application of electrical power to a transducer for normal operation.

Expansion Board   Plug-in circuit board that increases the capabilities of a computer. A data-acquisition board is one example.

Expansion Slot   Actual space provided within a computer system for individual expansion board.

Flash A/D Converter   Extremely fast A/D conversion technique in which an array of $2^N - 1$ comparators perform the conversion in a single cycle. N equals the number of bits used to represent the converter output.

Floating-Point Numbers   Real numbers that contain decimal parts or are written in scientific notation.

Flow Nozzle   Device that measures fluid flow rate based on the principle of the Bernoulli equation applied to an area change in a conduit.

Frequency Output   Output of a transducer in the form of frequency that varies as a function of applied measurand.

Frequency Response   Variation of the output of a device with respect to the frequency of the input signal.

Full Bridge   Wheatstone-bridge configuration utilizing four active elements or strain gages.

Full Scale   Specified maximum input value of a measurement system.

Gage Factor   Measure of the ratio of the relative change of resistance to the relative change in strain applied to a strain gage.

Gage Pressure   Difference between absolute pressure and ambient pressure.

Gain   Ratio between the output and the input of an electronic device such as an amplifier. Frequently expressed in a logarithmic form in decibels (dB).

Gain Error   Error that results if the gain of a device is not as specified or expected.

Gantt Chart   Type of bar chart used in project scheduling.

General-Purpose Interface Bus (GPIB)   Standard bus used for the interface and control of programmable instruments. Also called the IEEE488 bus.

Half Bridge   Wheatstone bridge with two active elements or strain gages.

Hardware   Physically visible parts of a computer, such as CPU, disk drives, and so on.

Head   Pressure in a fluid system expressed in terms of the vertical height of a fluid column.

High-Level Language   Programming language that simplifies the creation of computer codes, such as BASIC, C, Pascal, or FORTRAN.

High-Pass Filter   Signal filter that attenuates low-frequency signal components.

Histogram   Graphical representation of a frequency distribution by a series of rectangles, where the width of the rectangle represents the range of a variable and the height represents the frequency of occurrence.

Humidity Ratio   Ratio of mass of water vapor to the mass of dry air in a given volume of moist air.

Hysteresis   Difference in output of the measurement system when the measurand is approached from higher and lower values.

Input Impedance   Impedance measured across the input terminals of an electrical device. Often frequency dependent.

Input/Output (I/O) Channels   Channels used for transferring data into and out of a computer system. I/O channels include communication ports, operator interface devices, and data-acquisition and control channels.

Instrumentation Amplifier   Specialized amplifier with high input impedance and high CMRR.

Integrated Circuit (IC)   Semiconductor circuit element that contains multiple, interconnected circuits.

Integrating A/D Converter   A/D conversion technique in which the analog input voltage is integrated over time.

Interface   Means by which electronic devices are connected and interact with each other.

Ionization Vacuum Gage   Device that measures vacuum pressures by creating ions of the gas and measuring a current flow of the ions.

Laminar Flowmeter   Device used to measure fluid flow rate based on the pressure drop of laminar flow through small passages.

Least Significant Bit (LSB)   Rightmost digit in a binary number.

Least-Squares Line   Line fitted to a series of test data points such that the sum of the squares of deviations of the data points from the line is a minimum.

Linear Measurement System   Measurement system whose behavior can be described by an ordinary time-dependent linear differential equation.

Linearity   Closeness to which a calibration curve approximates a specified straight line. Linearity error is expressed as the maximum deviation of any portion of the calibration curve (average of upscale and downscale values) from a specified straight line.

Linearity Error   *See* Linearity.

Linear Variable Differential Transformer (LVDT)   Device used to measure linear displacement by modifying and sensing the spatial distribution of an alternating magnetic field.

Load Impedance   Impedance applied to the output terminals of a device by the connected external circuitry.

Loading Error   Measurement error caused when the use of a measurement device changes the value of the measurement.

Lowpass Filter   Signal filter that attenuates high-frequency signal components.

Machine Language   Instructions written in binary code, which can be executed directly by a computer. The structure of the code depends on the computer.

Manometer   Device used to measure pressure by measuring the height of a column of liquid.

Mass Storage   Device such as hard disk or tape drive used to store large amounts of computer data.

McLeod Gage   Manometric device used to measure vacuum pressures.

Mean of a Population, μ   Population parameter obtained from Eqs. (6.18) and (6.14) for continuous and discrete random variables, respectively.

Mean of a Sample, $\bar{x}$ Value calculated by dividing the sum of measured values by their total number. Also called the average.

Measurand   Physical quantity, property, or condition that is measured.

Median of a Population   Value of the random variable, x, at which the cumulative distribution function $f(x)$ is 0.5.

Median of a Sample   Number in the middle, when all the observations are ranked in their order of magnitude.

Microbending   Bending of an optical fiber, which results in attenuation of light intensity passed through the fiber. It is used as a sensing method in fiber-optic sensors.

Mode of a Population   Value of x corresponding to the peak value of the probability of occurrence of any given continuous or discrete distribution.

Mode of a Sample   Value of x corresponding to the maximum frequency on the distribution curve, or the highest probability in the case of a discrete random variable.

Monitor   Display screen used with a computer system.

Most Significant Bit (MSB)   Leftmost bit of a binary number.

Multiplexer (MUX)   *See* Analog Multiplexer.

Natural Frequency   Frequency of free (not forced) oscillations of the sensing element of a fully assembled measurement system.

Noise   Undesirable electrical interference causing degradation of a signal from sources such as electric machinery, ac power lines, and radio-frequency transmitters.

Nonintrusive Measurement   Measurement that causes negligible loading error. Usually true for techniques that do not require use of a physical probe for measurement.

NO$_x$ Oxides of nitrogen. A term usually used to indicate the sum of oxides of nitrogen (primarily nitric oxide, NO, and nitrogen dioxide, NO$_2$) in exhaust gases.

Nyquist Frequency   In sampling a continuous signal at discrete times, the Nyquist frequency is one-half the sampling frequency and represents the maximum signal frequency that can be sampled without aliasing.

Nyquist Theorem   *See* Sampling Rate Theorem.

Operational Amplifier   Electronic amplifier with a high input impedance, low output impedance, and a very high low-frequency gain.

Order of Measurement System   Order of the differential equation representing the dynamic behavior of the system.

Orifice Meter   Device that measures fluid flow rate based on the principle of the Bernoulli equation applied to an area change in a conduit.

Output Impedance   Impedance as measured on the output terminals of a device such as a transducer or amplifier.

Parallel A/D Converter   *See* Flash A/D Converter.

Parameter   Numerical value defining some property of a population.

Parts per Million (ppm)   Number of particles per million particles. Often used in air pollution terminology.

Phase Angle   Change in phase of an output with respect to a sinusoidal input.

Piezoelectric   Property of certain materials (such as quartz) that generate electric charge when deformed, and vice versa.

Piezoresistive   Property of certain materials that change resistance when deformed.

Pirani Vacuum Gage   Device used to determine vacuum pressures by sensing the gas thermal conductivity.

Pitot-Static Probe   Device used to measure fluid velocity based on the principle of the Bernoulli equation.

Poisson's Ratio   Negative of the ratio of the lateral strain to the axial strain.

Population   Entire collection of items from which a sample is drawn for test purposes.

Potentiometer   Device in which the resistance varies as a function of linear or angular displacement.

Precision Error   *See* Random Error.

Probability Distribution Function   Function representing the frequency of occurrence of a random variable versus the value of the variable.

Probability of Occurrence   Number of successful occurrences divided by the total number of trials.

Process Control   Field in which instruments and computers are used to monitor and control industrial processes.

Proving Ring   Simple ring of metal that can be used to determine force by measuring the diametrical deformation.

Psychrometric Chart   Chart from which the properties of moist air can be determined. It represents the relationship between dry-bulb temperature, wet-bulb temperature, humidity, and other properties of moist air.

Pyrometer   Device that measures temperature by measuring thermal radiation (usually, high-temperature values such as in a furnace environment).

Random Access Memory (RAM)   Semiconductor devices in a computer that can store digital data in a retrievable form. Data can be written to a particular location without having to sequence through previous locations. RAM is volatile, so all data are lost when the power is discontinued.

Random Variable   Variable that can assume any real value in a certain domain.

Range   Measurand values, which a measurement device is intended to measure, specified by the upper and lower limits.

Random Error   Same as precision error, a nonrepeatable error due to unknown or uncontrollable factors influencing the measurement.

Reference Junction   Means of compensating for ambient temperature variations and standardizing the readout of thermocouple circuits.

Refractive Index   Ratio of the speed of light in a vacuum to the speed of light in a transparent medium.

Regression Analysis   Method for establishing an explicit mathematical relationship between a dependent variable and one or more independent variable(s).

Relative Humidity   Ratio of actual water vapor mass to the saturated water vapor mass at a given temperature and volume.

Repeatability   Instrument's ability to produce the same output repeatedly under identical conditions.

Replication   Process of repeated independent observations made under identical test conditions.

Resolution   Smallest detectable change in a measurement.

Resolution Error   Error due to imperfect resolution. Usually taken as one-half of the resolution.

Resonant Frequency   Input frequency at which a device responds with maximum output amplitude.

Response Time   Length of the time required for the output of a measurement system to reach a specified percentage of its final value in response to a step change in the measurand.

Ringing Frequency   Frequency of an input producing maximum measurement system output corresponding to the damped natural frequency of an underdamped system.

RS-232   Digital communication standard for connecting modems, terminals, printers, and other digital devices.

Sample   Random selection of items from a population, usually made for evaluating the characteristics of the population.

Sample-and-Hold Device (S/H)   Circuit used to acquire and store an analog voltage rapidly and accurately for use by other devices.

Sampling Rate   Rate at which a signal is sampled (read) by a digital data-acquisition system. Expressed in Hz.

Sampling Rate Theorem   Theorem which states that if a continuous signal is sampled with at least twice the frequency of its highest-frequency component, it can be fully recovered from the sampled data.

Seebeck Effect   Principle that describes how a low-level voltage is formed in a circuit by the junctions of two dissimilar metals when the junctions are held at different temperatures. It is the basis for the operation of thermocouples.

Sensing Element   Part of the transducer that reacts directly in response to the measurand.

Sensitivity   Ratio of change in the output to the change in the input of a measurement device.

Sensor   Device that produces an electrical output that corresponds to a physical input (temperature, pressure, light, etc.).

Settling Time   Time period elapsed from the application of a step input into a measurement device to the time that the output enters and remains within a specified error band around its final value.

Shielded Cable   Transmitting line that has a protective foil or other sheathing around it to guard against electromagnetic interference.

Shielding   Protective covering that isolates a circuit or an electrical transmitting line from unwanted electromagnetic or radio-frequency interference.

Signal Conditioner   Device that modifies an electrical signal to make it more acceptable as an input to other devices.

Signal-to-Noise Ratio   Ratio of signal level to noise level. Often expressed as a logarithmic ratio in decibels (dB).

Simultaneous Sample/Hold   Data-acquisition subsystem in which several sample-and-hold circuits sample multiple analog channels simultaneously. Each channel requires one sample-and-hold device.

Software   Computer programs, operating systems, high-level languages, and so on, that are held in some type of storage medium and are loaded into RAM for execution when needed.

Span   Algebraic difference between the upper and lower range limits of the input or output of a measurement device.

Spatial Error   Measurement error that occurs when a measurand (such as temperature) varies in a spatial region and an attempt is made to represent that variation with measurement in a limited number of locations.

Spectral Analysis   Process of determining the component frequencies of a signal.

Stability   Ability of a measurement device to maintain a constant output when a constant input is applied.

Standard Deviation of a Population, σ   Square root of the variance of the population.

Standard Deviation of a Sample   Square root of the variance of a sample; a measure of the scatter of test data about the mean.

Static Measurement   Measurement of a measurand that does not vary with time or to which the measurement device has an instantaneous response.

Static Sensitivity   Ratio of the output to the input of a measurement system under static loading.

Statistic   Numerical attribute of a sample.

Stiffness   Force required to deform an elastic medium a unit of length. It is the inverse of compliance.

Strain   Ratio of the change in length due to application of a stress to the initial unstressed length.

Strain Gage   Sensor that converts a strain into an electrical signal. The most common type causes a change in resistance. Used as a sensor in pressure, force, and other measuring systems.

Strain-Gage Rosette   Three suitably arranged (usually in 45° or 60° angles) strain gages for measuring biaxial strains on a surface.

Successive Approximation A/D Converter   A/D converter that uses an interval halving method to determine each bit of the digital output.

Systematic Error   Repeatable error in a measurement that does not vary with replication. The same as bias error. It is estimated by the difference between the true value and average of a large number of readings.

Tachometer   Device for measuring the angular velocity of shafts.

Thermistor   Semiconductor temperature sensor that exhibits a large change in resistance resulting from a small change in temperature.

Thermocouple   Temperature sensor consisting of the junction of two dissimilar metals. The output voltage produced is a function of the difference in temperature between the hot and cold junctions of the two metals.

Time Constant   Usually used for first-order dynamic systems. It is a characteristic time of a system indicating how fast the system will reach a steady state when subjected to a step change in input. It is defined as the time that the output change reaches 63.2% of its final value.

Transducer   Device that converts physical parameters such as temperature, pressure, motion, and so on, into an electrical output.

Uncertainty Estimation of error in a measurement or in a result, usually determined with a certain level of confidence (often 95%).

Undamped Natural Frequency Angular frequency at which a system will oscillate in the absence of damping.

Unipolar Signal that includes values of only one sign (e.g., always positive or always negative).

USB Universal Serial Bus Port A connection on personal computers for digital communication with peripheral devices such as printers and scanners.

Variance of a Population, σ^2 Defined for a random variable x by Eqs. (6.5) and (6.7) for discrete random variables, and Eq. (6.19) for continuous random variables.

Variance of a Sample, S^2 Sum of squares of the differences between each observation and the sample mean divided by the sample size minus 1 [Eqs. (6.6) and (6.7)].

Venturi Tube Device that measures fluid flow rate based on the principle of the Bernoulli equation applied to an area change in a conduit.

Wet-Bulb Temperature Temperature of an air-temperature-measuring sensor that is covered with a wet wick and moved at a significant velocity (over 3 m/s) relative to the air. Together with the dry-bulb temperature, used to determine humidity from a psychrometric chart.

Wheatstone Bridge Network of four resistances, a voltage source, and a voltmeter used to measure the change in resistance of a sensor. Sometimes includes capacitors and is used to measure the change in capacitance of a sensor.

Young's Modulus Ratio of the normal stress to the strain in a material.

Zero Drift Change in the zero or null setting of an instrument over time.

Zero Offset Deviation of the output of a device at the zero or null point from the correct value.

Index